ORGANIC CHEMISTRY 1 PRIMER 2019

BY RHETT C. SMITH, ANDREW G. TENNYSON, AND TANIA HOUJEIRY

AF822212

Marketed by Proton Guru

Find additional online resources and guides at protonguru.com!

There is a lot of online video content to accompany this book at the Proton Guru YouTube Channel! Just go to YouTube and search "Proton Guru Channel" to easily find our content.

Correlating topics with your course: The homepage at protonguru.com provides citations to popular text books for further reading on each topic in this book so that you can follow along using this book in any course using one of these texts.

Instructors: Free PowerPoint lecture slides to accompany this text can be obtained by emailing IQ@proton.guru from your accredited institution email account. The homepage at protonguru.com provides a link to citations to popular text books for further reading on each Lesson topic in this primer.

Executive Editor: Rhett C. Smith, Ph.D. You can reach him through our office at: IQ@protonguru.com

Printed in the United States of America

10 9 8 7 6 5 4 3 2 1

ISBN 978-1-7332325-9-3

Organic Chemistry 1 Primer 2019

Rhett C. Smith, Ph.D.
Andrew G. Tennyson, Ph.D.
Tania Houjeiry, Ph.D.

Table of Contents

Preface to the Primer

This brief, plain language Primer is meant to provide the most basic principles and concepts of organic chemistry as simply as possible. It is meant to be an excellent *primer* to read *before* lecture so that once you get to lecture you will already have some knowledge of what will be discussed. It is also meant to be an excellent pre-exam review. Especially in cases where you are preparing for a final exam – which may be cumulative and cover material that you have not specifically studied for weeks or months – it should be useful as a more compact review than will be offered by a text book or other study guides.

Online Content, Reaction Guide and Practice Problems

There are many resources to support the material covered in this book. Many of these resources are found on protonguru.com (launched in July 2018 and still evolving with new content!).

The Proton Guru website includes:

- Links to video content including solved practice problem videos and lectures for important topics
- References to chapters in popular text books used at universities so that you can use this book and all the associated YouTube video content easily in any course using those text books

Proton Guru also has a reaction and practice problem guide that can be used with this Primer, "Organic Chemistry 1 Reactions and Practice Problems" by Rhett C. Smith. This book can be found on Amazon.com and includes:

- Brief review of each reaction with bullet point lists of key features of each reaction and a quick self test/answers for each reaction
- Problem Sets for important topics, with problems like you might see on class or standardized exams
- Progress Checks: longer practice exams with solutions

PART I. Structure, Stability and Conventions

Lesson I.1. Coulombic Forces and Representation of Structure

Lesson 1.1. Coulombic Forces

Coulombic forces are the forces that exist between charged particles. Particles may be either negatively charged or positively charged. Negative charges repel other negative charges, and positive charges repel other positive charges. Such repulsive forces tend to push like-charged entities away from one another as much as possible. If you hold two like-charged entities in close proximity, you will have to apply force, thus *exerting energy* to keep them close together. This creates a *strain*, and the interaction of like-charged particles is said to be **unfavorable** or **destabilizing.**

Oppositely-charged particles attract one another. This means that if two oppositely-charged particles come in close proximity, they will **spontaneously** pull towards one another; their interaction is said to be **favorable**, and arrangements that maximize such attractive forces are **stabilizing**.

The fundamental particle of negative charge is the electron (sometimes written as "e^-"). The fundamental particle of positive charge is the proton (sometimes written as "H^+"). In the familiar picture of an atom, the positive charge is located in the nucleus while the negatively-charged electrons are held in an "orbit" or "cloud" around the nucleus by attractive Coulombic forces. In an atom, the number of electrons is equal to the number of protons, so that the atom is overall neutral. In general chemistry, you learned that an atom can lose one or more electrons, leaving an excess of positive charge. The resultant species, an ion, is charged. A positively-charged ion is called a **cation** (pronounced "CAT-eye-on"). If an atom gains one electron or more, there is an excess of negative charge, thus forming a negatively-charged ion, which is called an **anion** (pronounced "ANN-eye-on").

Example I.1.1.

Provide the charge of the ion formed when each of the following species undergoes the indicated change:

a. $Cl \xrightarrow{\text{gain one electron}} ?$

b. $[CH_3]^+ \xrightarrow{\text{gain one electron}} ?$

c. $O^- \xrightarrow{\text{gain one electron}} ?$

d. $Br \xrightarrow{\text{lose one electron}} ?$

e. $O^- \xrightarrow{\text{lose one electron}} ?$

f. $Cu^+ \xrightarrow{\text{lose one electron}} ?$

Solution I.1.1.

Adding an electron (in a-c) leads to one additional unit of negative charge. So, in a. the neutral Cl becomes Cl^-. For b., the +1 charge of the $[CH_3]^+$ cation becomes (+1–1) = 0, giving $[CH_3]$. For c., O^- has a –1 charge so after adding another unit of negative charge it becomes O^{-2}. If an electron is removed from a species, it gains one more unit of positive charge. So, in d. the Br becomes Br^+, while in c. the O^- becomes a neutral O atom, and in f. Cu^+ becomes Cu^{+2}.

Lesson I.1.2. Representation of Structure

We learned in general chemistry (or even high school chemistry) how to draw Lewis dot structures to represent molecular structures, and how to interpret these representations. If you do not recall how to write Lewis structures, you will need to review this to succeed in organic chemistry. In the Lewis structures, one can use a dot to represent one electron and a line to represent two electrons. It is vital to your ability to understand organic chemistry that you recognize that a line or a dot represents negatively-charged electrons in the outer shell of an atom (the valence electrons). The atomic symbol (the letter(s) you use for the atom in a structure) represents the positively-charged nucleus, along with any inner-shell electrons (which typically do not interact with other atoms).

In your study of the valence shell electron pair repulsion (VSEPR) theory, you likely also encountered a way to draw bonds in order to represent a three-dimensional shape on a flat surface by using wedge or hashed lines. In this convention, wedges represent bonds coming out of the surface (e.g., chalkboard, whiteboard, paper, computer screen, etc.), hashed lines represent bonds going into the surface, and simple lines represent bonds that lie within the plane of the surface:

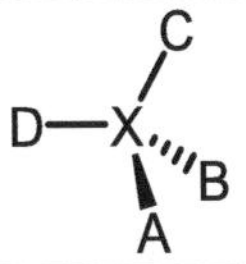

C-X and D-X bonds: in the plane of the page
A-X bond: comes out of page, with A closer to the viewer
B-X bond: goes into the page, with B farther from the viewer

In addition to the familiar Lewis structures and wedge/hashed lines, organic chemists use the *line-bond* notation to represent organic structures. The line-bond notation is especially helpful for organic compounds having multiple carbon atoms in chains or rings. An example of how the line bond is used to abbreviate a five-carbon-long chain is shown below:

Lewis Structure

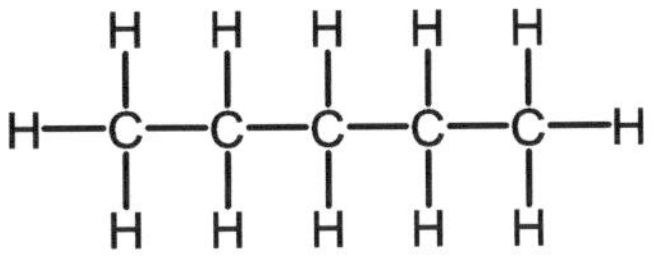

Condensed Formula

$CH_3CH_2CH_2CH_2CH_3$
or
$CH_3(CH_2)_3CH_3$

Line-Bond Structure

In the line-bond notation, a line represents a bond, just as it does in the Lewis structure notation. However, in the line-bond notation, each bend in a line, end of a line, or change in the number of parallel lines (e.g., single bond to triple bond) represents a carbon atom. In this notation, one assumes the correct number of H atoms on each carbon for the given charge. A neutral carbon atom has four bonds, so any bonds not explicitly shown in the line-bond notation are assumed to be to H; one does not need to draw out the H atoms on carbon atoms whose symbols are not drawn out as a C. All other non-C atoms, as well as any H atoms attached to non-carbon atoms *must* be shown. Also, if you draw out a 'C' letter for a carbon atom by choice (even though you do not have to), you would need to show all the H atoms on that C, for example:

For C_3H_6:

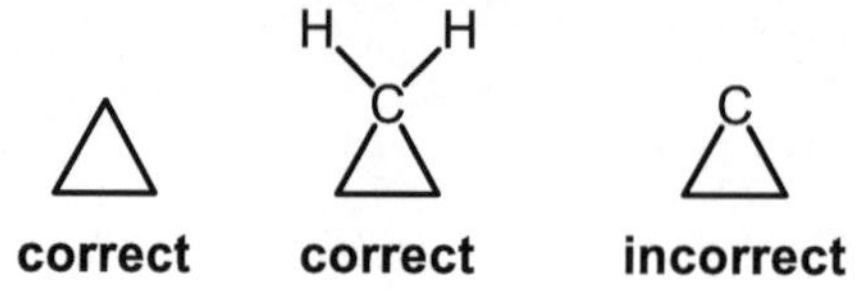

Here are a few examples of compounds shown in both Lewis structure and line-bind notation. One can readily see how the line bond notation can save a lot of time in drawing a structure as well as provide a simplified, compact picture of the molecule:

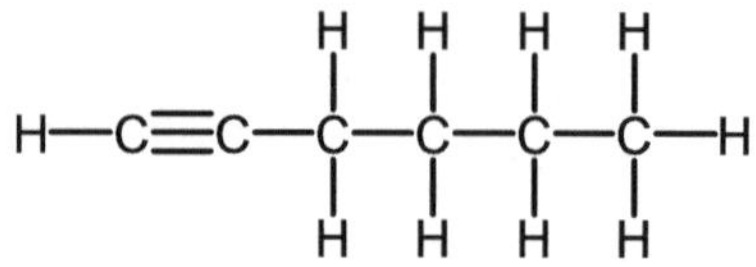

represents the same molecule as:

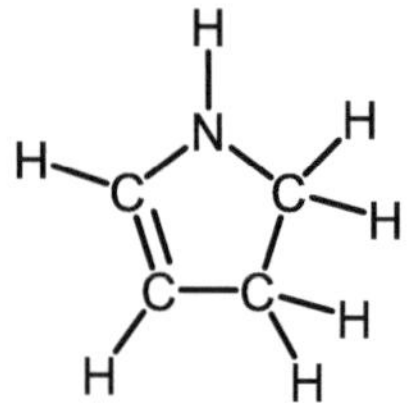

represents the same molecule as:

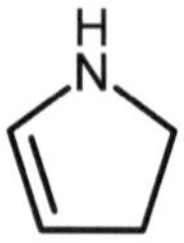

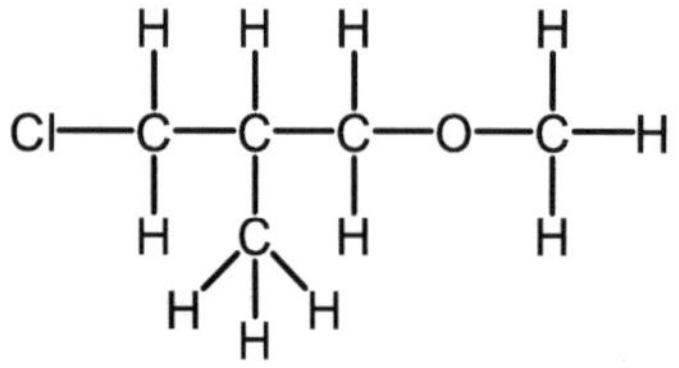

represents the same molecule as:

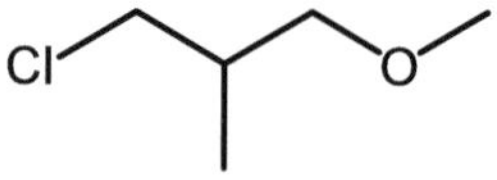

Example I.1.2.

Provide the line-bond structure for each of these Lewis structures:

A) 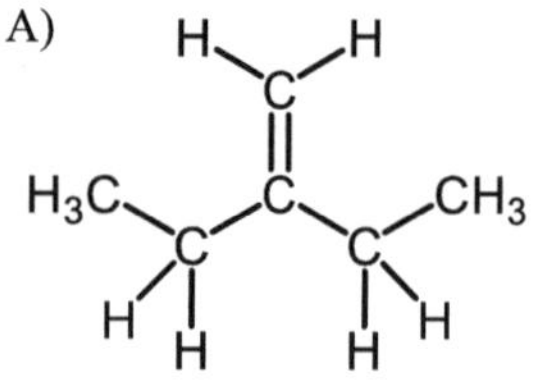

B) 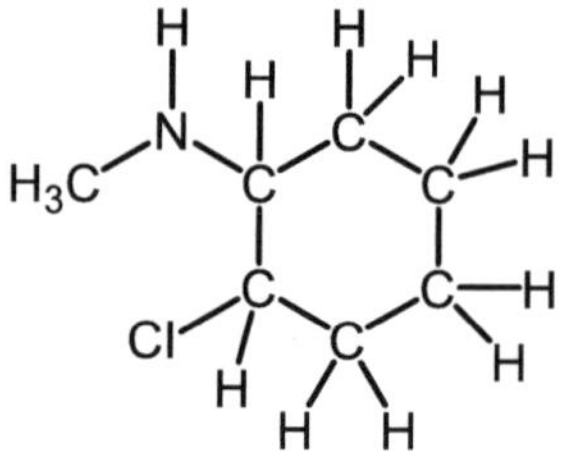

C) 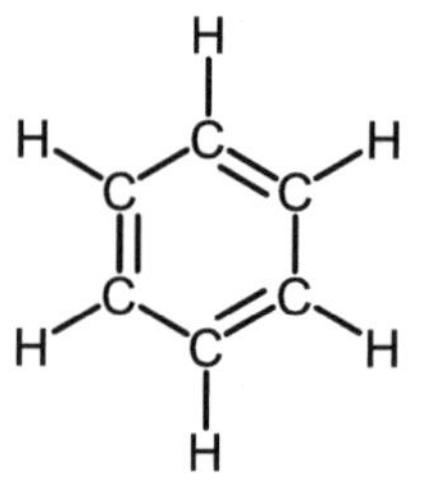

Solution I.1.2.

A) 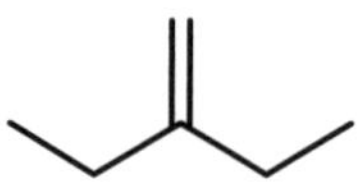

B) 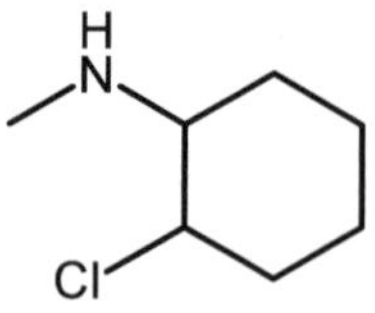

C) 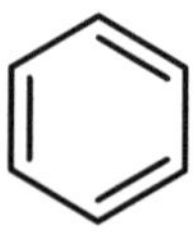

Lesson I.2. Polarity, Dipole Moments and Formal Charge

Lesson I.2.1. Polarity

In your prior chemistry courses, you learned about electronegativity and its influence on polarity, dipole moments, etc. We will briefly review how some of these concepts apply specifically to organic chemistry. Electronegativity is the pull an atom exerts on electrons in a bond. A higher value of electronegativity indicates a greater pull for electrons. Electronegativity of an element increases towards the top and towards the right side of the periodic table, with F being the most electronegative element.

In a bond between two atoms of equal electronegativity, the electrons in the bond will be held exactly in between the two nuclei in a symmetric distribution of charge, so there are no regions of unusually high negative or positive charge, and thus the bond is **nonpolar**. If the electronegativity values for two atoms are quite similar, then those bonds are not significantly polar. **An important example of nonpolar bonds in organic chemistry are C–H bonds.**

Examples of Nonpolar Bonds

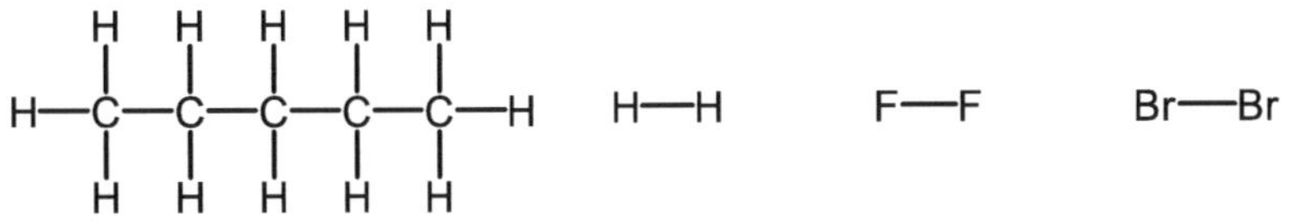

ALL the bonds in this molecule!

If one of the atoms in a bond has a significantly higher electronegativity, it pulls the electrons more closely towards it, so that there is more negative charge on its end of the bond, and less on the other end from which the electrons were pulled, creating a *polar bond*:

Arrows Point to Polar Bonds in these Examples

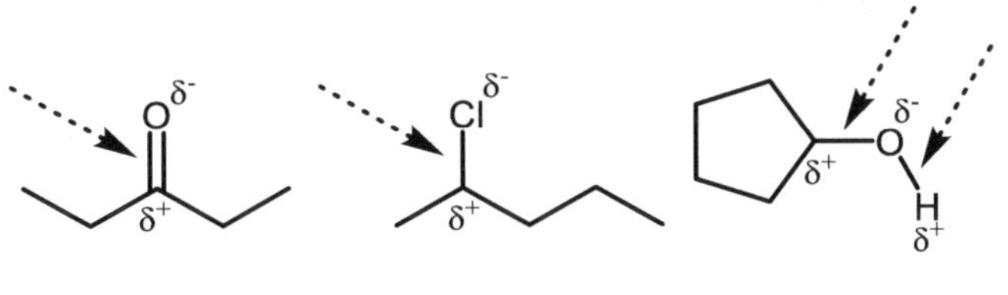

In the picture above, you may notice two new symbols. The δ^- symbol indicates **the partial negative charge** present at the negative pole of a **polar bond**, while the δ^+ symbol indicates the **partial positive charge** present at the positive pole of the polar bond. The use of these symbols to indicate what sites on a molecule have an excess of charge will become very important to your ability to solve problems as we move through organic chemistry. The greater the difference in electronegativity between the two bonded atoms, the more polar the bond will be, and thus there will be greater amounts of positive and negative partial charges on each side of the polar bond.

The polarity of a bond may also be represented by a **dipole moment** arrow. These arrows are vector arrows that have a plus sign at one end and the arrow points to where electrons are pulled:

Using Dipole Arrows to Indicate Polar Bonds

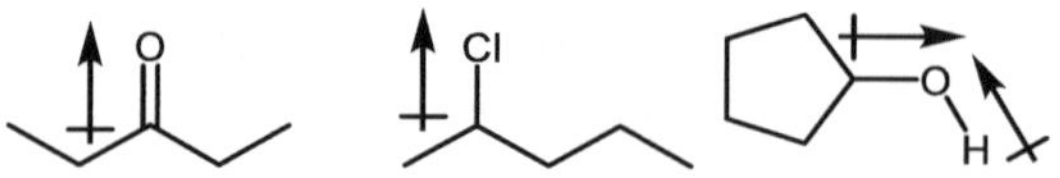

Example I.2.1.

Fill in the δ+ and δ– symbols for any polar bonds in the structures below.

A) 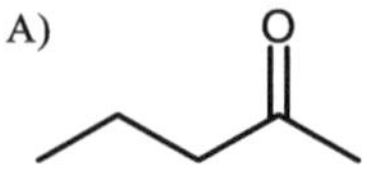 B) 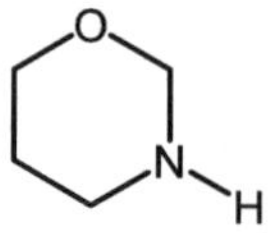 C) 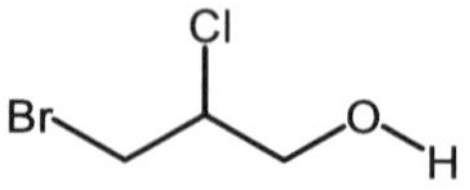

Solution I.2.1.

A) 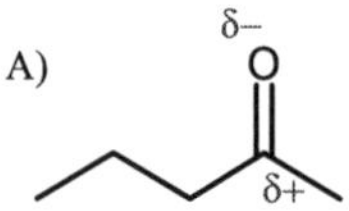 B) 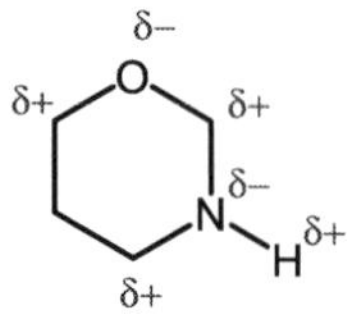 C) 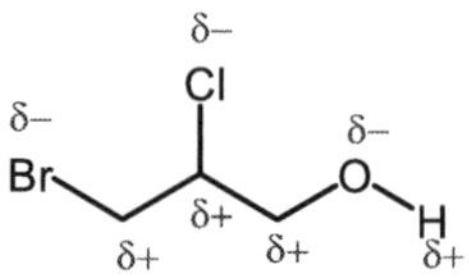

Lesson I.2.2. Dipole Moments of Molecules

One or more polar bonds may be present in a molecule, and each of the polar bonds that is present may pull electrons in a different direction. The sum of all these vectors representing electron-pulling forces in their respective directions is equal to the **dipole moment of that entire molecule**. If there are two bonds of equal polarity, but the two are oriented such that they pull electrons in opposite directions, then the molecule as a whole is not polar. An example is carbon dioxide (CO_2):

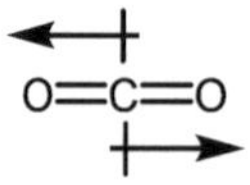

The vector arrows have equal magnitude and point in opposite directions. They therefore cancel one another out. This is not a polar molecule, even though there are polar bonds present.

In cases where there are two dipole vectors pointing in non-opposing directions, the sum of the vectors can result in a dipole moment for the molecule that points in a direction in which neither of the polar bonds is pointing. An example is in water:

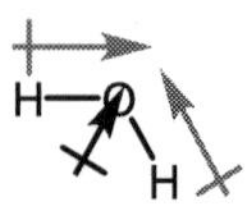

The small black vector is the dipole moment for the molecule. The larger grey arrows represent the dipole for each bond. The sum of bond vectors is the dipole moment of the molecule.

Example I.2.2.

Fill in the molecular dipole moment for each molecule:

A) 

B) 

C) 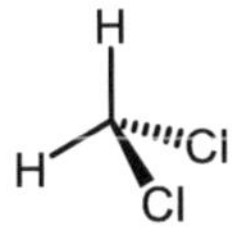

D) 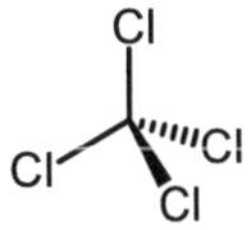

Solution I.2.2.

A) 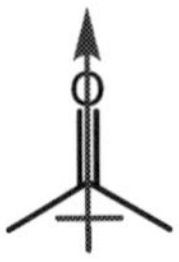

B) 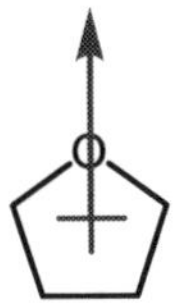

C) 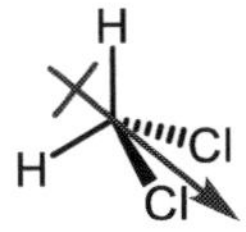

D) 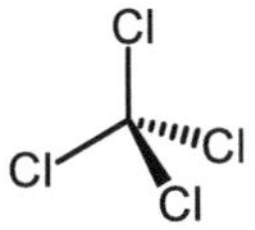

No molecular dipole despite polar bonds; the pull in all four directions cancel out!

Lesson I.2.3. Formal Charge

In addition to the partial charges discussed above, some atoms can take on a full positive or negative charge, as you have no doubt seen in ions in your prior courses in chemistry. It is important to know how to assign formal charges in organic chemistry. The formula for formal charge of an atom in a structure can be given as follows:

Formal charge =
[# of valence e^- in neutral atom] – [# nonbonding e^- on the atom] – [number of bonds to that atom]

Consider carbon in several different structures:

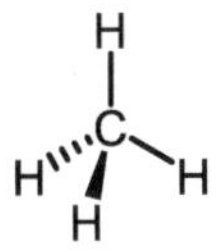 (valence – nonbonding electrons – bonds) (4 – 0 – 4) = formal charge of 0 (zero) on C	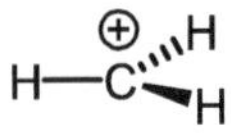 (valence – nonbonding electrons – bonds) (4 – 0 – 3) = formal charge of +1 on C
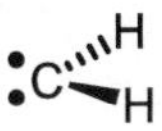 (valence – nonbonding electrons – bonds) (4 – 2 – 2) = formal charge of 0 (zero) on C	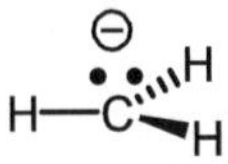 (valence – nonbonding electrons – bonds) (4 – 2 – 3) = formal charge of -1 on C

In the structures above, all the bonds and non-bonding electrons are shown. A neutral carbon atom has four valence electrons. One simply plugs the values into the equation and the formal charge is readily determined. It is vitally important to determine which atoms in a molecule have partial or formal charges, so that we can begin to predict the properties and reactions of molecules as a result of Coulombic attraction or repulsion. Learning the ability to make such predictions forms the majority of an introductory organic chemistry course.

<u>Example I.2.3.</u>

Provide all non-zero formal charges for atoms in these structures:

A) 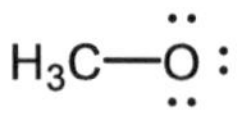

B) 

C) 

D) 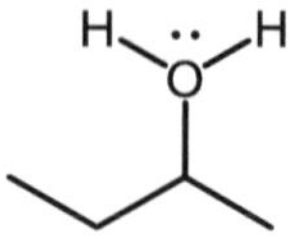

<u>Solution I.2.3.</u>

A) 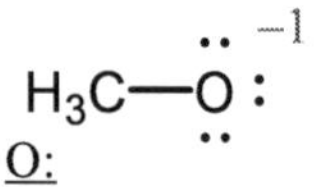

<u>For O:</u>
6 valence – 6 lone pair electrons – 1 bond
= F.C. of –1

B) 

<u>For C:</u>
4 valence – 2 lone pair electrons – 3 bonds
= F.C. of –1

C) 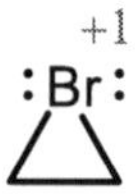

<u>For Br:</u>
7 valence – 4 lone pair electrons – 2 bonds
= F.C. of +1

D) 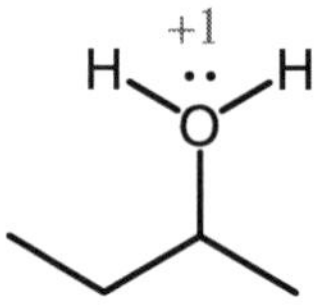

<u>For O:</u>
6 valence – 2 lone pair electrons – 3 bonds
= F.C. of +1

Lesson I.3. Intermolecular Forces, Boiling Point/Melting Point and Ionic vs. Covalent

Lesson I.3.1. Intermolecular Forces

There are three types of intermolecular forces on which we will focus in this book:

i. Hydrogen Bonding
ii. Dipole–Dipole interactions
iii. van der Waals interactions (London dispersion forces).

The **strongest** of these intermolecular forces is **hydrogen bonding** (sometimes abbreviated "H-bonding"). Hydrogen bonding is a specific type of Coulombic attractive force between the partial negative charge of one polar bond and the partial positive charge on an H atom (induced by the H atom being in a very polar bond):

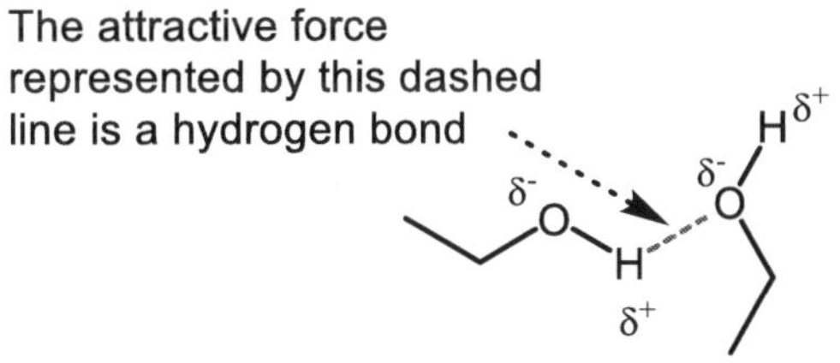

The greater the charges on H and X in the molecule, the greater the Coulombic attraction will be, so the two molecules will be more strongly attracted to one another. The elements with the three highest electronegativity values are F > O > N. Consequently, the most polar H–element bonds occur in molecules in which hydrogen is bonded to one of these elements. Molecules capable of the strongest H-bonds all feature H–F, H–O and H–N bonds. For the purposes of organic chemistry courses, you will typically see compounds with H–O and H–N bonds engaging in hydrogen bonding.

The **second strongest** of the intermolecular forces we will cover are **dipole–dipole interactions**. Dipole–dipole interactions are simply the attractive forces between a partial positive charge on an atom in one molecule and the partial negative charge on an atom in another molecule. The greater the partial charges, the stronger the Coulombic attraction between the two:

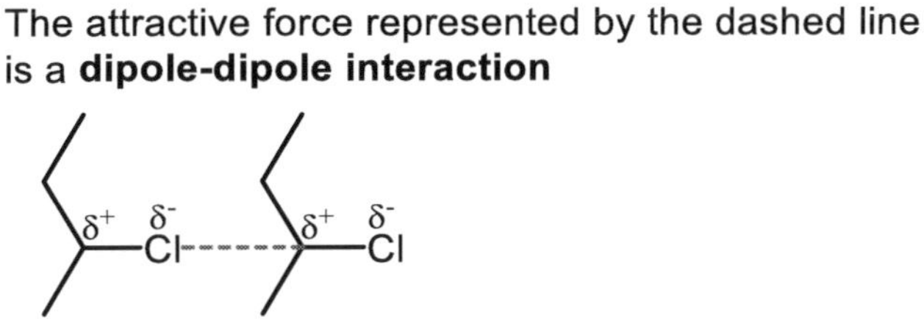

The weakest of the intermolecular forces we will cover are van der Waals interactions, sometimes called London Dispersion Forces. The van der Waals interactions arise from temporary dipoles that form in molecules that lack polar bonds. When two such molecules come in close proximity, the electron clouds on the atoms in one molecule repel the electron clouds on the atoms in the other molecule. This will, very briefly, cause a very weak 'induced dipole' on the molecules whose electrons were repelled. These forces, however, are far weaker than either dipole–dipole interactions or H-bonds.

Lesson I.3.2. Using Intermolecular Forces to assess Relative Melting and Boiling Points

In a solid, it is the intermolecular forces that hold molecules close together with enough force to allow the solid to maintain its shape. A solid will only melt into liquid form when the intermolecular forces are disrupted enough to allow the molecules to flow past one another. To disrupt intermolecular forces, one needs to add energy, generally by heating the sample to its melting point. *The stronger the intermolecular forces, the more heat (higher melting point) is needed* to melt the sample. This observation allows us to predict the relative melting points of several samples simply by comparing their structures.

The boiling point of a liquid can be estimated in a similar way. In a liquid, the molecules remain close to one another even though the molecules can flow past one another. When enough energy is added to boil the liquid, the intermolecular forces are completely overcome, allowing the molecules to break free into the gaseous state, whereupon the molecules too far apart to experience intermolecular forces between one other. *The stronger the intermolecular forces, the more heat (higher boiling point) is needed* to boil the sample.

Example I.3.1.

List the strongest intermolecular force that is present between molecules in a sample of each of the following and rank the compounds 1–4 in terms of boiling point, 1 being highest.

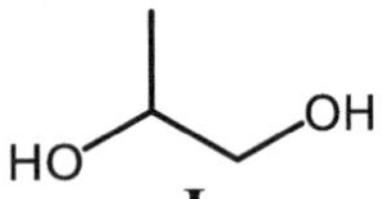

I

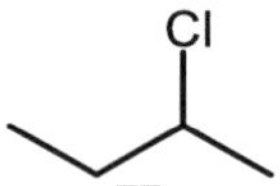

II

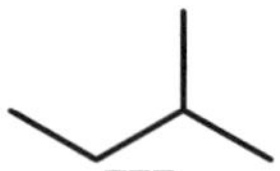
III

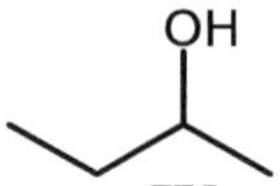

IV

Solution I.3.1.

The strongest intermolecular force in each is:

I: H-bonding (two sites)

II: dipole-dipole

III: van der Waals interactions (London dispersion forces)

IV: H-bonding (one site)

The boiling point increases as the strength of the force increases, and compound I has *two* H-bonding units whereas compound IV only has one. So, the order of boiling point is **I** > **IV** > **II** > **III**, where compound **I** has the highest boiling point.

Lesson I.3.3. Effect of Molecular Weight and Branching on Boiling and Melting Point

We have seen that the type of intermolecular forces in a sample has a dramatic effect on the boiling and melting points. What if we have to compare two samples that have the same type of intermolecular forces? For example, hydrocarbons are nonpolar, so they only have dispersion forces. To compare the b.p. or m.p. of two hydrocarbons, we need additional rules. First, **the greater the molecular weight** of the molecule, **the higher the boiling point (b.p.) or melting point**:

Molecule	b.p. (°C)
C_4H_{10} 	**–0.5**
C_5H_{12} 	**36.1**
C_6H_{14} 	**68.7**
C_7H_{16} 	**98.4**
C_8H_{18} 	**125.7**

Second, **the larger the surface area the higher the boiling point and melting point**. This means that the linear molecules have higher b.p./m/p. compared to branched molecules of the same mass:

C_5H_{12}		
	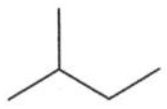	
b.p =36.1°C	b.p =27.9°C	b.p =9.5°C

I.3.4. Ionic versus Covalent Compounds

Sometimes the difference in the electronegativity of two atoms bound to one another is so large that when the compound is placed in a solution one of the atoms pulls all of the electrons away from the other. The result is an anion (X^-) and a cation (M^+):

$$MX \longrightarrow M^+ + X^-$$

Such compounds are called ionic compounds. You probably learned the difference between ionic and covalent compounds in a previous chemistry course, so here we just review how this is relevant to organic chemistry. It is relatively easy to identify most of the ionic compounds you will see in organic chemistry because **ionic compounds have a metal-nonmetal bond**. In a first semester organic chemistry course, the **metals are usually Li, Na and K**. So, all of these formulae represent ionic compounds that will dissociate into ions in solutions: NaOH, NaH, $KOCH_3$, KI, KCN, $LiOC_2H_5$, $NaNH_2$, $NaOOCCH_3$, etc. These will all dissociate as indicated below. Note that it is the nonmetal that is directly attached to the metal that gets the negative charge when the compound dissociates:

$$NaOH \longrightarrow Na^+ + {}^-OH$$

$$NaH \longrightarrow Na^+ + H^-$$

$$KOCH_3 \longrightarrow K^+ + {}^-OCH_3$$

$$KI \longrightarrow K^+ + I^-$$

$$LiOC_2H_5 \longrightarrow Li^+ + {}^-OC_2H_5$$

$$KOCH_3 \longrightarrow K^+ + {}^-OCH_3$$

$$NaNH_2 \longrightarrow Na^+ + {}^-NH_2$$

$$NaOOCCH_3 \longrightarrow Na^+ + {}^-OOCCH_3$$

acetate

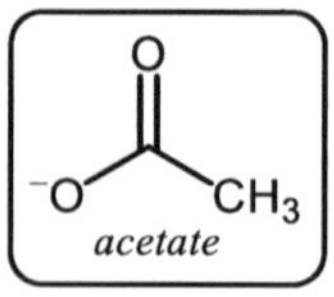

Strong acids will also dissociate into ions, and the strong acids you will predominantly see in organic chemistry will be H_2SO_4 and HX (X = Cl, Br, I). On the other hand, **other nonmetal-nonmetal bonds are covalent and will not dissociate to form any ions in solution.** Examples of covalent bonds include all of the bonds in the following molecules:

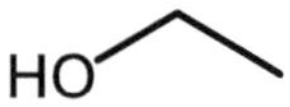

$HOCH_3$

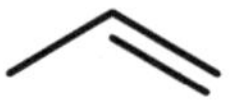

H_2O

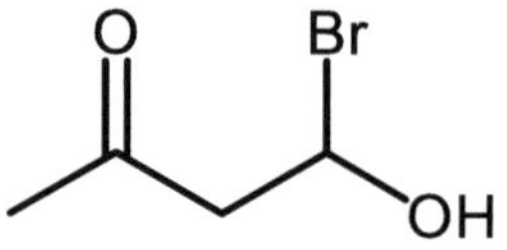

Lesson I.4. Nucleophiles, Electrophiles and Functional Groups

Lesson I.4.1. Nucleophiles and Electrophiles

Thus far, we have seen that intermolecular forces are capable of influencing properties of a pure compound. However, there can also be attractive forces between two *different* chemical species. If this force is strong enough and circumstances are correct, the two species may undergo a **chemical reaction: a rearrangement of atoms and electrons** (possibly bonding electrons) to create one or more new species. How can we begin to understand the chemical changes we see among organic compounds? How can we predict what reactions might happen between two species? These questions are complex, but understanding Coulombic forces is a good first step in our quest to answer them. To begin this task, it is useful to divide chemical species into two classes: **nucleophiles** and **electrophiles**.

A nucleophile (Greek for "nucleus-loving") will be attracted to positive charges, such as nuclei, cations, or the partial positive-charged end of a polar bond. A nucleophile will thus tend to be something with a full or partial negative charge on it, or it may simply have loosely-held electrons like a lone pair. Because a double bond consists of four electrons and a triple bond consists of six electrons, **C=C and C≡C bonds also have an excess of electrons in one spot can be nucleophiles** (we will see a more precise explanation in Lesson I.6). Some examples are shown here:

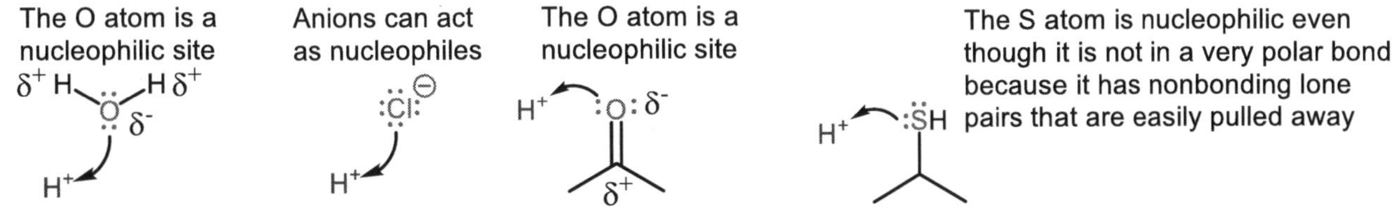

The H nucleus, a proton (H^+) is used as an example of an electron acceptor
Curved arrows show from what site the electrons would be taken to give them to the acceptor.

If we were to compare two nucleophiles to determine which is more nucleophilic (more quickly gives its electrons away), some common sense will guide us. First, an anionic atom gives electrons away more quickly than a neutral atom. Second, a less electronegative atom gives electrons away more quickly than a more electronegative atom. So, in the examples above, the Cl^- is the best nucleophile because it is an anion, but the other nucleophiles are neutral. The sulfur (S) atom is the next-best nucleophile because it is less electronegative than oxygen (O). We do not yet know how to decide which of the neutral O atom examples is more nucleophilic, but we will learn this later on.

An electrophile (Greek for "electron-loving") will be attracted to negative charges such as anions, lone pair electrons, or the partial negative-charged end of a polar bond. An electrophile will thus tend to be something with a full or partial positive charge on it, or something with an atom that has less than an octet of electrons in its valence shell. Some examples are shown here:

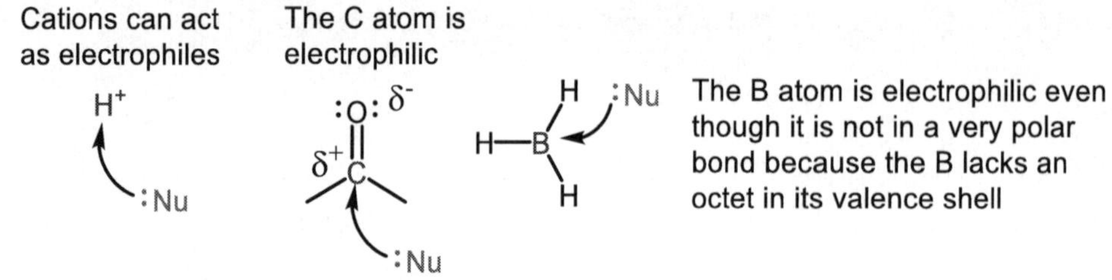

The symbol "Nu" is used to represent a nucleophile.
Curved arrows point to the electrophilic site to which electrons could be donated.

By filling in the partial charges on the atoms in a molecule, we identify the likely sites to which an electrophile or a nucleophile might be attracted. An example of this type of analysis is provided in Example I.4.1. We will do many of these assessments throughout the course.

Example I.4.1.

Identify each molecule as a potential nucleophile, electrophile, neither, or both. Draw arrows pointing towards electrophilic sites and draw arrows pointing away from nucleophilic sites.

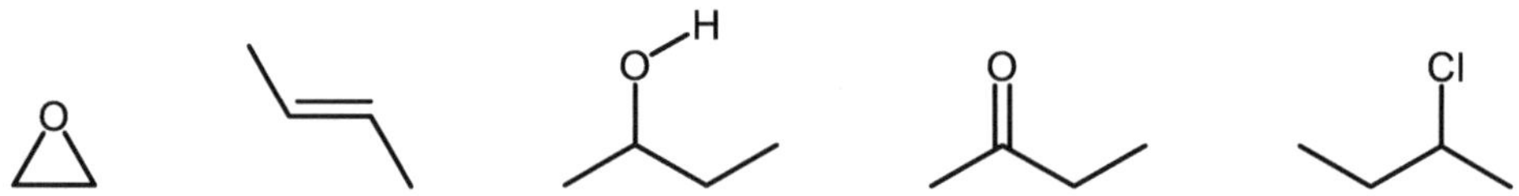

Solution I.4.1.

Each atom with a lone pair may act as a nucleophile. Each atom that is the partial positive end of a polar bond is electrophilic. It is helpful to highlight the partial charges, C=C and C≡C bonds, and fill in the lone pairs so that we do not miss any possibilities for electrophilic or nucleophilic sites:

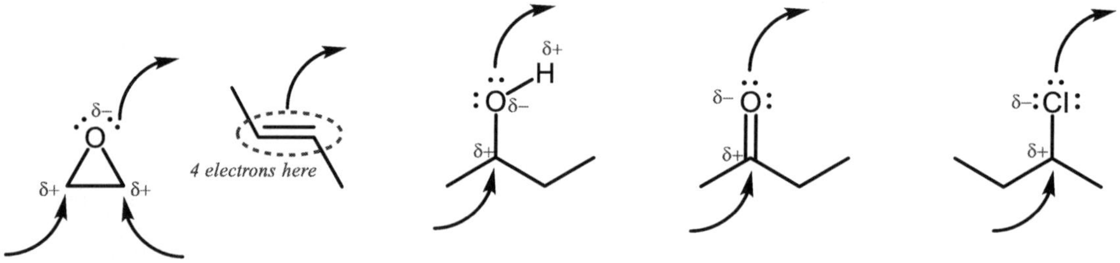

Lesson I.4.2. Functional Groups

We now have the ability to identify sites in molecules between which there may be attractive forces, and we have also begun to rationalize how these attractive forces could lead to a chemical reaction. With these skills, we can understand that a C=O bond might generally interact with a nucleophile in a consistent way, regardless of which specific molecule contains that C=O bond. You would expect the partial positive charge on the C of the C=O unit to attract nucleophiles, for example. You might also expect the partial negative on the O in the O–H bond to attract electrophiles, regardless of the molecule on which the OH group resides. We can make generalizations about how a particular **group** of atoms might **function**. Organic chemists have identified specific groups of atoms that do react in predictable manners within a wide range of molecules. These predictably-reacting groups are called **functional groups**. The structures and names of the most important functional groups are shown in the following diagram. The "R" in these generic structures may be any chain or ring composed of C and H atoms, attached to where the "R" group is placed.

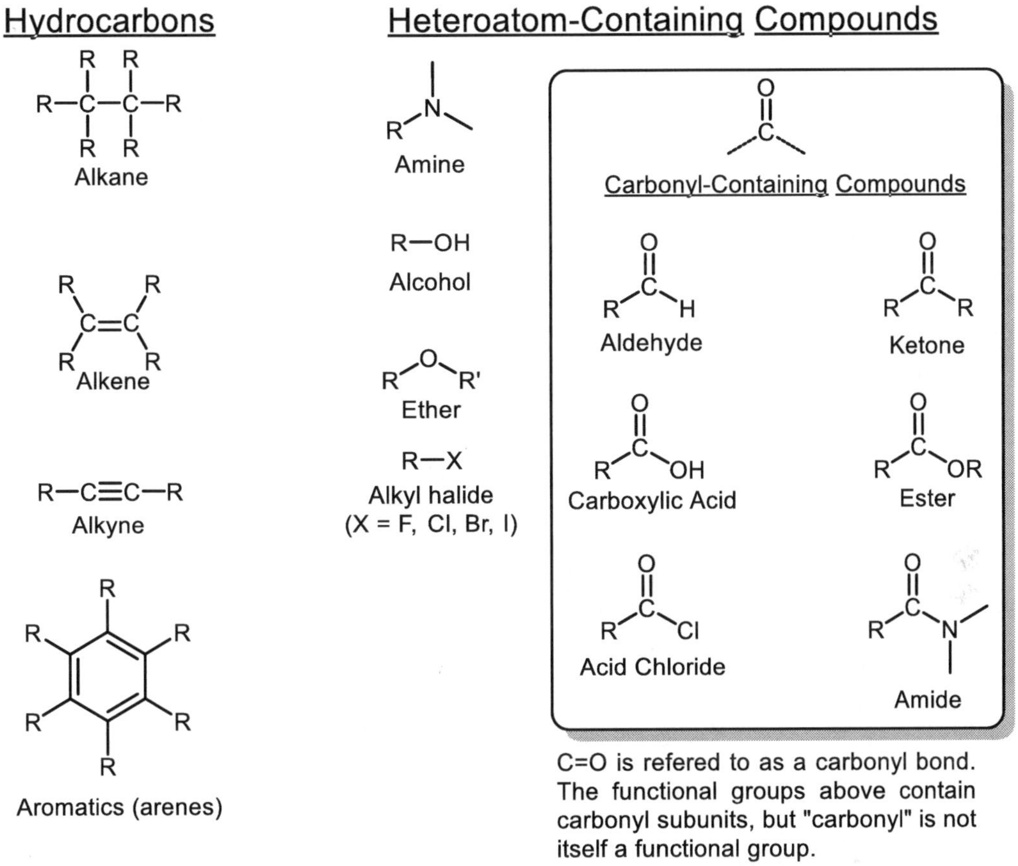

C=O is refered to as a carbonyl bond. The functional groups above contain carbonyl subunits, but "carbonyl" is not itself a functional group.

It is important to be able to look at a complex molecule and rapidly identify all such functional groups that are present. Below is an example of an anti-cancer drug with its functional groups labeled:

Taxol (a complex cancer drug) without (left) and with (right) functional groups labeled

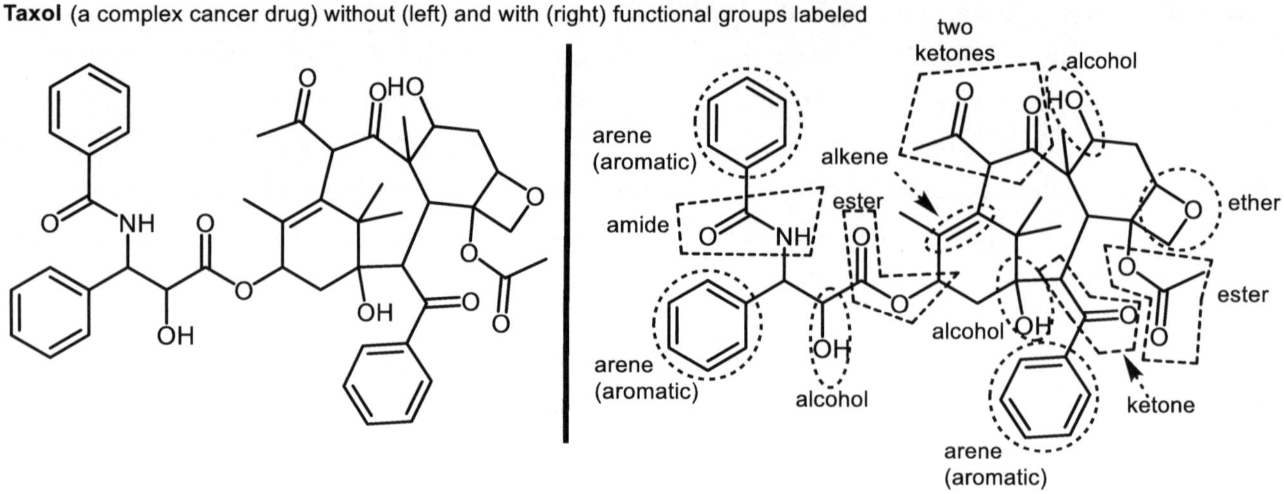

Example I.4.2.

Identify all non-alkane functional groups in this molecule by circling them and writing in the group's name beside each.

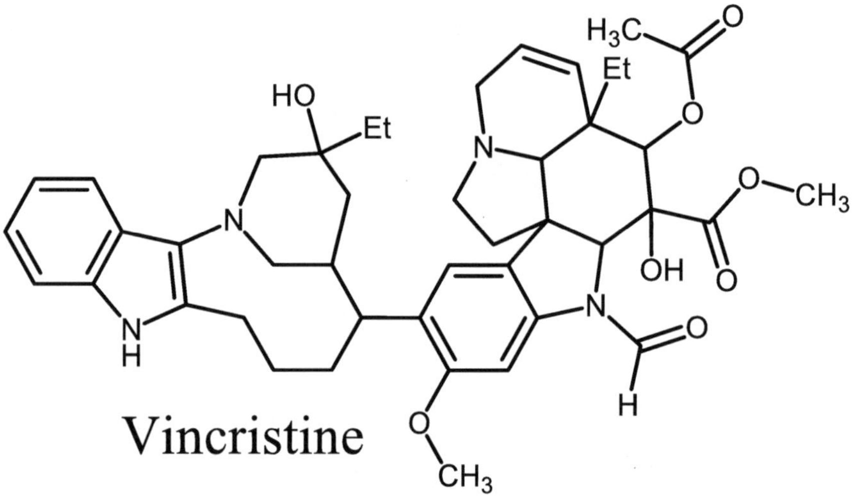

Vincristine

Solution I.4.2.

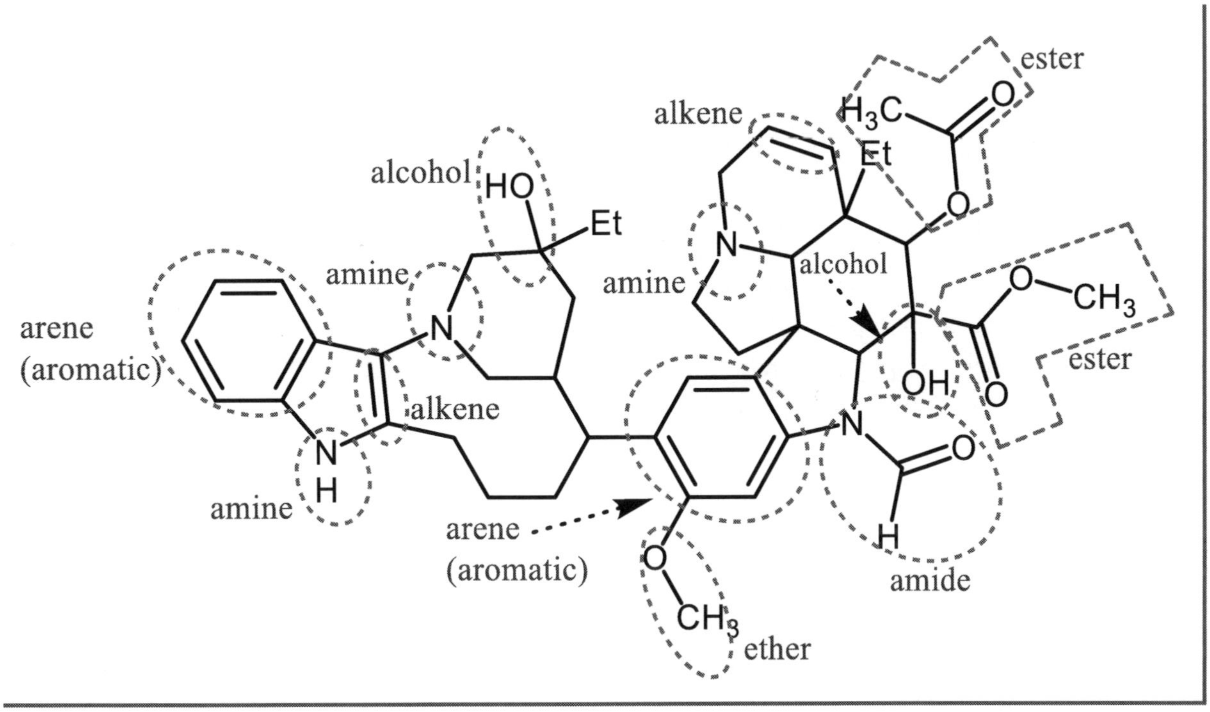

Lesson I.5. Arrow-Pushing Formalism

Lesson I.5.1. Curved Arrows

Each covalent bond is made of two electrons. In a chemical reaction, bonds are broken and/or made. Sometimes several bonds are broken/made simultaneously, such that a complex rearrangement of atoms and electrons takes place all at once. To represent these rearrangements, chemists have developed the **Curved Arrow Formalism**. In this convention, one uses arrows to point to where the electrons go. Most reactions in organic chemistry involve the movement of pairs of electrons. A curved arrow thus represents movement of a pair of electrons to wherever the arrow points. These arrows must be curved, to distinguish them from the straight arrows used in chemical equations (e.g., A + B → C + D). Here are a few examples of how the curved arrow formalism is used:

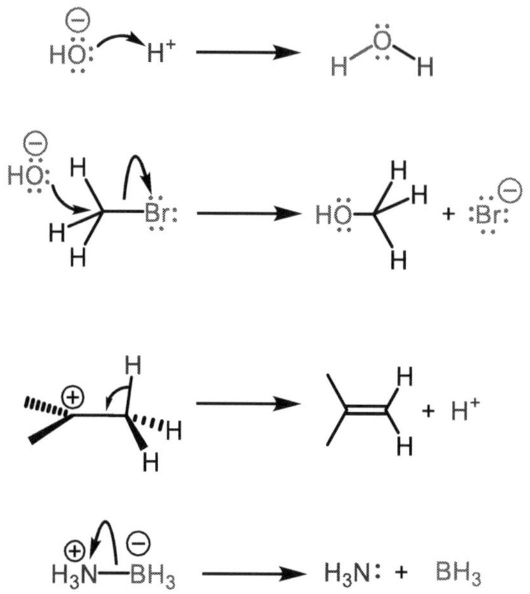

If only one electron moves, we instead use a "fishhook arrow", which has only half an arrow head to point to where the electron goes. Here are a couple examples using this fishhook arrow notation:

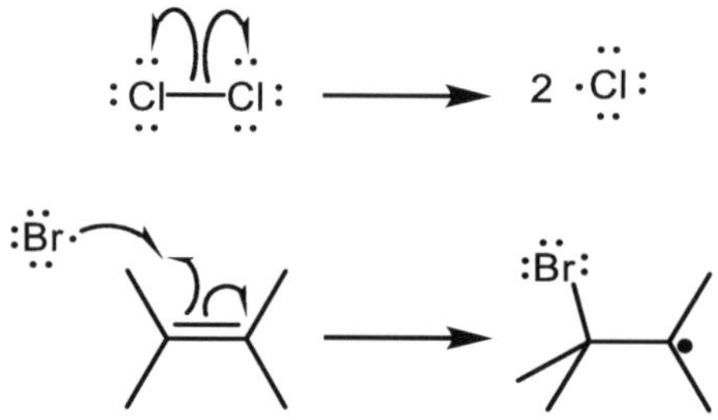

Soon, we will study specific patterns for electron movement and eventually be able to make predictions about what reactions happen. For now, focus on understanding how the arrows work.

Lesson I.5.2. Representing Reactions

We now have a considerable amount of knowledge that we can put together:

1) the curved arrow formalism allows us to represent movement of electrons

2) the ability to assign partial charges and formal charges allows us to identify sites to which electrons might be attracted and from what sites they may be pulled

3) we can identify species as being electrophilic (pull electrons towards them) or nucleophiles (they give electrons away)

Consider the observation that an acid, represented as H^+, and a base, represented as HO^-, react with one another to form water. Here are the Lewis structures of the reactants and the product:

We see that the product of this reaction is a combination of the OH and the H, with the new bond between the O and the H. The positive charge on H^+ tells us that it is electrophilic and will pull electrons towards it (our curved arrow should point to it). Conversely, the negative charge on HO^- suggests that it is nucleophilic and will donate electrons. The movement of electrons in the reaction is represented as:

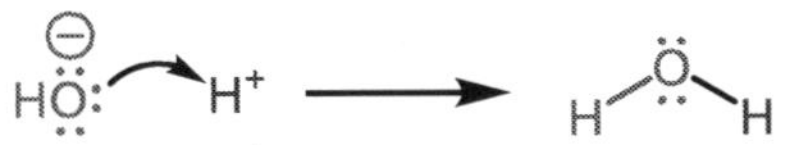

From the above representation of the reaction, we can say that the proton (H^+) acts as an electrophile, accepting electrons from the nucleophilic hydroxide (HO^-). When the O atom of the nucleophile shares electrons with the H atom in the electrophile, a new covalent bond between O and H results. The formal charges on the O (–1 in the reactant) and the proton (+1 in the reactant) change to 0 in the.

For the more complex example shown below, we again start by identifying the bonds made/broken:

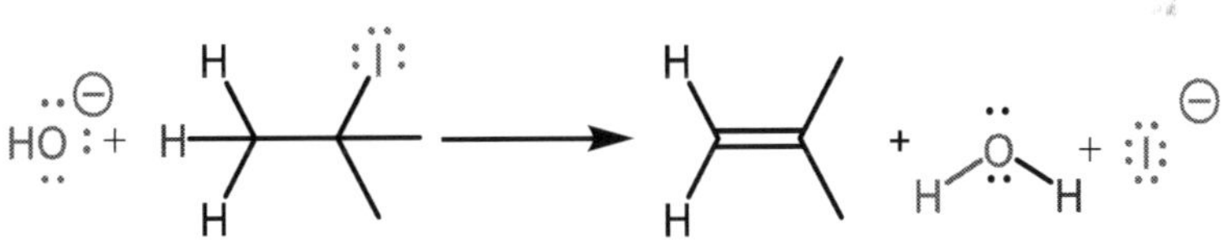

Bonds made: HO to H (to get water), second bond between two carbons.

Bonds broken: H–C bond (left side), C–I bond.

Next, evaluate changes in charge. The O in HO^- starts negative and ends neutral, so it must donate electrons away. The iodine begins neutral and ends negative, so electrons must be pushed onto it. The most reasonable way to accomplish all of these things in a single step is to move the electrons as follows:

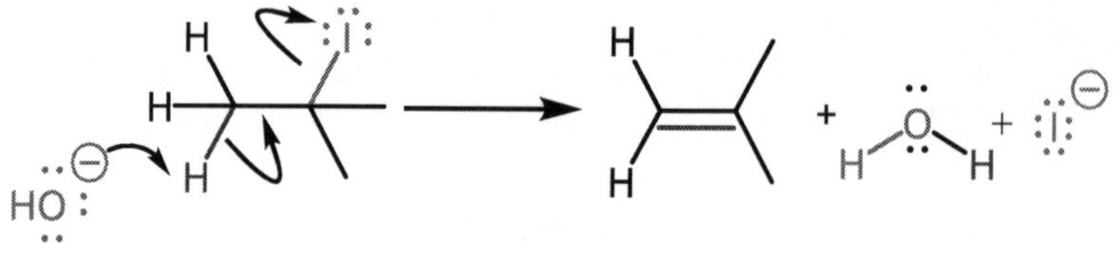

In the two examples above, we were given a set of reactants with a known product, and from that information we determined how the electrons would have to flow in order to accomplish the reaction. We can also predict products if we know how the electrons move. Consider these reactants and the accompanying curved arrows:

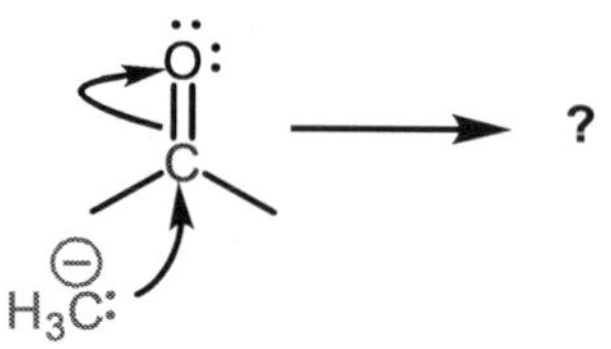

In this case, we can predict the products by filling in any new bonds formed, removing any broken bonds, and noting any changes in formal charge. The lower arrow indicates that the CH_3 will share a lone pair with the C of the C=O bond, leading to a new C–C bond. The upper arrow shows that the two electrons in one of the C=O bonds gets localized onto the O, so the O goes from having two lone pairs to three, and it acquires a formal chare of –1.

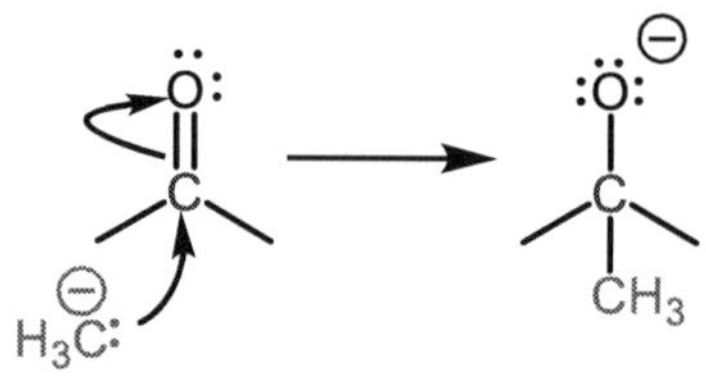

Note that, at this point, we are simply representing how electrons pairs would move about to yield a given molecule or, if we are told how the electrons will move, we can provide the products. We are not yet armed with the tools we need to *predict* what reaction will take place or to decide whether a given hypothetical reaction is reasonable. These skills will come in the following lessons.

Example I.5.1.

For 1–5, fill in the missing products of the indicated electron movement. For reactions 6–11, fill in arrows to show the flow of electrons needed to change reactants to the indicated product(s).

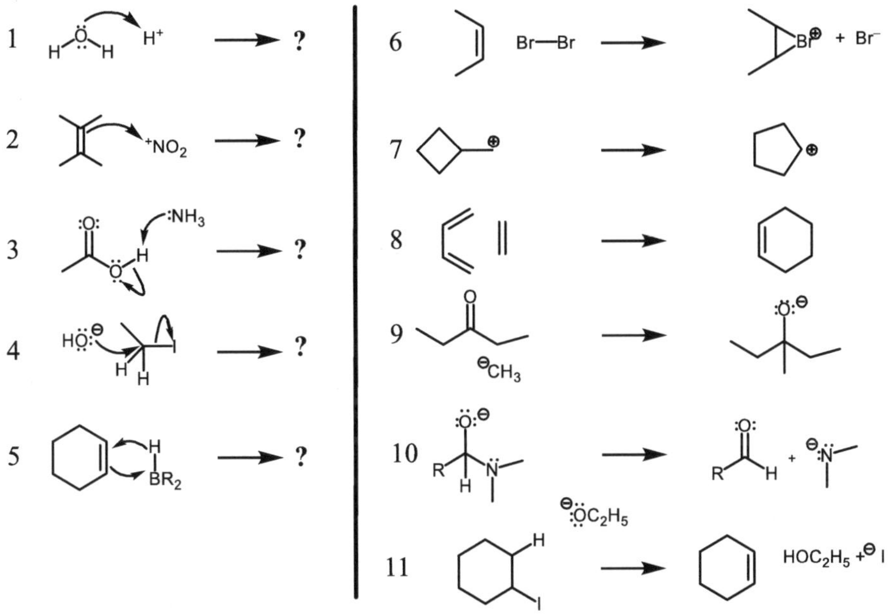

Solution I.5.1.

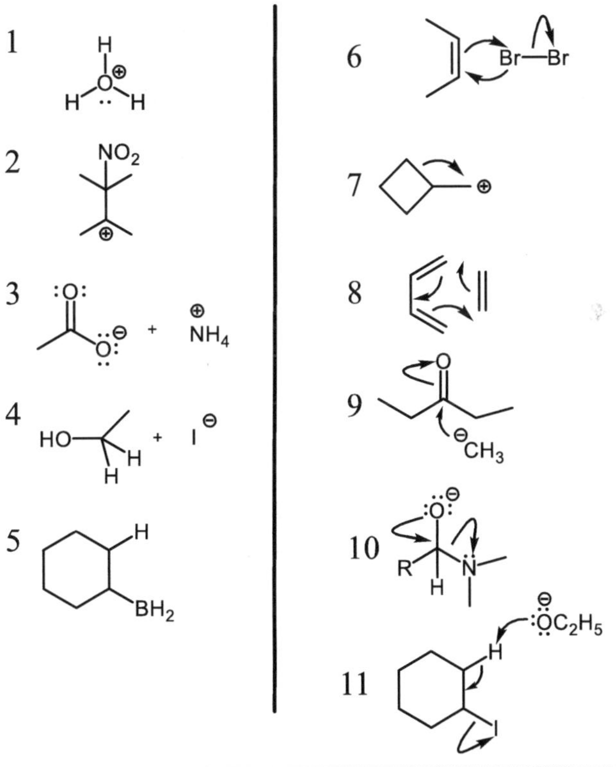

Lesson I.6. Hybridization, Sigma and Pi Bonds, Lone Pairs and Bond Geometry

Lesson I.6.1. Hybridization

Each element in the second row of the periodic table has one 2*s* orbital and three 2*p* orbitals in its valence shell. The valence shell electrons are the only ones we show in Lewis structures because they are the only ones involved in bonding and chemical reactions. To accomplish the most effective bonding and minimize repulsion between bonding pairs, the *s*- and *p*-orbitals present on *individual atoms* mix to form hybrid orbitals in *molecules*. The *s* orbital is always mixed into hybrid orbitals. The number of *p*-orbitals that mix with the *s*-orbital on a given atom depends on the sum of lone pairs and atoms attached to that atom as follows:

If (lone pairs + atoms attached) = 2 **then** hybridization = *sp*

If (lone pairs + atoms attached) = 3 **then** hybridization = sp^2

If (lone pairs + atoms attached) = 4 **then** hybridization = sp^3

Note that the total number of orbitals ($s + p$) that mix to form the hybrid orbitals is equal to (lone pairs + atoms attached).

An *sp*-hybridized atom has two *sp*-orbitals on it. To minimize repulsion between the electrons in these orbitals, the orbitals are oriented 180° apart, in a linear geometry:

The two *p*-orbitals that were not incorporated into the hybrid are still on the C:

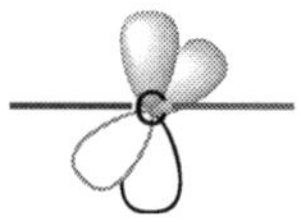

The hybrid orbitals may hold lone pairs or form bonds. The type of bond formed by a hybrid orbital involves electron density directly between the two atoms involved in the bond. These are called σ-bonds (σ is the lowercase Greek letter "sigma"). There can be no more than one sigma bond between any two given atoms, and each single bond is a σ bond.

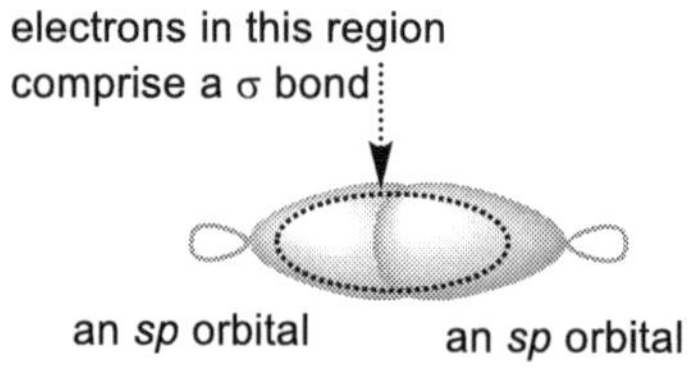

Each *p*-orbitals that does not mix into the hybrid orbital may interact with a *p*-orbital on an adjacent atom, forming a bond with electron density above and below the internuclear axis. Such a bond is called a π-bond (π is the lowercase Greek letter "pi"). Each bond beyond the first bond between two atoms (which is a σ bond) is a π bond. So, a double bond is made of one σ- and one π-bond

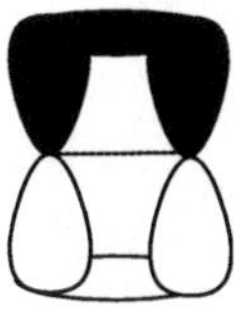

The *sp*-hybridized atom in particular has two *p*-orbitals on it, so it can make two π-bonds.

Molecules with sp hybridized atoms (bold):

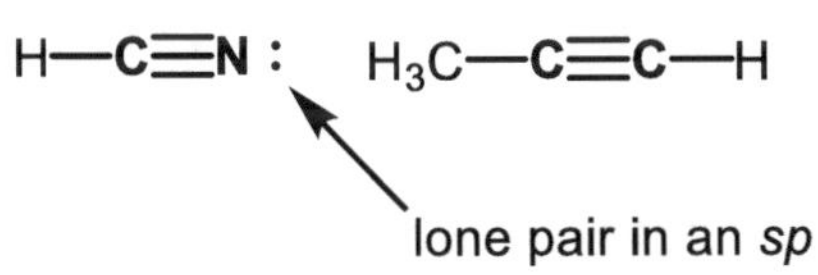

each triple bond is made of one σ bond involving an *sp* orbital and two π bonds using remaining *p* orbitals

It is important to know whether electrons are in a lone pair, σ-bond or π-bond because, for a given atom, lone pair electrons are held less strongly (attraction to only one nucleus) than bonding electrons (attracted to two nuclei). Furthermore, the π-bond electrons are held less tightly than the σ-bonding electrons. This knowledge will help us predict which electrons might be the easiest to pull away from a molecule to interact with an electrophile. Thus, for a given atom**, lone pair electrons would be easiest to pull away, followed by π-bonding electrons, and σ-bonding electrons would be most difficult to remove**. We would therefore predict that, if H^+ (an electrophile) interacts with an alcohol, it would most likely interact with the lone pair electrons on O. If H^+ interacted with an alkene, the electrophile would pull the π-bond electrons from the alkene. If H^+ interacted with an alkane, the σ-bond electrons might be too tightly held to be pulled away by the H^+. These simple predictions all turn out to be true!

An *sp*^{2}-hybridized carbon atom has three *sp*^{2}-orbitals on it. To minimize repulsion between electrons in these orbitals, the orbitals are 120° apart, in a trigonal planar geometry:

There will be one unhybridized *p*-orbital, so an *sp*^{2}-hybridized atom can make three σ-bonds and one π-bond. Here are some examples of molecules which incorporate *sp*^{2}-hybridized atoms:

Molecules with *sp*^{2} hybridized atoms (bold):

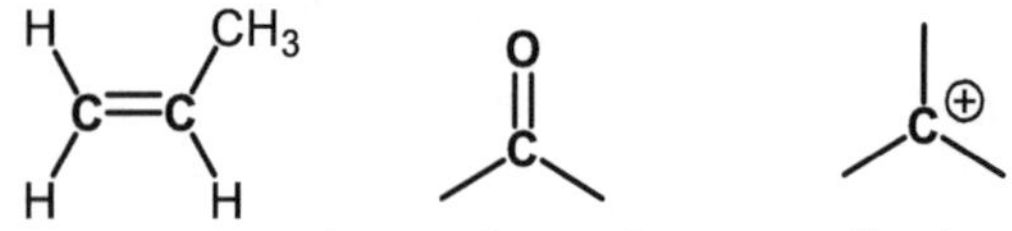

lone pairs on O are in *sp*^{2} orbitals

cation has an empty *p* orbital

each double bond is made of one σ bond involving an *sp*^{2} orbital and one π bond using the unhybridized *p* orbital

An *sp*^{3}-hybridized carbon atom has four *sp*^{3}-orbitals on it. To minimize repulsion between the electrons in each orbital, the orbitals are 109.5° apart, in a tetrahedral geometry:

There are no unhybridized *p*-orbitals on an *sp*^{3}-hybridized atom, so it cannot form any π-bonds. Example I.6.1 illustrates how atom hybridization states are identified in more complex molecules.

Example I.6.1.

Identify the hybridization for each atom to which an arrow is pointing:

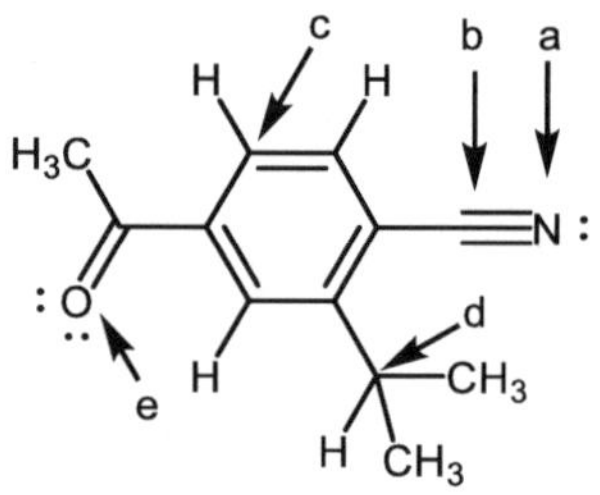

Solution I.6.1.

Nitrogen atom (a) has a lone pair and is bound to one other atom; (lone pairs + atoms attached) = 2, so hybridization = *sp.*

Carbon atom (b) is bound to two other atoms and has no lone pairs; (lone pairs + atoms attached) = 2, so hybridization = *sp.*

Carbon atom (c) is bound to three other atoms and has no lone pairs; (lone pairs + atoms attached) = 3, so hybridization = *sp*^{2}

Carbon atom (d) is bound to four other atoms and has no lone pairs; (lone pairs + atoms attached) = 4, so hybridization = *sp*^{3}

Oxygen atom (e) is bound to one other atom and has two lone pairs; (lone pairs + atoms attached) = 3, so hybridization = *sp*^{2}

Note that lone pairs are not always shown on atoms. Likewise, the H atoms on C are not always shown in the line-bond notation. The structure of the molecule given in Example I.6.1 could also have been provided as:

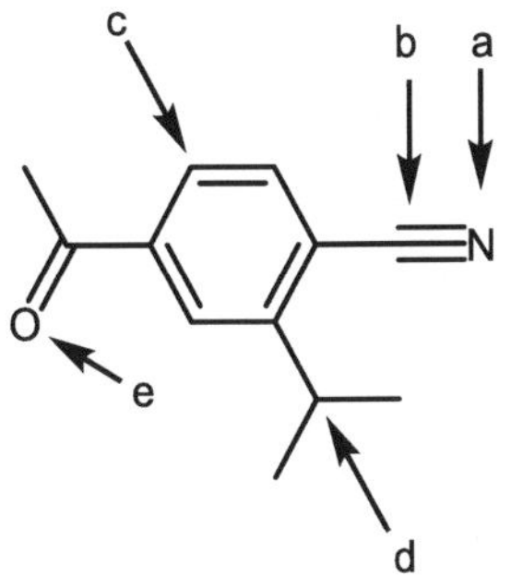

In such a case, you would be expected to use the periodic table and your knowledge of formal charge to fill in the proper number of electrons and H atoms, much like you would if you were trying to draw a Lewis structure from scratch.

Lesson I.6.2. Influence of Hybridization on Electronegativity

We know that electronegativity increases towards the top of the periodic table, where the electrons are in orbitals closer to the nucleus and thus have a stronger attraction to the nucleus. We also know that electrons in an *s*-orbital are closer to the nucleus than electrons in a *p*-orbital, and this is why the *s*-orbital is filled first in the electronic configuration of an atom. These facts make it clear that the orbital holding electrons influences how strongly the nucleus pulls on those electrons. The electronegativity of an atom in a molecule is thus influenced by what orbitals hold the electrons in its valence shell. A hybrid orbital has some fraction of *s*-orbital and some fraction of *p*-orbital. An *sp*-orbital is made of an *s*-orbital and a *p*-orbital, so this hybrid orbital has 50% *s*-character. Only one of the three orbitals mixed to make an sp^2-orbital is *s*, so it has 33% *s*-character. Only one of the four orbitals comprising the sp^3-orbital is *s*, so an sp^3-orbital has 25% *s*-character. Because *s*-orbitals are closer to the nucleus, the greater the "*s*-character", the closer the electrons in that hybrid orbital are to the nucleus, and the more electronegative that atom will be. A good rule of thumb is that an *sp*-hybridized C is similar in electronegativity to an sp^3-hybridized N. We will study the influence of this effect in later Lessons.

Lesson I.7. Resonance and Delocalization Energy

Lesson I.7.1. Representing Resonance using Curved Arrows

Some molecules can be drawn with two or more valid Lewis dot structures, in which the atoms are all connected in the same arrangement, with only the placement of electrons differing. These different forms of the same molecule are called *resonance forms* or *resonance contributors*. As an example, consider the task of drawing a Lewis structure for the anion, $[C_2H_3O]^-$. You could come up with two possible structures in which each non-H atom has its octet filled, the proper number of electrons has been placed, and the necessary overall charge is –1 has been reached:

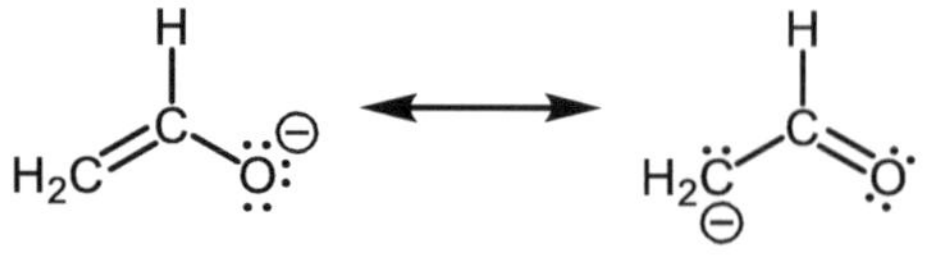

Note that the double-headed arrow used in the above graphic is only used to represent interconversion between two resonance forms of the same structure. This is *not* a chemical reaction, and it is *not* an equilibrium (which is represented using *two* half-headed arrows, each one pointing in opposite directions). Note, also, that the σ-bond framework remains the same in both structures: *only non-bonding and π-bonding electrons can be changed between resonance forms. No σ-bonds break when resonance forms interconvert.* We can use the curved arrow formalism to represent how these two resonance forms can interconvert:

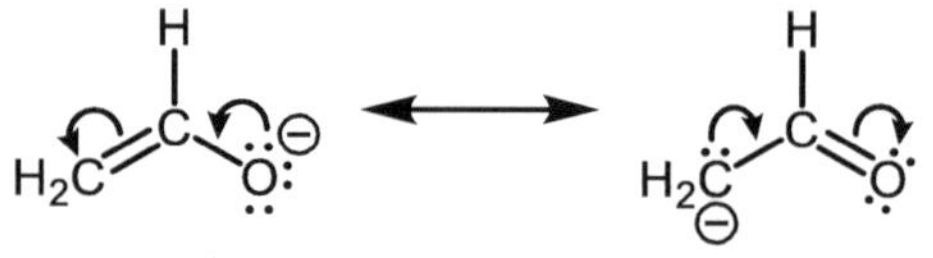

Lesson I.7.2. Resonance Hybrids and Delocalization Energy

So, which of the two possible resonance contributors for this anion is correct? Well, the actual structure of this anion will be some combination of both resonance contributors. The negative charge is spread out (i.e., *delocalized*) over more than one atom in this anion. The real structure is better represented by what is called a *resonance hybrid* of all of the contributors. To draw a resonance hybrid, you have to represent a blend of all possible contributors. To illustrate how this is done, consider the properties of the two contributors for this anion:

i. The σ-bonding skeleton is the same in both contributors, so this skeleton will be the same in the hybrid, and that is what we draw first:

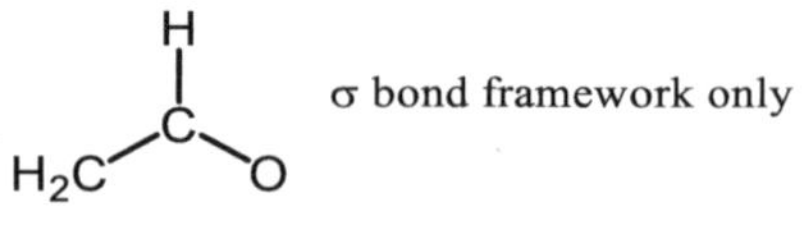

ii. In addition to the σ-bond between C and O, there is a π-bond in contributor on the right. In the hybrid, we *use a dashed line to represent the π-bond that is only in some, but not all, of the resonance contributors* to indicate a partial bond in that position:

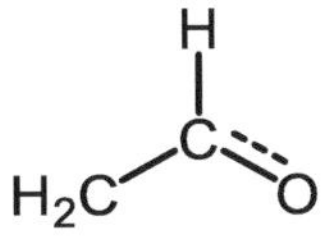

first partial π bond added

iii. In addition to the σ-bond between C and C, there is a π-bond in the contributor on the left, so in hybrid, we again use a dashed line to indicate partial multiple bonding:

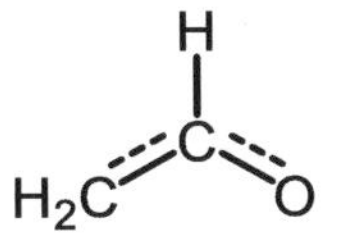

second partial π bond added

iv. Finally, in one contributor the negative charge is on the C and in the other contributor the negative charge is on the O. We therefore use the δ– symbol to show the *partial* occupancy of the negative Coulombic charge on each atom:

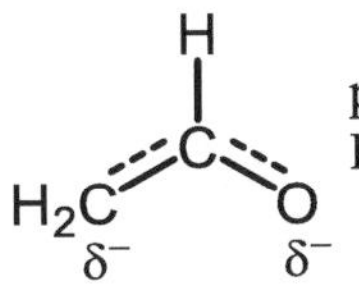

partial charges added;
RESONANCE HYBRID

v. This is one way to represent the *resonance hybrid* of this anion. You may also see resonance hybrids drawn with a curved line drawn near all the atoms involved in the "spread-out" multiple bonding, with the net charge drawn near the center of the curved line. The resonance hybrid for this anion could thus also be drawn like this:

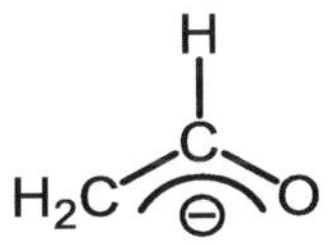

alternative representation of
the same resonance hybrid

This delocalization provides additional stability to a molecule by lowering repulsive forces and/or increasing attractive forces. A pair of electrons that is localized within a smaller volume experiences more Coulombic repulsion than a pair of electrons that is delocalized over a larger volume. The extra stability that is afforded to molecules that have multiple resonance forms is called the *delocalization energy* or *resonance energy*. We will see specific examples of this stabilization in lessons on anion, cation and radical stability later in this book.

Example I.7.1.

Provide TWO additional resonance contributors and the resonance hybrid structure for this cation:

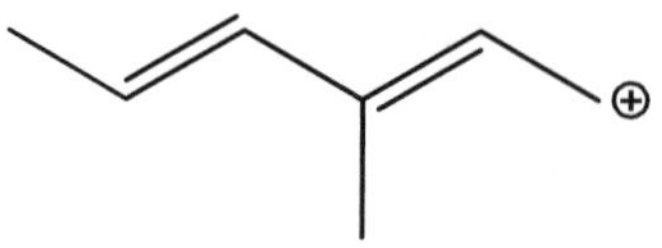

Solution I.7.1.

The resonance contributors (starting with the given structure) are:

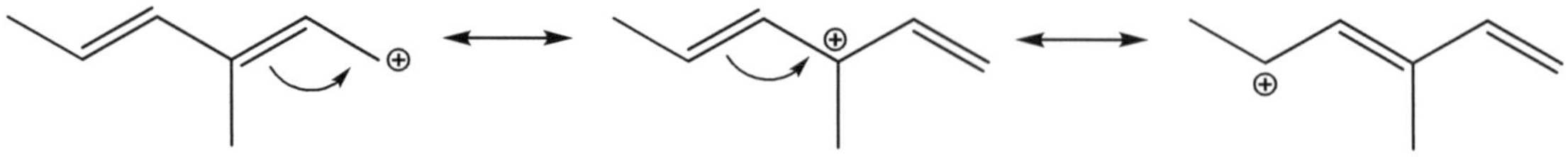

The hybrid will have partial positive charge at each site where there is positive charge in a contributor. Likewise, the hybrid will have π-bond character wherever there is a π-bond in any of the contributors. So, the resonance hybrid will be:

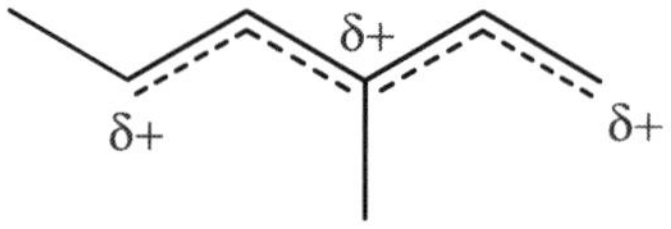

Lesson I.7.3. Influence of Resonance on Hybridization

We now know that, although lone pair electrons generally occupy hybrid orbitals, an exception occurs when a lone pair participates in the resonance delocalization of a π-system. In the case of such delocalization, the lone pair electrons are better represented as occupying an unhybridized *p*-orbital. This will change the assignment of hybridization in some cases. Consider the N atom in ammonia versus the N atom in vinyl amine:

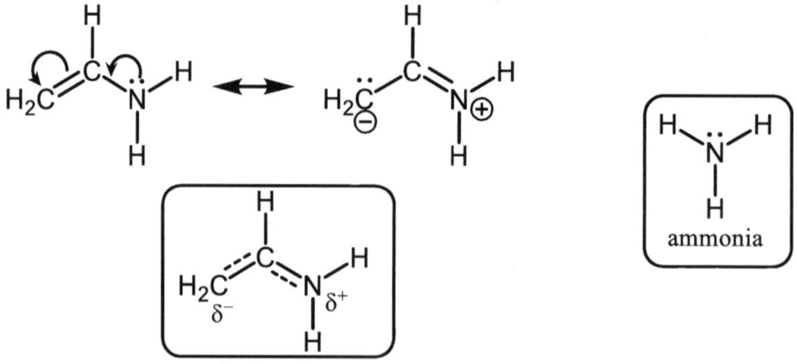

vinyl amine
resonance hybrid

In ammonia, the three pairs of electrons comprising the σ-bonds and the lone pair each occupy a hybrid orbital. There are four pairs of electrons, so four orbitals must hybridize together. The N in ammonia is thus sp^3-hybridized. The contribution of the N lone pair to resonance delocalization into the π-bond in vinyl amine, however, means that these lone pair electrons must occupy an unhybridized *p*-orbital (otherwise they could not be used to make a π-bond!). As a result, only the three pairs of electrons comprising the N–H and N–C bonds will reside in hybrid orbitals, and **the N in vinyl amine is thus better represented as sp^2-hybridized**. From this example, you can see that you must consider possible resonance contributions to the "real" structure prior to doing something as "simple" as assigning hybridization.

Example I.7.2.

Label each C atom in this structure as being *sp*, sp^2 or sp^3-hybridized.

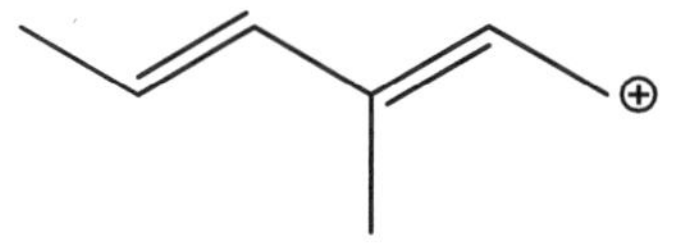

Solution I.7.2.

If we refer back to the solution to Example I.7.2, we will see that the resonance hybrid structure is:

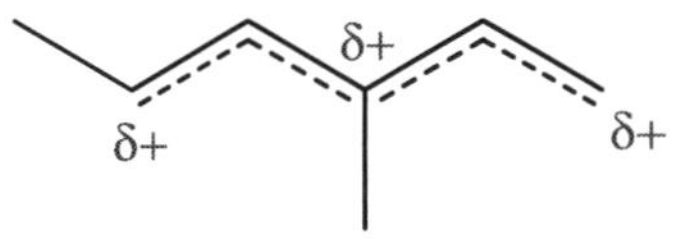

A carbon atom that has exactly three (lone pairs plus σ bonds) is sp^2-hybridized. A carbon atom that exactly two (σ bonds plus lone pairs) is *sp*-hybridized. A carbon atom with exactly four (σ bonds plus lone pairs) is sp^3-hybridized. This analysis leads us to the following assignments (treat dashed lines as half of a π bond):

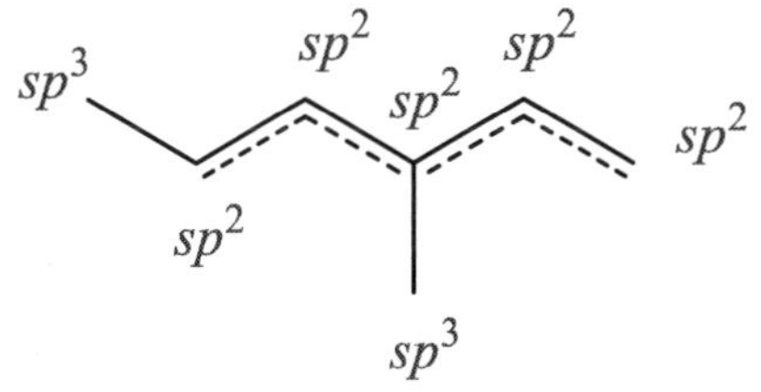

Lesson I.8. Applying the Arrow-Pushing Formalism

Lesson I.8.1. Identifying Improper Arrow-Pushing

As you go through organic chemistry, you will be faced with problems in which you must propose or evaluate mechanisms. A good skill to have as you approach such problems is to be able to recognize impossible arrow-pushing steps. For example, when an **arrow points from a positively-charged atom or towards a negatively charged atom, you should automatically say "that's impossible"**:

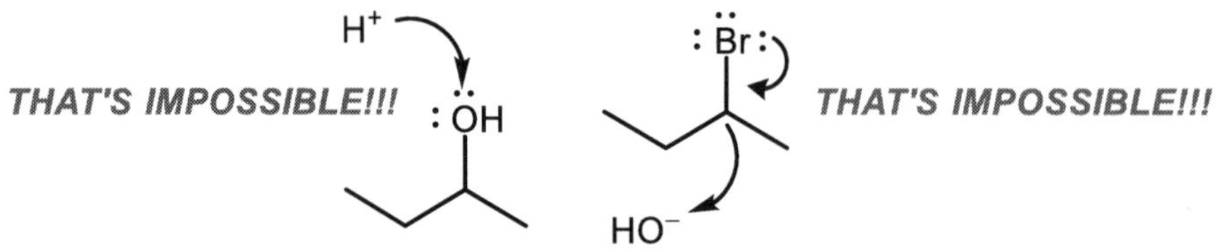

Another common mistake is forgetting about the hydrogen atoms that, although present in the molecule, you are not required to draw in a line-bond representation. Consider this **impossible arrow-pushing**:

The above mechanisms might look fine at first, but you should always **draw in the H atoms on any atom where you are making or breaking bonds**. If we do that in this case, it is obvious that **the proposed arrow pushing is impossible**:

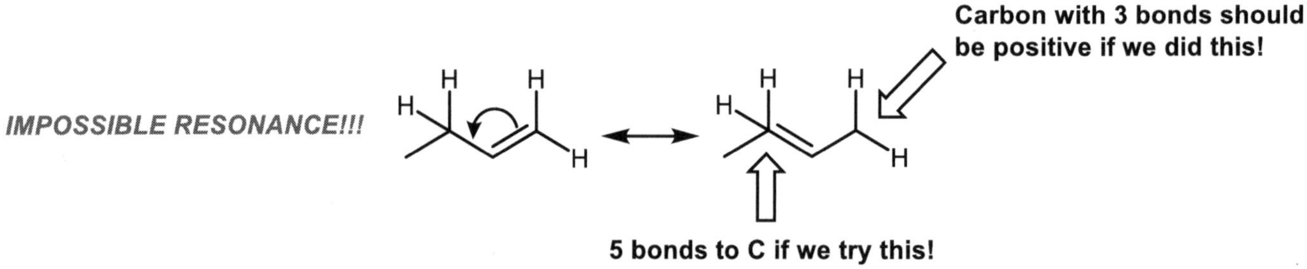

Lesson I.8.2. Common Elementary Steps of Organic Reactions

With our understanding of chemical structure, the curved arrow formalism, and the types of bonds found in typical organic molecules, we are now equipped to investigate chemical reactions in more detail. Organic reactions can involve many arrow-pushing steps along the path from starting materials to products and may include multiple intermediates along the way. Fortunately, a great many organic reactions, even those that occur over many steps, typically involve only a handful of elementary steps. Most such steps can take place either in the forward or reverse direction. As we continue our studies, we will begin to learn how to identify the conditions under which a given step is favored and likely to occur. In this lesson, however, we have only two goals: 1) introduce the names for the most common elementary steps, and 2) acquire the ability to identify which of these steps is occurring if we are shown a potential set of reagents and products.

Six of the basic steps of many organic reactions in introductory organic classes are as follows:

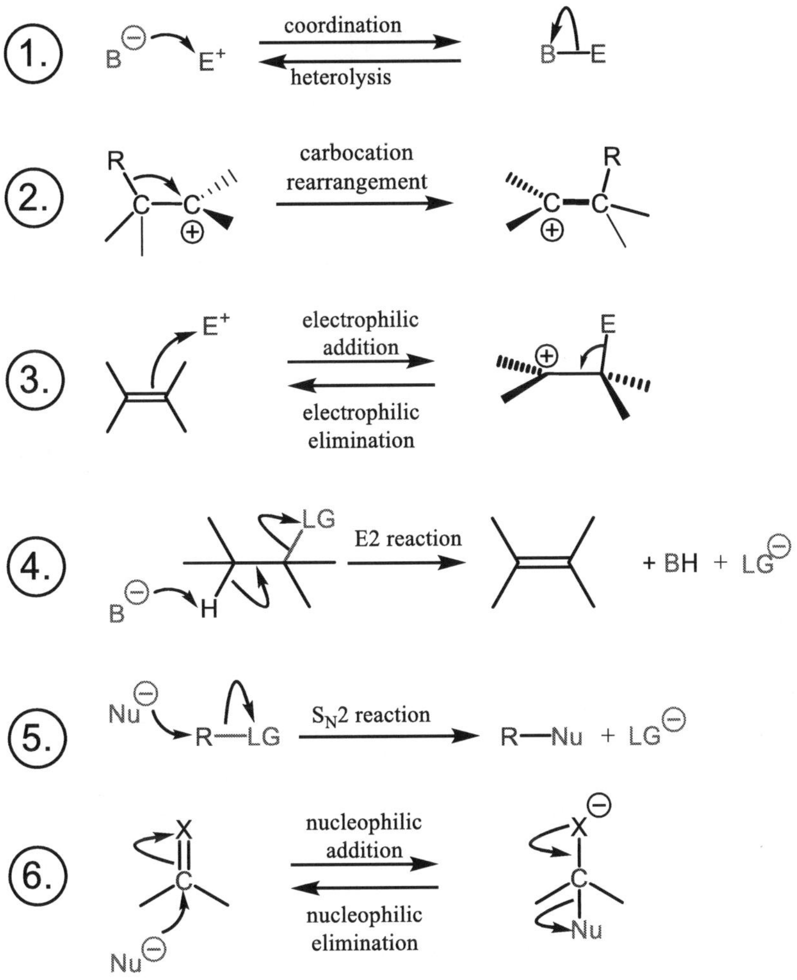

Lesson I.8.3. Identifying Elementary Steps in Multi-Step Mechanistic Pathways

Now that we have a list of some important elementary steps of organic reactions, our next goal is to gain the ability to identify which of these steps is occurring given a reaction scheme. It is helpful to tabulate the types of reagents needed to accomplish each step, as well as the type and number of bonds made/broken in the course of each elementary step. The table below summarizes these observations.

Step	Bonds Broken		Bonds Made		Notes
	σ	π	σ	π	
Coordination	0	0	1	0	One atom shares a pair of electrons with another to form a single bond
Heterolysis	1	0	0	0	Both electrons from the bond go to one atom
Carbocation Rearrangement	1	0	1	0	It must be a carbocation both before and after this step. Only one species involved.
Electrophilic Addition	0	1	1	0	Electrons are pulled **from** the π-bond **towards** an electrophile
Electrophilic Elimination	1	0	0	1	Electrons to make the π-bond come from a σ-bond
E2	2	0	1	1	Three simultaneous curved arrows required!
S_N2	1	0	1	0	Make and break σ-bond simultaneously
Nucleophilic Addition	0	1	1	0	Electrons are pushed **towards** the π-bonded atom **from** a nucleophile
Nucleophilic Elimination	1	0	0	1	Electrons to make the π-bond come from a lone pair, the broken σ-bond electrons push onto a leaving group

Example I.8.1

Identify which elementary step is represented in each of the following reactions:

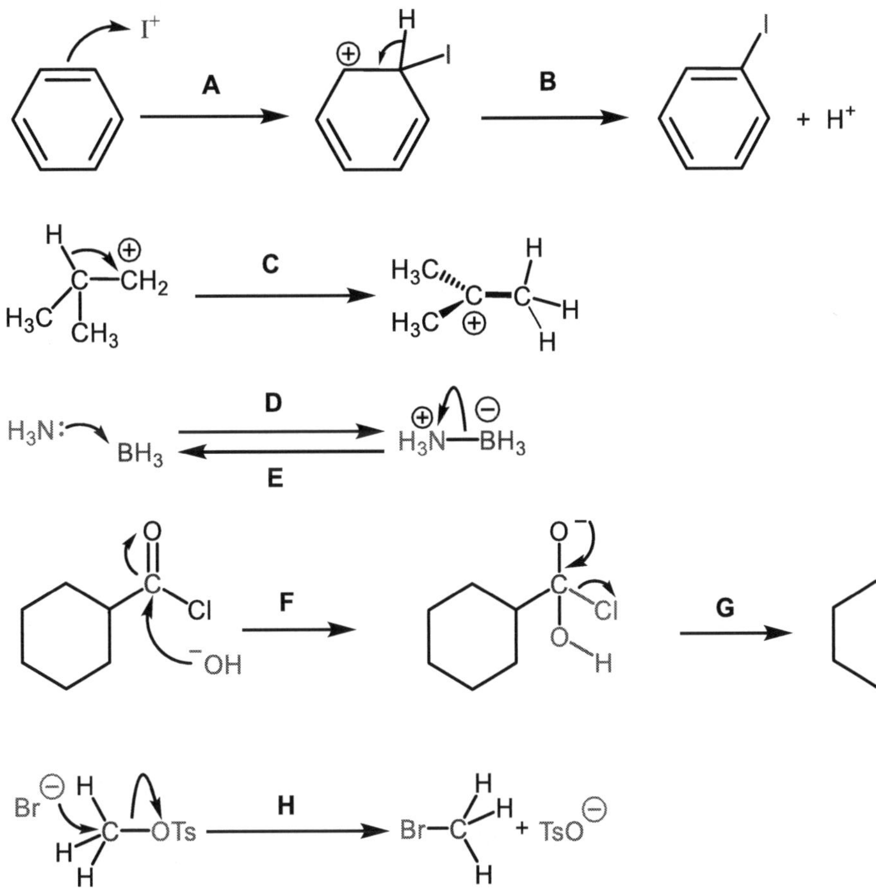

Solution I.8.2

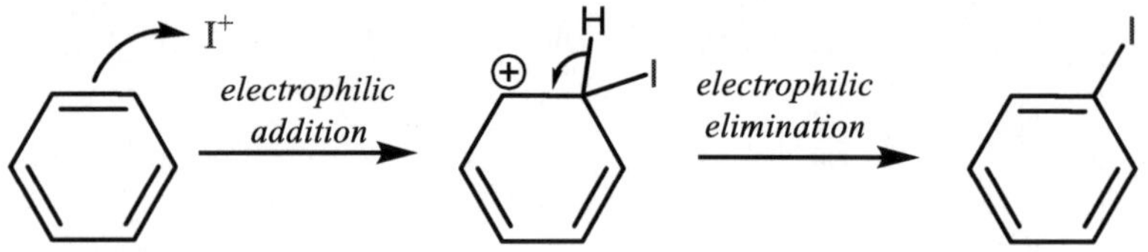

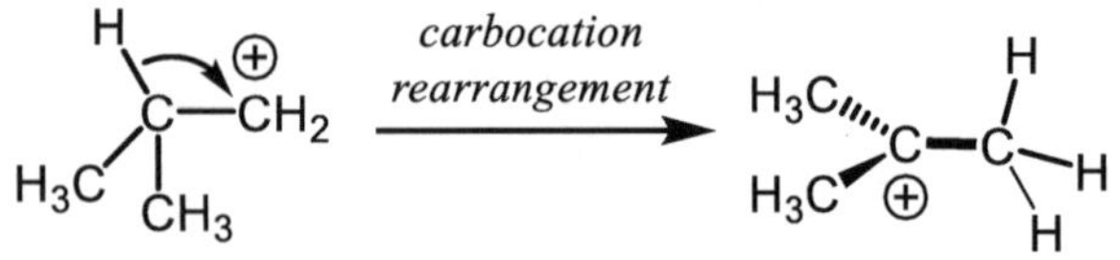

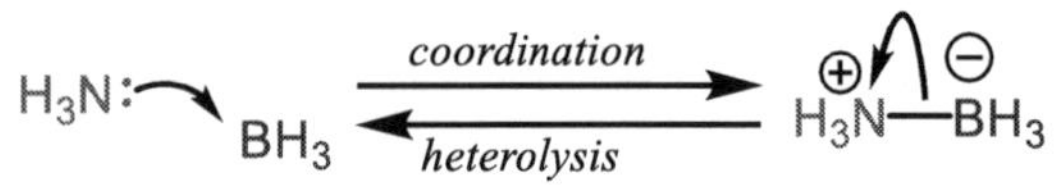

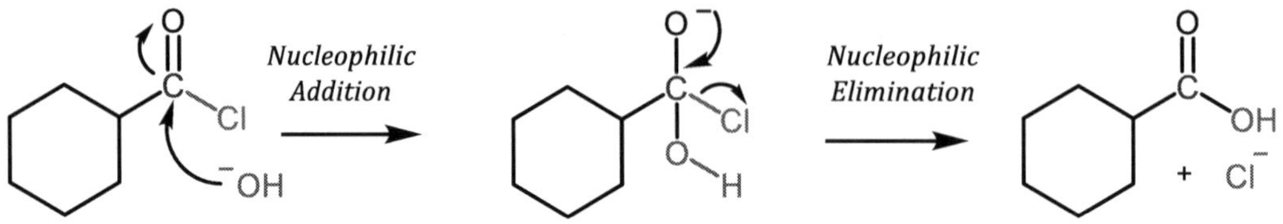

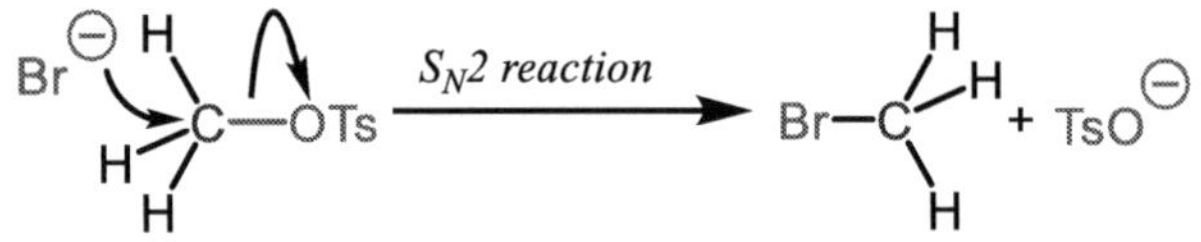

Lesson I.9. Definitions and Concepts related to Acids and Bases

Lesson I.9.1. Arrhenius and Brønsted-Lowry Definitions

There are several definitions of acids and bases used by chemists, and you have undoubtedly studied some of these in prior classes. We will briefly review some of those definitions here. The **Arrhenius definition** is that an acid is a substance that produces H^+ when dissolved in water. We know that H^+ in water is, in reality, the hydronium ion, H_3O^+, so you may see either H^+ or H_3O^+ written, where they are meant to represent the same thing. Examples of Arrhenius acids could be acetic acid (CH_3COOH), HCl, H_2SO_4, etc. The Arrhenius definition is that a base is a compound that makes hydroxide (HO^-) when dissolved in water. Examples of Arrhenius bases are NaOH, $[NH_4][OH]$, etc.

The **Brønsted-Lowry definition** considers an acid to be a proton donor – similar to the Arrhenius definition – but expanded upon the definition of what could be a base. In the Brønsted-Lowry definition, a base is a proton acceptor. All of the hydroxide-producing species defined as bases under the Arrhenius definition are still bases under the Brønsted-Lowry definition, because HO^- reacts with H^+ to make H_2O. However, the Brønsted-Lowry definition incorporates species other than HO^- that likewise react with H^+. For example, ammonia, pyridine and methoxide are all bases under the Brønsted-Lowry definition:

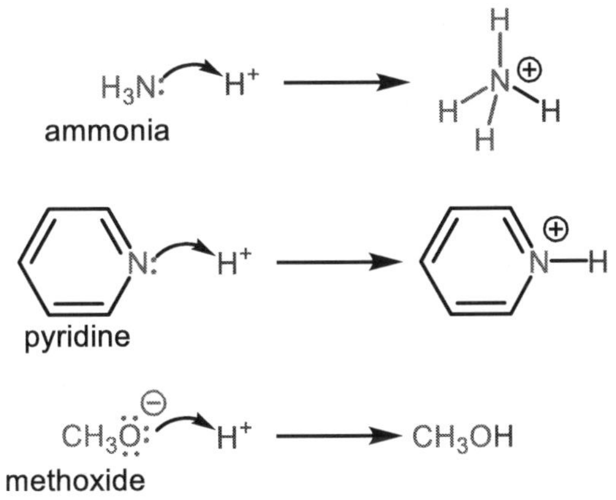

Regardless of the type of acid, scientists use pK_a values to quantify acid strength. A lower pK_a value corresponds to a stronger acid. We also know from prior courses that a stronger acid has a weaker conjugate base and that a stronger base has a weaker conjugate acid. This is because chemical reactions are more favorable when more thermodynamically stable species are produced, a theme that we will use to predict reactions throughout organic chemistry. We can conclude that **the more stable the conjugate base anion, the stronger the acid**. Conversely, **the less stable the conjugate base anion the stronger a base it is** because there is greater favorability for it to be neutralized by an acid.

Lesson I.9.2. Lewis Definitions

The final definition of acids and bases we will cover is the **Lewis definition**. In this definition, an acid is an electron pair acceptor and a base is an electron pair donor. In the Lewis definition, H^+ is still an acid and HO^- is still as a base, which is apparent in the following mechanism:

Likewise, ammonia fits the definition of a base, much as it did in the Brønsted-Lowry definition:

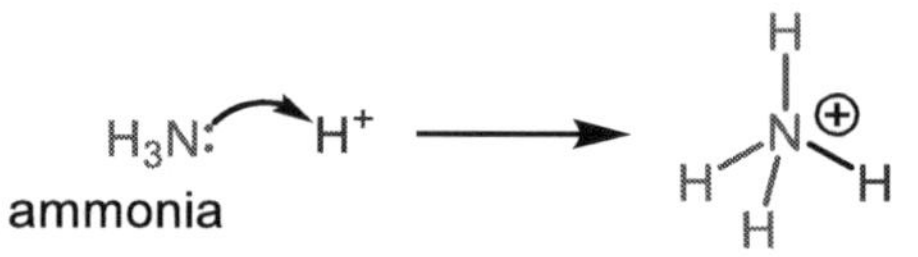

Unlike the Arrhenius or Brønsted-Lowry definitions, however, the Lewis definition allows for species other than H^+ donors to be acids. This is logical, because if we think of an acid as something that neutralizes a base, and we know that things other than H^+ can neutralize base, then these other base-neutralizing species are clearly functioning as acids! A classic example is borane (BH_3), which is a good electron pair acceptor because the B atom has only six valence electrons and accepting a pair of electrons allows it to achieve an octet. Borane is a good Lewis acid that can react with any of the bases we have seen in this lesson:

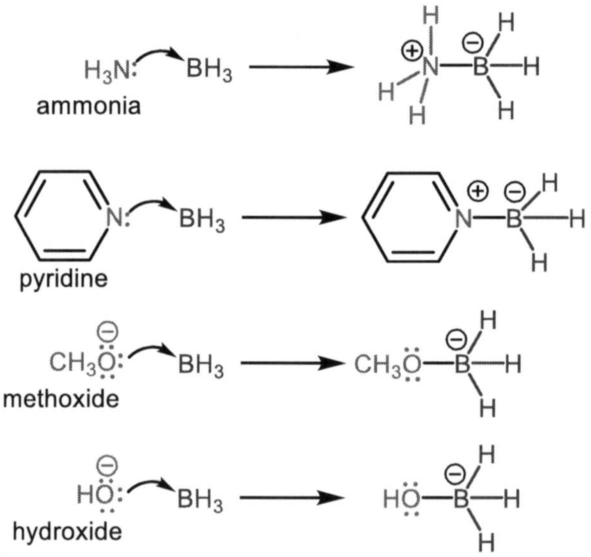

In Lesson I.8, we learned the coordination elementary reaction step. This is a Lewis acid–base reaction because one of the two species donates an electron pair (the base) and the other accepts this electron pair (the acid), and no other electron pair movements occur.

Lesson I.9.3. Using pK_a Values and Identifying Predominant Species in Solution

You may recall from your past chemistry courses that the strength of an acid can be quantified using the pK_a value and that a pH value can be used to quantify the acidity of a solution. The lower the pK_a value, the stronger the acid. Likewise, the lower the pH of a solution, the more acidic it is. One important skill to develop in organic chemistry is to be able to determine the predominant protonation state of a species in a solution of a certain pH. As a general rule of thumb, **a site will keep its proton until the pH of the solution is higher than the pK_a for that site**. Consider phosphoric acid (H_3PO_4), which has pK_{a1} = 2.2, pK_{a2} = 7.2, pK_{a3} = 12.3. This means that if we put phosphoric acid into a solution, it will keep all three of its protons until the pH is raised to above pH = 2.2. At a pH of higher than 2.2 (but not yet at 7.2), phosphoric acid will exist predominantly as $[H_2PO_4]^-$, until the next "checkpoint" pH of 7.2 is reached. Once we get above pH = 7.2, that solution contains enough base to take a second proton away. At that point, the predominant species is $[HPO_4]^{2-}$. Only above a pH of 12.3 will the last of the three protons be removed in the predominant form in solution. To summarize, the predominant form of H_3PO_4 in solutions of various pH values will be as follows:

more basic solution →

pH 0		pH = pK_{a1} 2.2		pH = pK_{a2} 7.2		pH = pK_{a3} 12.3	
PREDOMINANT FORM:	H_3PO_4		$[H_2PO_4]^-$		$[HPO_4]^{2-}$		$[PO_4]^{3-}$
		above pK_{a1}: take 1st proton		above pK_{a2}: take 2nd proton		above pK_{a3}: take 3rd proton	

Example I.9.1.

What will be the protonation state of glutamic acid at A) pH = 10.5 and B) pH = 7.4, given its fully-protonated form and pK_a value for each site?

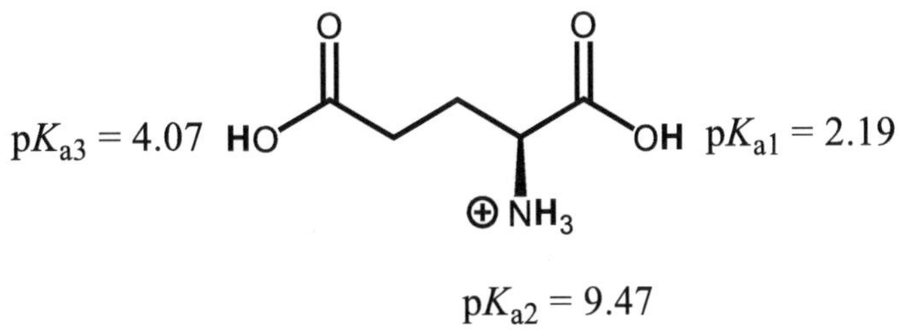

Solution I.9.1.

A) A pH of 10.2 is higher than any of the pK_a values for the three sites. This means the solution is sufficiently basic to take a proton from each of the three sites, to give:

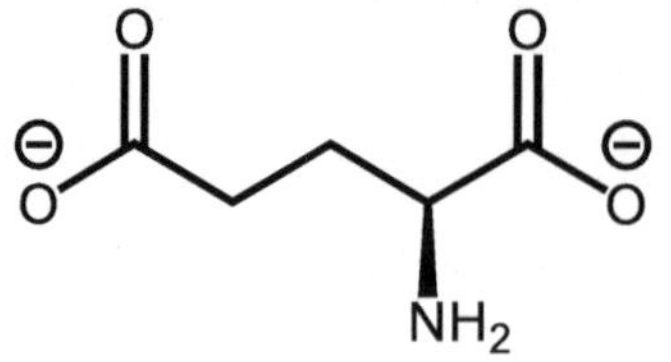

B) A pH of 7.4 is higher than the pK_a for two of the sites, but it is lower than the pK_a value for the ammonium site (ammonium is a positively-charged nitrogen). So, we would expect the two carboxylic acid sites to be deprotonated:

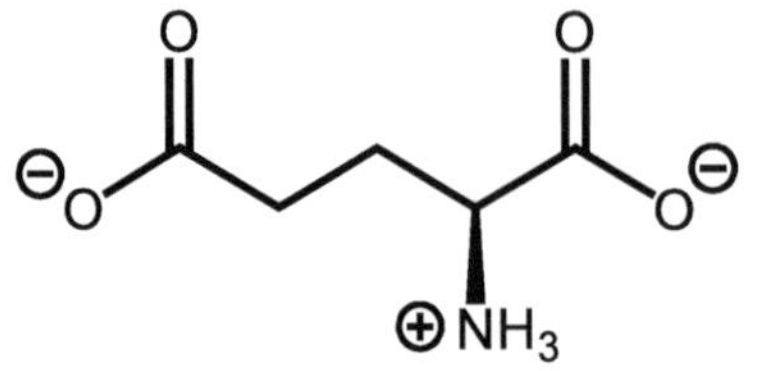

Lesson I.10. Relating Structure to the Strength of an Acid or Base

In the previous lesson, we observed that the more stable the conjugate base anion, the stronger the acid that produces it. The focus of this lesson is to use the chemical structure of a compound to *make relative predictions about acid strength.* We will consider each structural effect individually, in the order of generally most influential to least influential on stability. We will refer to these effects as the **SERI effects** (standing for size, electronegativity, resonance and inductive effects).

Lesson I.10.1. Influence of Deprotonated Atom's Size on Acidity

As we saw in Lesson I.10.1, allowing electrons to spread out over a larger volume via resonance alleviated Coulombic repulsion. For the same reason, **when the negative charge is on a larger atom, the anion is more stable** than an analogous species featuring the negative charge on a smaller atom. If the conjugate base anion is more stable, the acid producing that anion is a stronger acid. We conclude that, all else being equal, the stronger acid in a pair of molecules is the one in which the negative charge produced upon deprotonation resides on a larger atom. Consider H_2O vs. H_2S. If H_2O acts as an acid by losing H^+, a negative charge is on O in the conjugate base (^-OH). If H_2S acts as an acid by losing H^+, a negative charge is on a much larger S atom in the conjugate base (^-SH). We would predict that ^-SH is more stable than ^-OH, which would mean that H_2S is a stronger acid than H_2O. The pK_a values confirm our prediction.

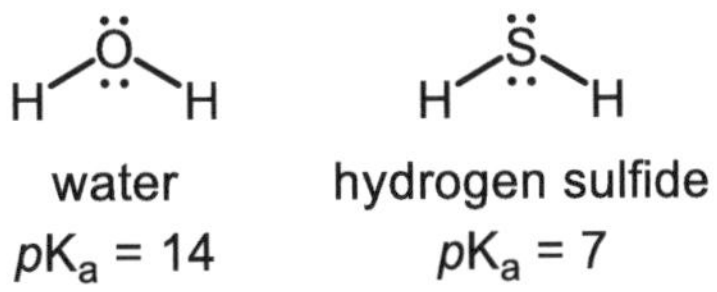

To assess anion stability, we do not look at the size of the entire molecule, just the size of the atom bearing the formal negative charge in the conjugate base. Larger atoms in the same molecule are irrelevant to anion stability if they never acquire any negative charge following deprotonation.

Lesson I.10.2. Influence of Deprotonated Atom's Electronegativity on Acidity

Another factor leading to increased anion stability is increased attraction between the electrons of the negative charge and the nucleus of the atom. This occurs when the positive charge on the nucleus increases, while the distance from the nucleus to the electrons remains relatively constant.

Atoms in the same row of the periodic table have the same valence orbitals, and consequently have *roughly similar sizes* (the small differences are unimportant to anion stability). Two atoms of the same size have the same distance between the valence electron and the nucleus. So, within a row of the periodic table, the attraction of the electrons to the nucleus increases as the charge on the nucleus increases. In fact, this trend gives rise to the electronegativity differences among the elements within a row of the periodic table. Going from left to right in the second row of the periodic table (i.e., B, C, N, O, F), for example, the charge on the nucleus increases and so does the electronegativity.

When we compare two anions, each of which has a formal negative charge on an atom in the same row, such as HO^- and H_2N^-, we see that **the stability of the anion increases as the electronegativity of the anionic atom increases**. In terms of relating this observation to acidity, we can state that **the stronger acid is the one that leads to deprotonation of the more electronegative atom** if those two atoms are in the same row. It is also important at this point to recall the effect of hybridization on electronegativity (Lesson I.6).

The effect of electronegativity is generally not as influential as the ability to spread charge out via resonance delocalization or atom size. Therefore, it is best to check first for resonance and consider the size of the atom being deprotonated, and then, only if the two possible deprotonation sites are atoms in the same row of the periodic table which have no possibility for resonance, to check the electronegativity.

<u>Example I.10.1</u>

Which is the strongest acid?

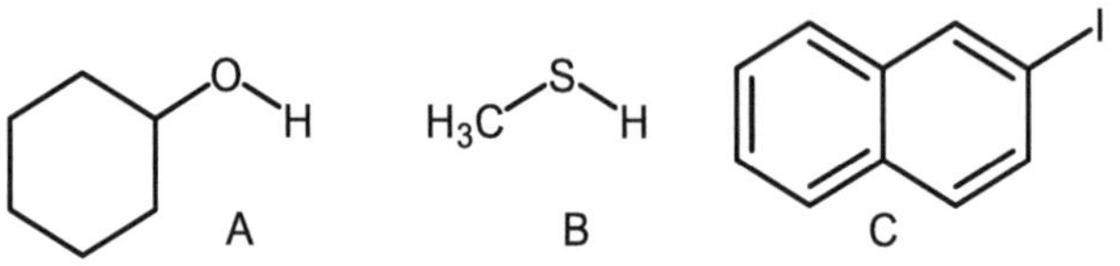

A B C

<u>Solution I.10.1</u>

First, we identify the atom to be deprotonated in each case. In A, it will be the O because it is more electronegative than C. In B, it is S because S is larger than C. In C, we do not have a choice because only carbon atoms have protons in this case. Thus, B is the strongest acid and C is the weakest acid.

Lesson I.10.3. Influence of Conjugate Base Resonance Stabilization on Acidity

Repulsive forces destabilize a chemical species. Attractive forces stabilize a chemical species. One factor that increases anion stability is to decrease repulsion between electrons in the anion. Repulsion can be alleviated when the negative charge is allowed to spread out over a larger volume. In Lesson I.7, we learned that charge can be spread out (delocalized) through resonance.

If the conjugate base anion is more stable, the acid producing that anion is a stronger acid. We conclude that, all else being equal, the stronger acid in a pair of molecules will be the one that has more resonance delocalization of the negative charge on the conjugate base anion.

Example I.10.2

Which is the stronger acid?

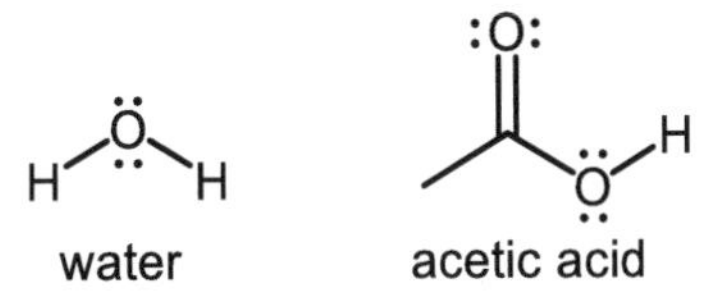

Solution I.10.2

The way to answer these types of questions is to draw the conjugate bases. We note that the conjugate base for acetic acid has resonance stabilization. We conclude that it is the more stable (weaker) conjugate base. We therefore conclude that acetic acid is the stronger acid.

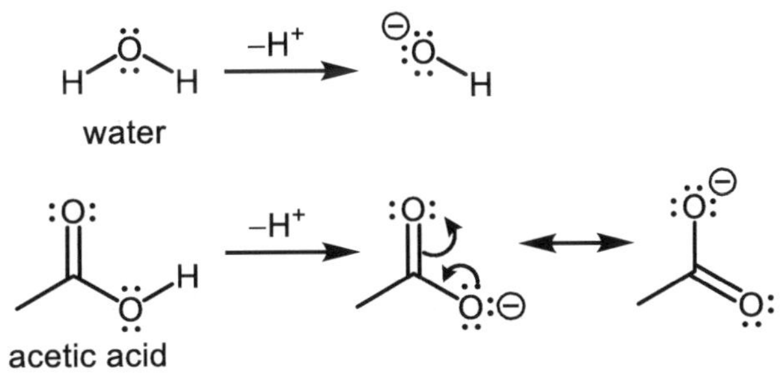

Lesson I.10.4. Influence of Inductive Effects on Acidity

"Inductive effect" is a general term for any effect *induced* by *nearby* groups. Inductive effects can be divided into two types: **stabilizing** and **destabilizing**. In the context of anion stability, stabilizing inductive effects are observed when groups *near* the anionic atom have some attraction for the negative charge. Conversely, destabilizing inductive effects are observed when some nearby group has a repulsive interaction with the anionic atom's negative charge. Furthermore, the magnitude of the effect, whether stabilizing or destabilizing, is greater the closer the attractive/repulsive unit is to the anionic atom. An example of a stabilizing inductive effect is an attractive force between the anionic atom and a nearby partial positive end of a polar bond:

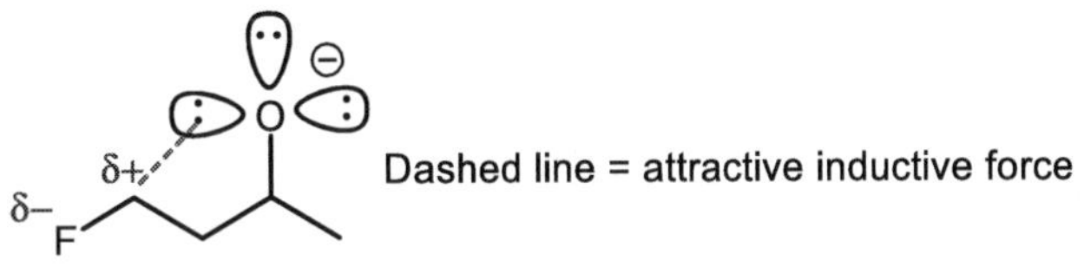

An example of a destabilizing inductive effect would result from repulsion between the negative charge and electrons in nearby bonds:

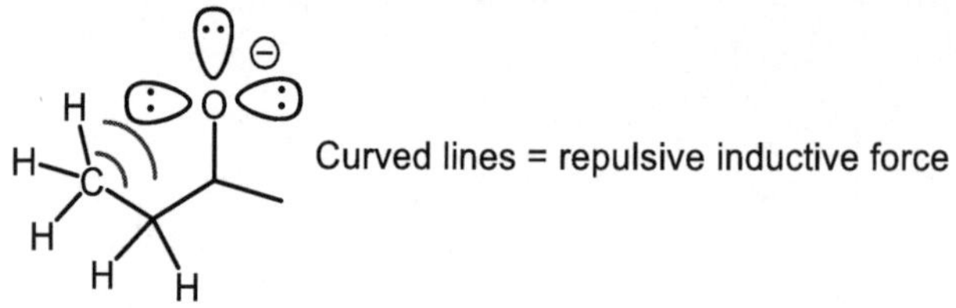

Inductive effects generally have less influence on stability than do the effects of resonance, size and electronegativity.

Example I.10.3

Which is the strongest acid?

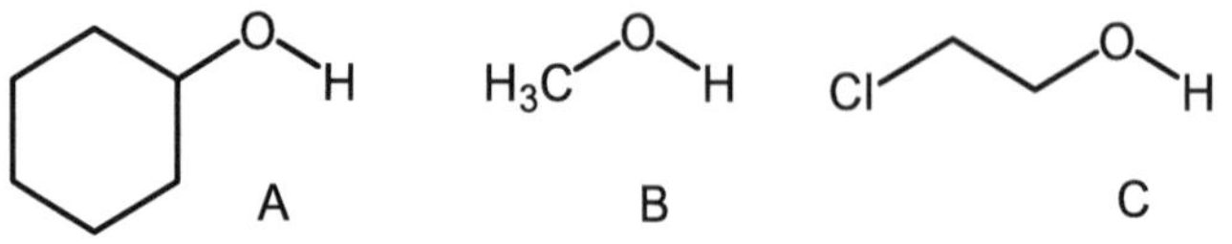

Solution I.10.3

All of these species will be deprotonated at the OH group to leave a negative charge on O. In B, there are no strong attractive or repulsive forces between the methyl group and the O. In contrast, the larger cyclohexyl group in A will have an inductive repulsion for a negatively-charged O. So, A is a weaker acid than is B. This leaves us with C. The conjugate base of C will experience an attractive inductive force:

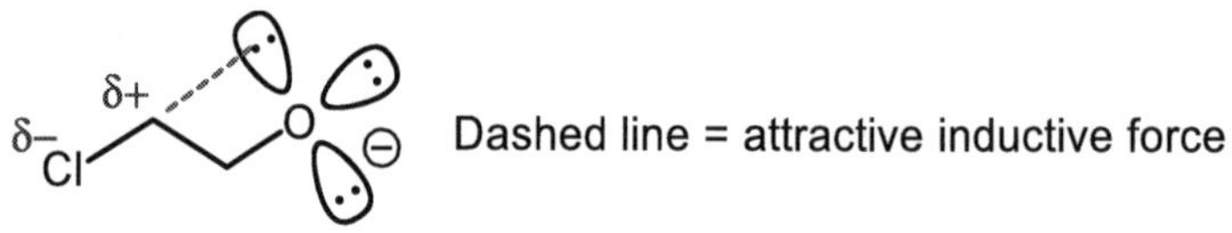

For these reasons, C is the strongest acid, followed by B, and A is the weakest acid.

Lesson I.11. Stability of Carbocations and Alkenes

Lesson I.11.1. Influence of Resonance/Inductive Effects on carbocation Stability

In Lesson I.10, we examined how various structural features influence the stability of anions. We will now study how such structural features influence the stability of cations. We will, however, limit our discussion to carbocations (cations having the positive charge on C):

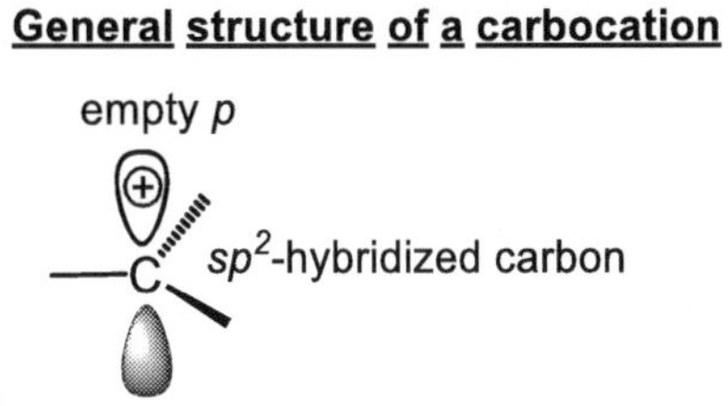

Because we are limiting our discussion to species with a positive charge on C, we do not need to consider size or electronegativity effects like we did in Lesson I.10. We will only consider inductive effects and resonance.

For most carbocations encountered in first-semester organic chemistry, the inductive effect trend comes into play. In the Lesson I.10, we saw that σ-bonding electron pairs repel negative charge, so having more "branches" near an anion was destabilizing. Because electrons *attract* positive charge, a σ-bonding electron pair will be attracted to the empty *p*-orbital on the carbocation, which is a *stabilizing* interaction. This specific interaction is called **hyperconjugation.**

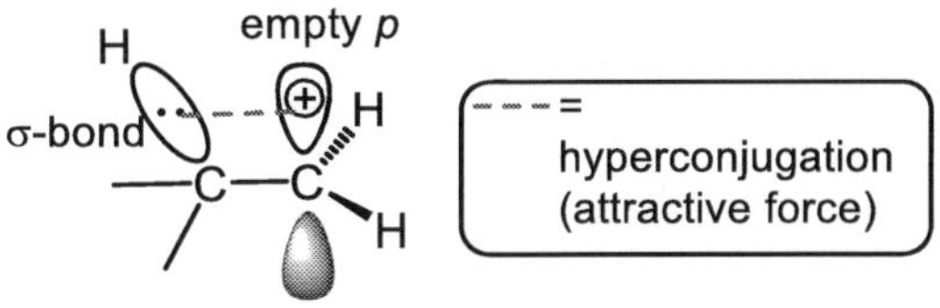

Due to hyperconjugation, the more non-H substituents on the cationic C, the more stable it is. For convenience, chemists refer to a carbocation in which the positively-charged **carbon makes bonds to one carbon as a primary (1°)** carbocation**, with two carbons as secondary (2°), and with three carbons as tertiary (3°)**:

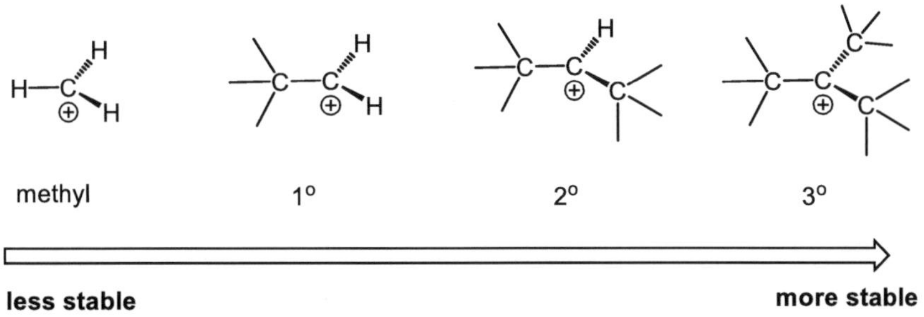

Resonance can also stabilize a cation by allowing the positive charge to spread out over a larger volume. A good rule of thumb is that stabilization afforded by the presence of additional resonance contributors is equivalent to having one additional alkyl chain. The trend in carbocation stability, with resonance factored in, is as follows:

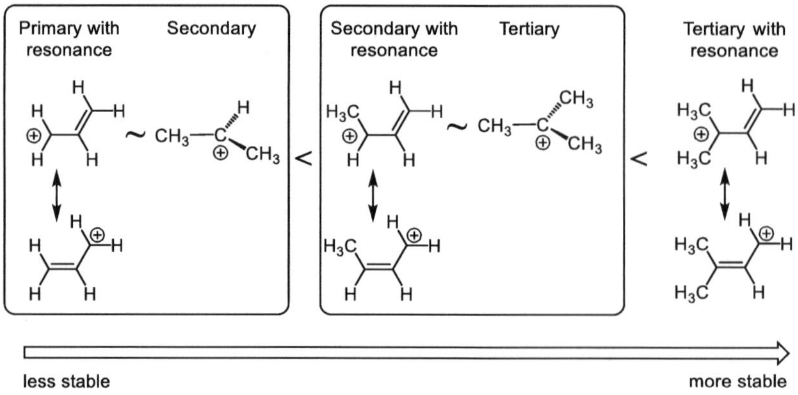

Example I.11.1

Which is the more stable species within each pair of molecules shown?

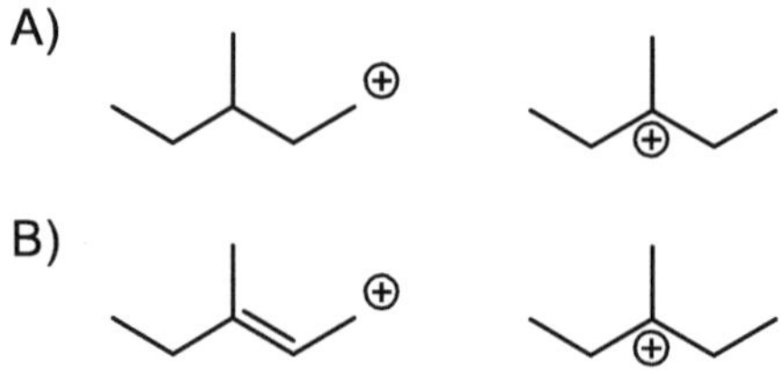

Solution I.11.1

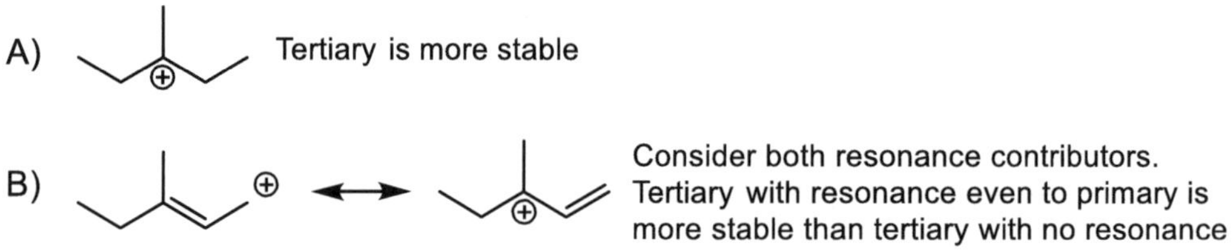

Hydrogenation of an alkene leads to addition of an H atom to each C atom in the C=C bond:

<u>Hydrogenation</u>

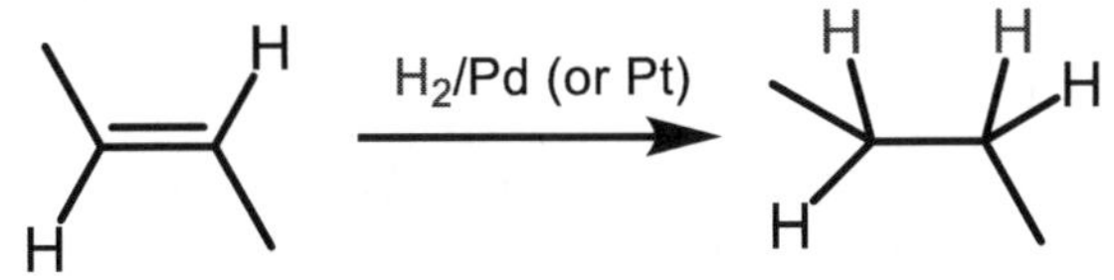

The resulting σ-bonds are stronger than the π-bond that was replaced. As a result, hydrogenation of an alkene to an alkane will be exothermic (i.e., give off heat), and this is referred to as the "heat of hydrogenation". Scientists have observed that, as the sp^2-hybridized carbons in the alkene become more electron deficient, the alkene will become less stable with respect to the alkane, and thus a greater heat of hydrogenation (ΔH_h) will be observed upon converting the alkene to the alkane.

In Lesson I.6, we saw that an sp^2-hybridized C (like the C atoms in a C=C bond) is more electronegative than is an sp^3 hybridized C. In Lesson I.11 we learned that a carbon atom that has a pull for electrons (for example a carbocation) is stabilized by electron donation from sp^3-hybridized substituents. In an analogous fashion, each alkyl branch on an alkene stabilizes it further by electron donation. Thus, the ranking of stability of simple alkenes is tetrasubstituted > trisubstituted > disubstituted > monosubstituted > ethylene. Ethylene is the least stable of the alkenes (i.e., highest in energy), so it will give off the most energy upon hydrogenation. As an alkene becomes progressively more stable, it will give off progressively less energy upon hydrogenation.

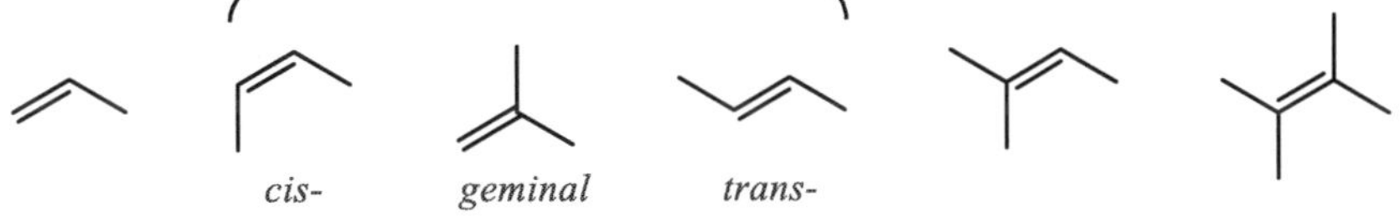

Degree of substitution:	none	mono	di	di	di	tri	tetra
	$H_2C{=}CH_2$		*cis-*	*geminal*	*trans-*		
ΔH_h: (kcal/mol)	–32.8	–30.1	–28.7	–28.5	–27.5	–27.0	–26.6

Disubstituted alkenes can have both substituents on the same carbon of the C=C (labeled *geminal* in the table above), the substituents may both point to the same side of the C=C unit, one off each C of the C=C (called the *cis*-isomer) or the substituents may point in opposite directions, one off each C of the C=C (called the *trans*-isomer). *cis*- > *geminal* > *trans*-. The reason that the *cis*-isomer is less stable than the *trans*-isomer is that the *cis*-isomer has unfavorable steric repulsions. **Steric repulsion is the term given to a case in which two groups physically "bump into" one another. When this happens, the electron clouds around the groups repel one another:**

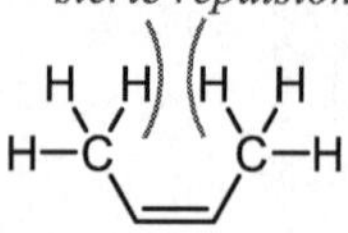

cis-isomer

It is not sterics that makes the geminal isomer higher in energy than the *trans*-isomer. Instead, it is an electronic reason. Remember from Lesson I.6.2 that the sp^2-hybridized C is more electronegative than an sp^3-hybridized C, so the C atoms in the C=C unit are stabilized by sp^3-carbon donors. One side of the C=C in 2-methylpropene has no donor groups, but both sides of the C=C bond have electron donors in the *trans*- isomer, making it more stable. The general trend for the heats of hydrogenation for alkenes is, in descending order: $CH_2{=}CH_2$ > monosubstituted > external disubstituted ~ *cis*-isomer of internal disubstituted > *trans*-isomer of internal substituted > trisubstituted > tetrasubstituted.

Example I.11.2

Which of the following alkenes would have the higher heat of hydrogenation?

Solution I.11.2

Both molecules are disubstituted alkenes, but in **A** the double bond has *cis* stereochemistry, whereas **B** has *trans* stereochemistry. **A** is less stable (i.e., higher in energy) than **B** due to steric interactions, and will thus release more energy upon hydrogenation of the C=C bond. **A** will have the higher heat of hydrogenation.

Lesson I.12. Predicting Reaction Spontaneity and Direction of Equilibria

Lesson I.12.1. Favorable Reactions

In Lesson I.10, we saw that the acidity of a species increases as the stability of the deprotonated species (conjugate base anion) increases. In acid–base reactions, and in all chemical reaction, **the formation of a more stable species is more favorable than the formation of a less stable species**. In the same line of reasoning, **consumption of less stable species is more favorable**. When chemists refer to something as being more favorable, they mean more energetically favorable (more thermodynamically favorable). A more favorable reaction will require less energy to carry out. Indeed, a **spontaneous reaction** is one in which the products are more stable than the reactants. You may know from prior chemistry courses that for spontaneous reactions, the Gibbs free energy (ΔG) for the reaction is negative, referred to as an exergonic reaction; the opposite is called an endergonic reaction. Often people ignore entropy and refer to spontaneous reactions as those in which the enthalpy (ΔH) for the reaction is negative, referred to as an exothermic reaction; the opposite is called an endothermic reaction. Spontaneous reactions typically give off heat as a result of reaction. An example would be burning gasoline. You do have to put in a small initial bit of energy (e.g., a spark) to get the reaction to begin, but overall you will get much more energy out upon combustion of the gasoline than you put in. So how can we predict when a reaction will occur spontaneously?

In Lessons I.10 and I.11, we discussed how to evaluate the attractive/repulsive Coulombic forces present within anions and cations to assess their relative stability. Now we can use these skills to predict relative spontaneity of a reaction. The questions "which is the more spontaneous reaction", "which is more thermodynamically favorable", "which is more energetically favorable", or simply "which is more favorable" are all asking the same thing: *which reaction consumes the less stable starting materials or produces the more stable products?*

Example I.12.1

Assuming that each process is mechanistically allowed, use your knowledge of stability trends to predict whether formation of product would be thermodynamically favorable (spontaneous):

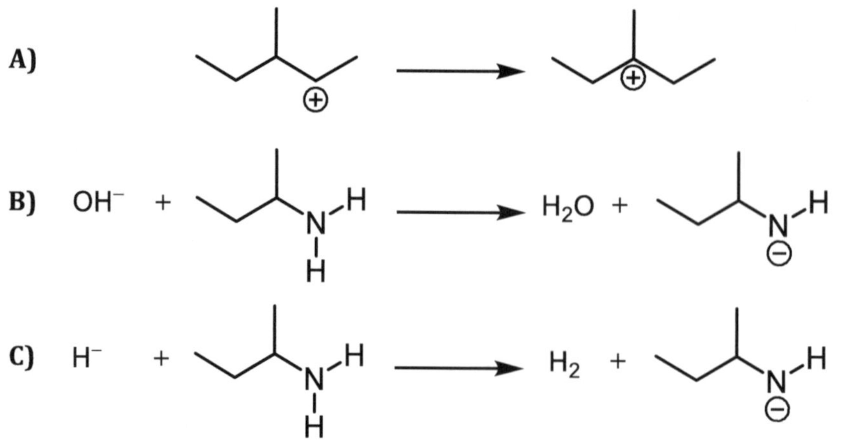

Solution I.12.1

In reaction A) the reactant is a 2° carbocation and the product is a 3° carbocation. We know from the Lesson I.11 that a 3° carbocation is more stable than a 2°, due to hyperconjugation. Reaction A) is therefore spontaneous.

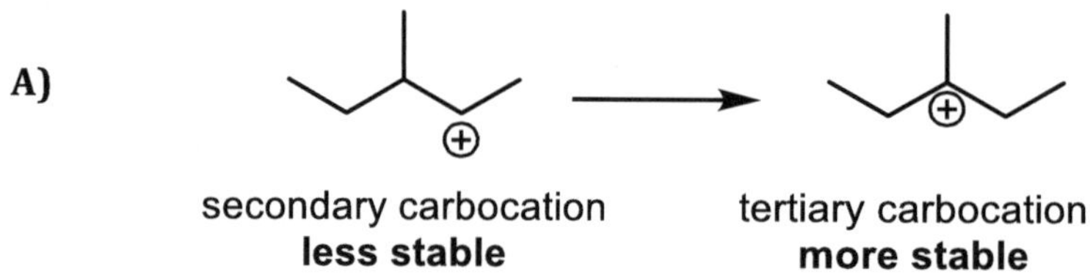

In reaction B) there are two reactants, one anion and one neutral species. There are likewise two products, also an anion and a neutral species. Which species do we focus on in our analysis of stability? Unless a neutral species has some significant strain (we will learn about these later), which is not present in reaction B), ionic species have the greatest contribution to stability/instability, because they have formal Coulombic charge. For reaction B), this means we compare HO^- to $[C_4H_9NH]^-$. From Lesson I.9, we know that an anion having the negative charge on a more electronegative atom (i.e., oxygen) is more stable than an anion having the negative charge on a less electronegative atom (i.e., nitrogen). So, in reaction B), the reactant anion is more stable than the product anion. The reaction is therefore **not** thermodynamically favorable in the forward direction:

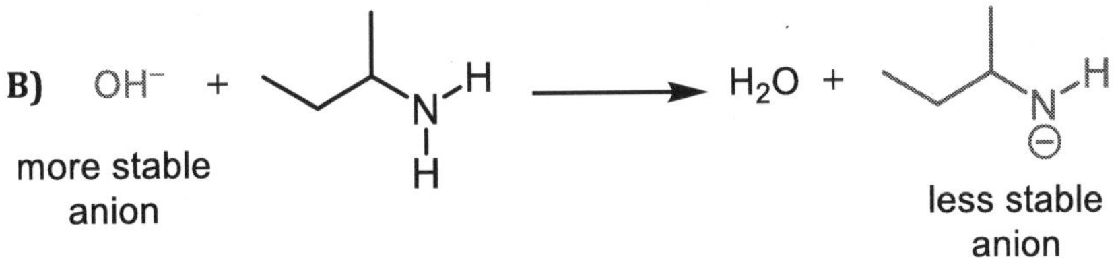

In reaction C), we again compare the anions. From Lesson I.9, we know that an anion having the negative charge on a larger, more electronegative atom like N (as in the product anion) is more stable than an anion having the negative charge on a smaller, less electronegative atom like H (as in the reactant anion). So, in reaction C), the reactant anion is less stable than the product anion. The reaction is thermodynamically favorable in the forward direction.

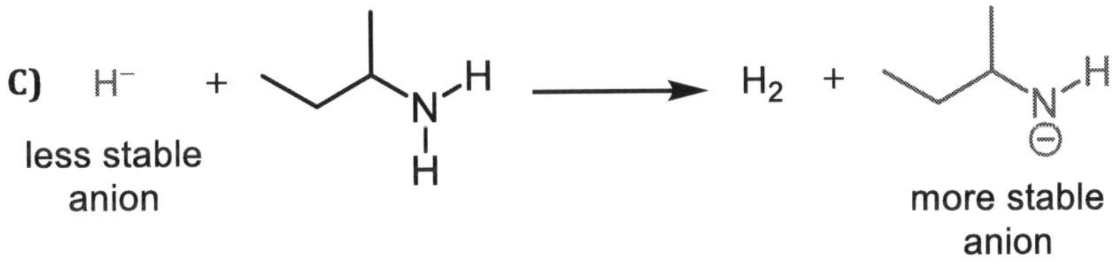

Lesson I.12.2. Equilibria

A state of equilibrium occurs when the rates of the forward and reverse reactions are identical so that the concentration of all species remains constant. The side of an equilibrium that has more stable species is the side that is favored. The product side is favored in cases where the products are more stable than the starting materials and vice versa. If the products and reactants are equally stable, then neither side is favored, which leads to an equal abundance of products and reactants. The equilibrium constant (K_{eq}) for any equilibrium (regardless of reactant/product stability) is given by:

$$K_{eq} = [\text{products}]/[\text{reactants}]$$

We can conclude that, for a favorable reaction, the equilibrium constant is greater than one – there will be a higher concentration of products than reactants, because the products are more stable. For an unfavorable reaction, the equilibrium constant will be less than one. In general, **the greater the value of the equilibrium constant, the more thermodynamically favorable the reaction is**. This fact allows us to quantify the relative spontaneity of reactions and, by extension, the relative stabilities of different species.

Lesson I.12.3. Using pK_a to Predict Reaction Direction

To predict the outcome of an acid-base reaction, note that the *more acidic* proton will be donated:

$$CH_3NH_2 \quad + \quad CH_3OH \rightleftharpoons CH_3\overset{\oplus}{N}H_3 \quad + \quad CH_3O^{\ominus}$$

pKa = 40 pKa = 15.5

To predict which side of the equilibrium is favored, remember that **the equilibrium always shifts from the stronger acid to the weaker acid.**

reactants favored

$$CH_3NH_2 \quad + \quad CH_3OH \xleftarrow{\hspace{2em}} \; \rightleftharpoons CH_3\overset{\oplus}{N}H_3 \quad + \quad CH_3O^{\ominus}$$

$pK_a = 15.5$ $pK_a = 10.7$

Lesson I.13. Reaction Coordinate Diagrams and Reaction Rate

Lesson I.13.1. Parts of a Reaction Coordinate Diagram

In Lesson I.12, we learned how to predict relative favorability of a reaction by comparing the relative stability of reactants and products across multiple reactions. It is often helpful to have a pictorial representation of the energy changes involved in a reaction. For this purpose, chemists use what is called a **reaction coordinate diagram**. A reaction coordinate diagram has energy on the *y*-axis and reaction progress (time) on the *x*-axis. Here is an example:

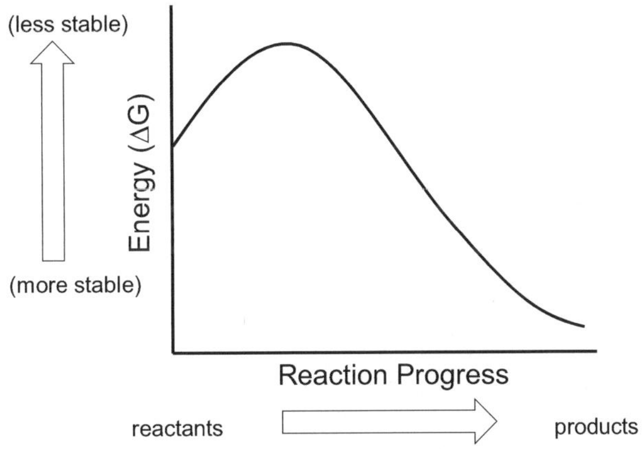

There are several critical pieces of information that we can gain from a simple analysis of a reaction coordinate diagram. First, we can tell whether the reaction is thermodynamically favorable or not:

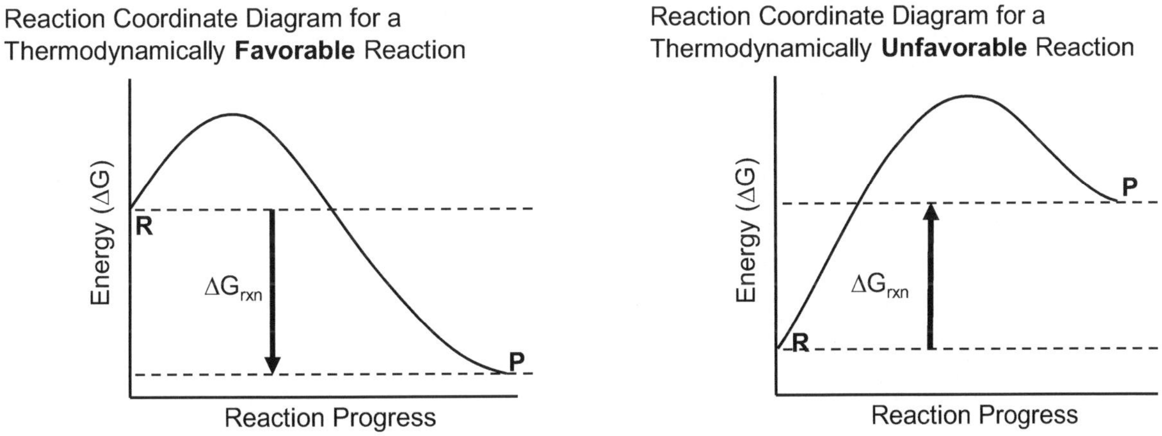

In a spontaneous reaction, the energy of products (**P**) will be lower in energy (i.e., more stable) than the reactants (**R**).

We can also tell how many steps, intermediates, and transition states are in the mechanistic path from reactants to products. Each peak represents a transition state (**T**) for each reaction step. Transition states are not isolable species, and only transiently exist at local energy maxima. Each valley (local energy minimum) corresponds to an intermediate (I). An intermediate is an isolable species, which serves as the starting point for another step. Consider the reaction coordinate diagrams below:

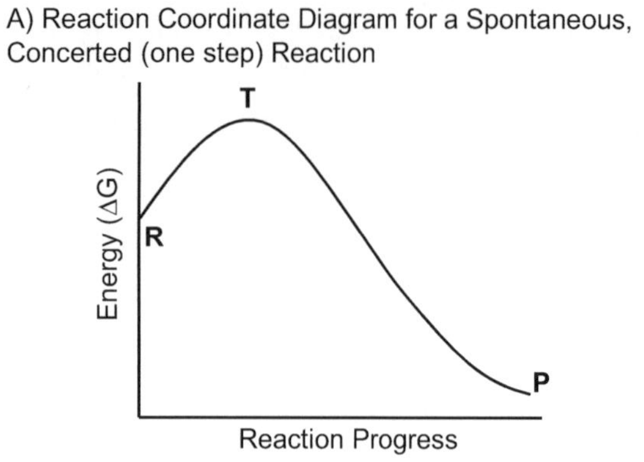

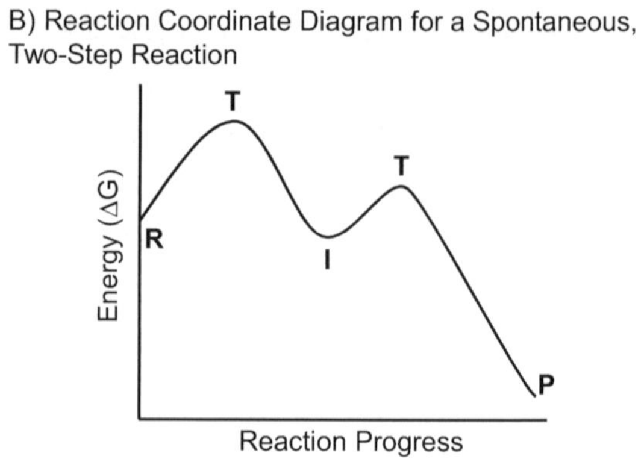

Note the obvious difference between the observed energy changes for a reaction that takes place via one mechanistic step (single step reactions are called **concerted** reactions, diagram A) and one that takes place via two steps (diagram B). If one records the energy changes during a reaction, mechanistic information can be found.

<u>Example I.13.1</u>

A scientist does a reaction and observes energy changes shown in this diagram:

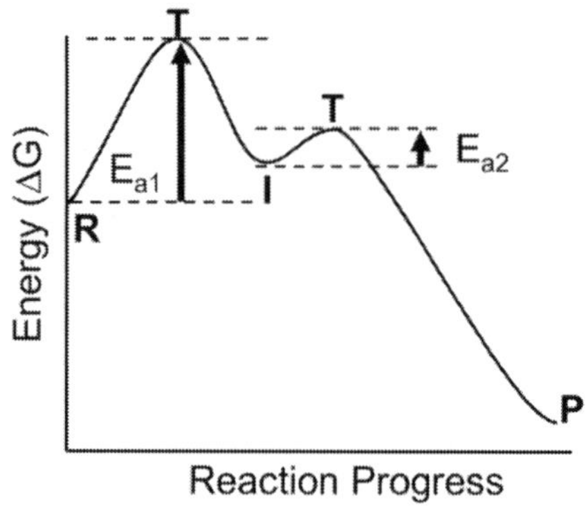

Which of the following mechanisms is consistent with this observation?

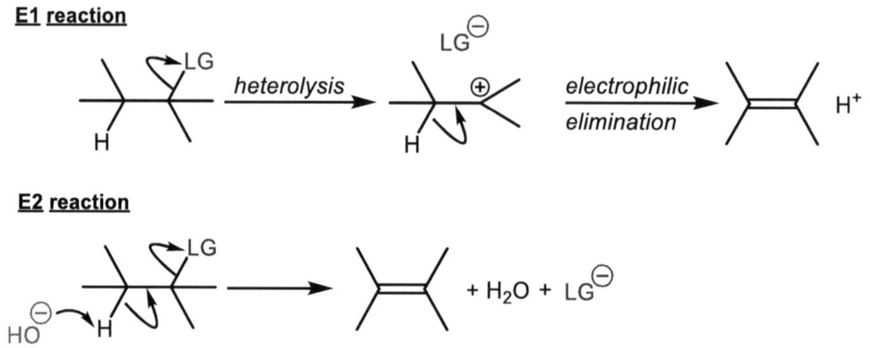

Solution I.13.1

The reaction coordinate diagram indicates that there are two steps in the mechanism. The E2 reaction choice shows only one step (an E2 step that we learned in Lesson I.8). The E1 reaction, however, consists of two elementary steps, heterolysis followed by electrophilic elimination (also from Lesson I.8). The E1 mechanism is consistent with the observed energy changes represented in reaction coordinate diagram.

The rate of a reaction can also be determined from a feature in the reaction coordinate diagram. You may recall from General Chemistry that the rate constant (k) for a reaction is proportional to the energy of activation (E_a) of a reaction, as outlined in the Arrhenius equation:

$$k = Ae^{-(E_a/RT)}$$

(A = Arrhenius constant, R = gas constant, T = temperature)

We are not going to use the equation here, but we can conclude **that a higher energy of activation leads to a slower reaction.** The energy of activation is shown on a reaction coordinate diagram below: it is the amount of energy needed to get "over the hump" of any given step of a reaction.

Energies of Activation for Step 1 (E_{a1}) and two (E_{a2})

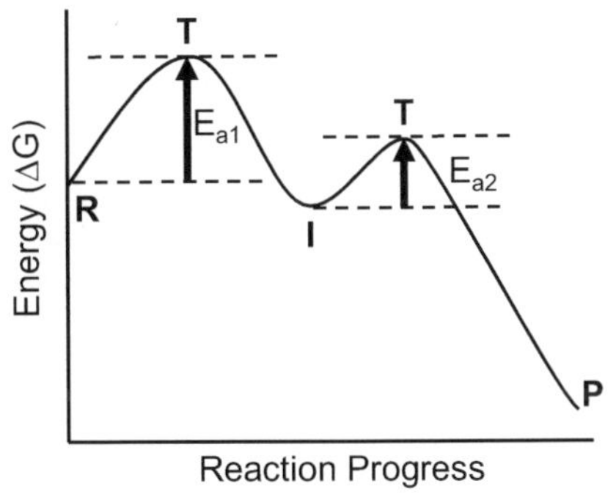

In the reaction coordinate diagram above, the first step has a higher energy of activation than does the second step, so the first step will be slower. This leads to another piece of information that we can gain from the diagram. The slowest step of a reaction is termed the **rate-limiting step**. Knowing the rate-limiting step helps one make predictions about reaction trends and products, as well as practical laboratory or production plant considerations, such as solvent selection or the amount of heat that must be supplied for a reaction.

Lesson I.13.2. Rate Law Expressions

The overall rate for a reaction is determined from its rate-limiting step, the slowest step of a reaction. If a reaction is concerted, there is only one step. If there is more than one step in a reaction (each transition state, or "hill" is a step), then the step with the highest E_a is the rate-limiting step. Once we know the rate-limiting step and the mechanism of that step, it is possible to write a rate law that expresses how the rate changes if we change the concentration of our reactants. Consider this rate-limiting step for a reaction:

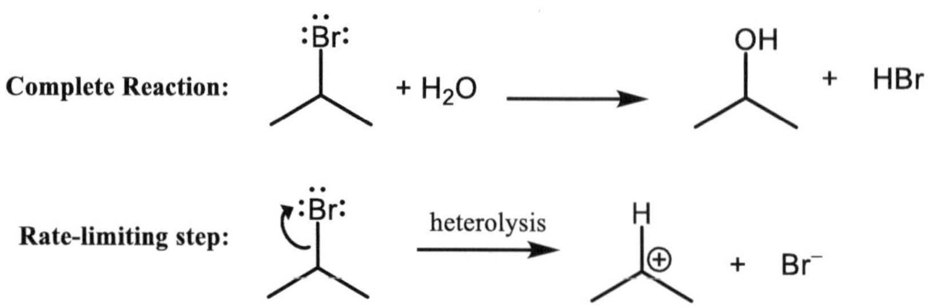

Although two species, an alkyl bromide and water, are needed for the reaction, only one of these, the alkyl bromide, is a starting material for the rate-limiting step. The rate law expression will only depend on the concentration of one species in such a case. Such reactions are called **unimolecular** reactions or **first-order** reactions. Remembering that we put a species in brackets to represent its concentration, the rate law for this reaction will be:

$$\text{Rate} = k[\text{alkyl bromide}]$$

Consider another case:

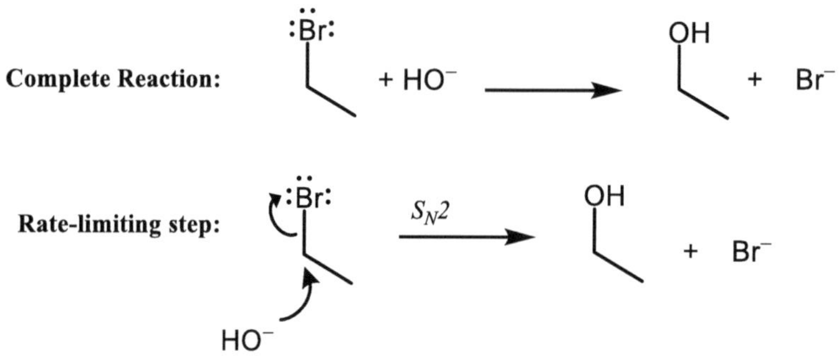

Here, the rate-limiting step features hydroxide anion and the alkyl bromide, so the rate law expression will depend on the concentration of two species. Such reactions are called **bimolecular** reactions or **second-order** reactions. Remembering that we put a species in brackets to represent its concentration, the rate law for this reaction will be:

$$\text{Rate} = k[\text{alkyl bromide}][\text{HO}^-]$$

Lesson I.14. Nomenclature I: Alkanes, Alkyl Halides, Alcohols and Cycloalkanes

Lesson I.14.1. Linear Alkanes and Substituents

In Lesson I.4, we saw examples of the functional groups typically covered in introductory organic chemistry. Alkanes are molecules consisting of only carbon and hydrogen atoms, with all the carbons interconnected by only single bonds. There is an infinite number of ways that C and H atoms could combine to form alkanes, so it is important to develop a naming system (a nomenclature convention) for alkanes to use in our studies. **Linear alkanes** have all carbons attached in a linear chain, with no C-containing branches or other **substituents** coming off of the chain other than hydrogen. Below are shown the linear alkanes, up to ten carbon atoms in length, along with their names. On the right is the name for substituent chains having the same number of C atoms as the corresponding linear alkane on the left. In addition to these alkyl substituents, halide substituents (fluoro for –F, chloro for –Cl, bromo for –Br and iodo for –I) can be used in naming alkyl halides following the same rules in Lesson I.14.2 for alkanes.

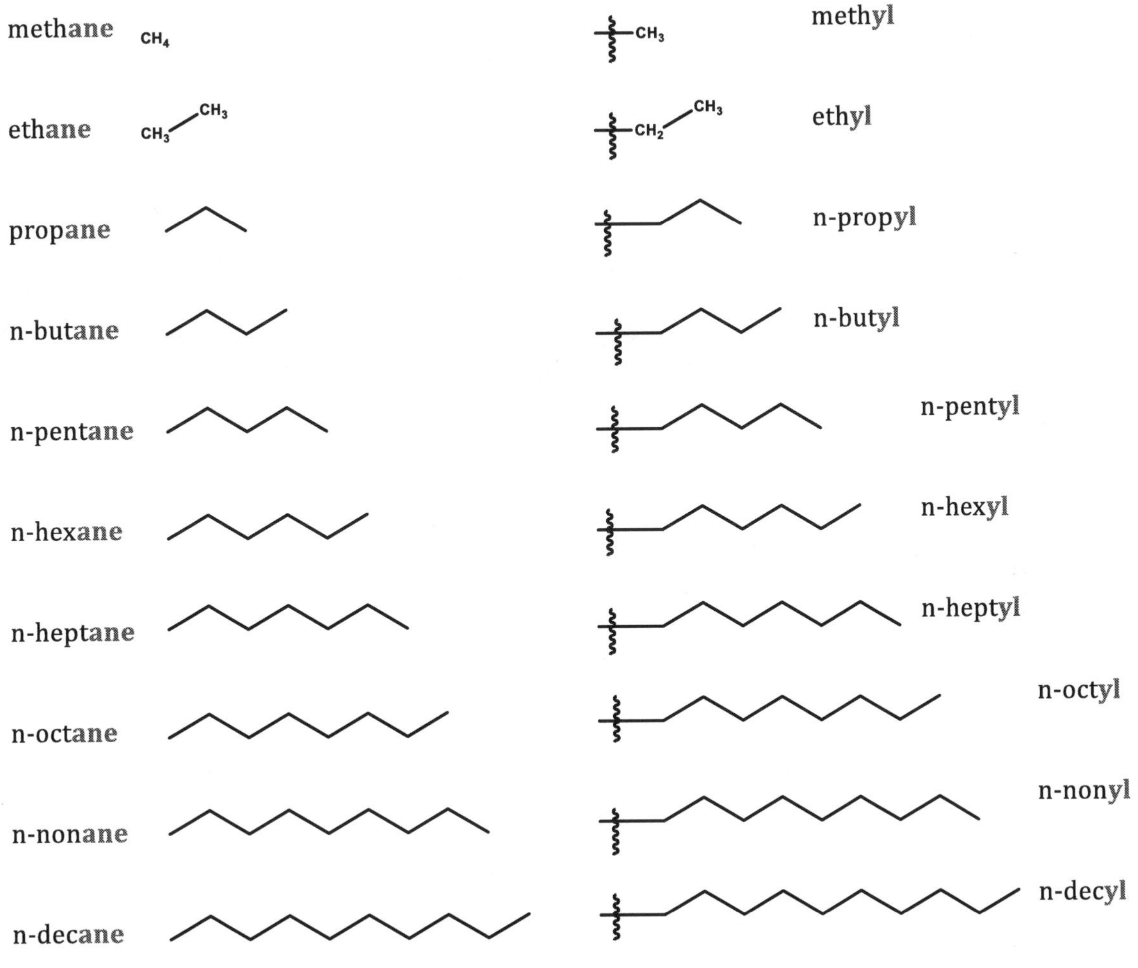

Note that some of the linear chains have an "*n*-" in front of their names. This '*n*-' stands for "normal", and **it is usually left off of the name and we assume that it is a "normal" linear chain**. On the other hand, some arrangement of hydrocarbon chains that are placed as substituents have specific names that use other designators in front of the part of the word telling us how many C are present. The more common ones are provided below for reference. To help remember which is which, notice that the "iso" prefix refers to species having the Y-shaped unit, the *sec*- prefix indicates attachment at a *sec*ondary carbon, and the *tert*- prefix refers to attachment at a *ter*tiary carbon.

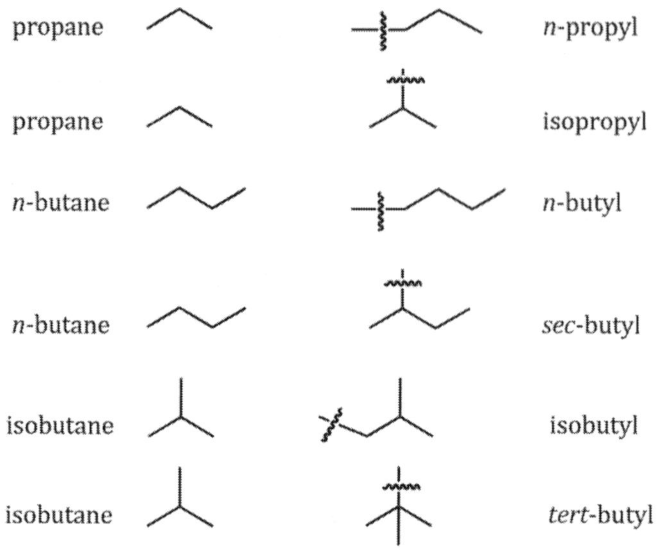

There are also different ways that the same set of carbon atoms can be arranged for a given molecular formula. If two molecules have identical molecular formulae but differ in their bond connectivity, we term these **constitutional isomers**. As an example, the possible constitutional isomers having formula C_5H_{12} are provided below:

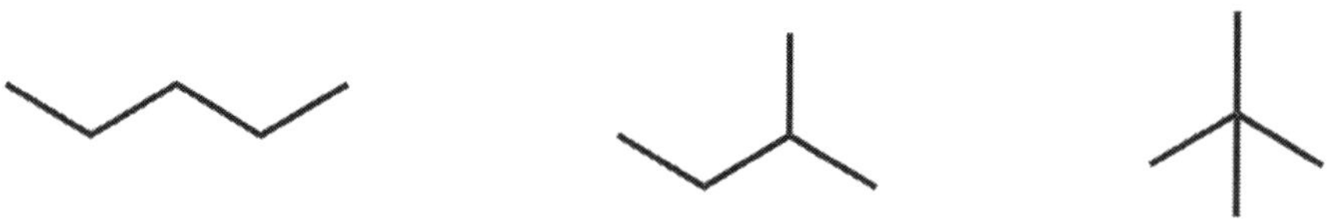

Lesson I.14.2. Naming Branched Alkanes and Alkyl Halides

We will now use a series of examples to illustrate how to use the following six-step guide for naming alkanes. Note that, for most alkanes, you will only need some of these rules.

1) Find the longest chain. This is the 'parent chain'; the other things coming off of the parent chain are its substituents.
2) Number the carbon atoms in the 'parent chain' in the way that gives the lowest number to the substituent closest to an end of the parent chain.

3) If more than one type of substituent is present, list them in alphabetical order in the name.
4) If more than one of the same substituent are present on your parent chain, use di, tri, tetra, etc., prefixes to denote this (these prefixes do not count when alphabetizing, though; neither do the *n*-, *sec*-, or *tert*- prefixes; however, the *iso* prefix DOES count)
5) If numbering leads to the same lowest number substituent in either direction, the correct numbering gives the lowest number to the substituent that is first alphabetically.
6) If you find two different possible parent chains of the same length, you choose the one with more substituents coming off of it.

Consider the molecule on the left. The longest chain (step 1) has five carbons, so the parent chain is pentane (highlighted on the right).

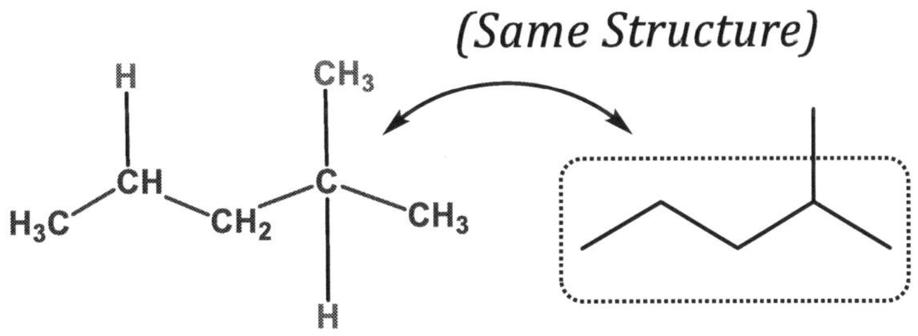

Having identified the parent chain, we move on to step 2, numbering the carbon atoms in the parent chain such that the substituent (here a methyl) has the lowest number:

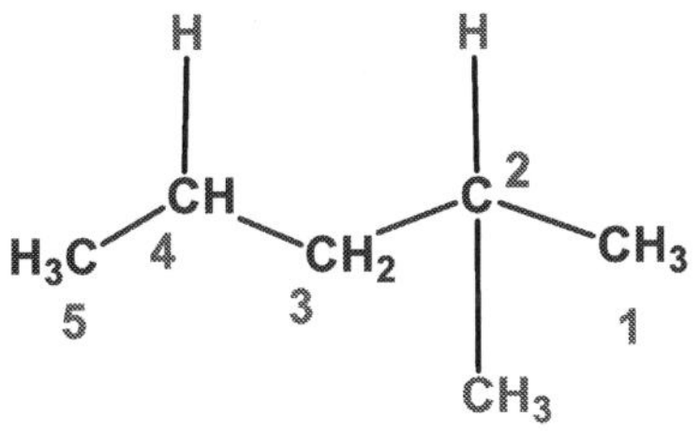

This molecule is thus called **2-methylpentane**. Note that there is always a dash between a numeral and the substituent to which it refers. Consider another example, in which we have already applied rules 1 and 2, and have found a heptane chain having a methyl substituent at position 2 and an ethyl substituent at position 4.

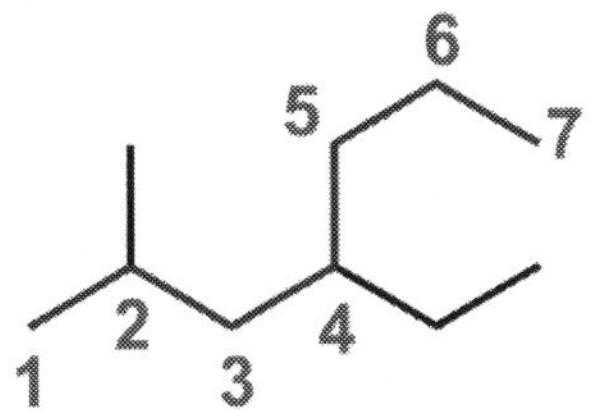

We now need to apply rule 3 (list substituents alphabetically) to compose the final name of **4-ethyl-2-methylheptane**.

In a case in which a parent is substituted by more than one of the same substituent, for example:

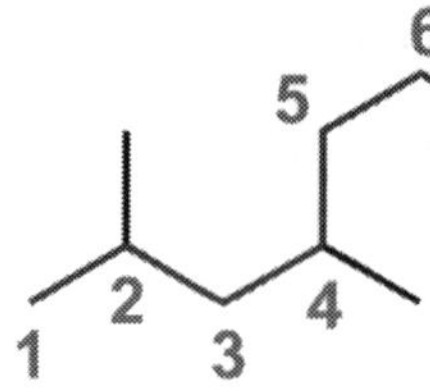

We will need rule 4 involving use of di-, tri-, etc. prefixes. The molecule above is thus properly called **2,4-dimethylheptane**.

If we encounter a case like this:

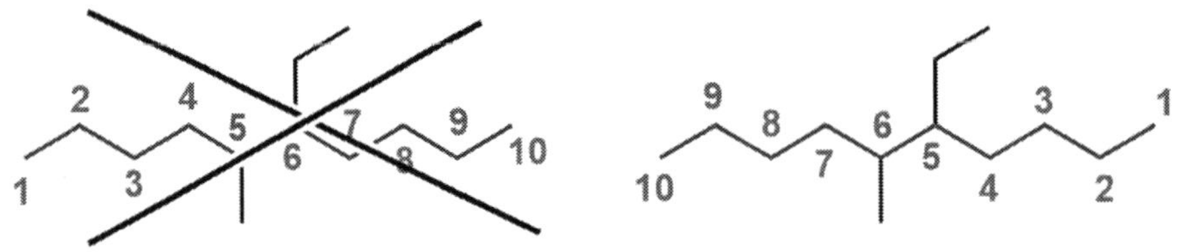

In which there are two possible ways to number the parent chain, we need to apply the fifth rule. This rule tells us that in a tie, give the alphabetically first substituent the lower number, as shown on the right, above. This approach allows us to properly name the molecule as **5-ethyl-6-methyldecane**.

We may encounter a molecule in which two possible longest parent chains are identified, as in the example below.

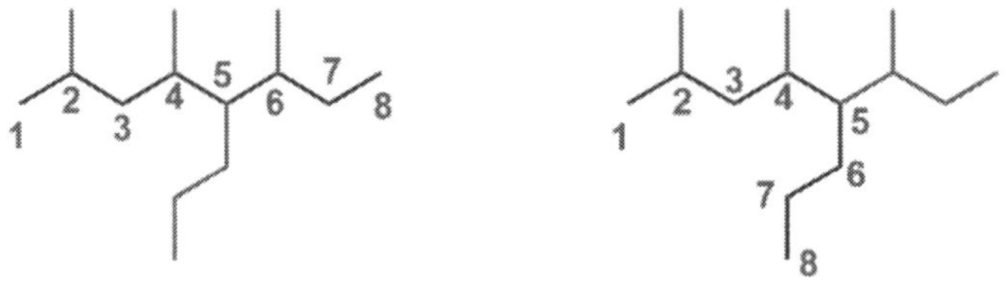

In these cases, rule 6 tells us to choose the parent chain having more substituents (the one on the left), so this molecule is properly called **2,4,6-trimethyl-5-propyloctane**.

Alkyl halides are named using the same set of rules as are alkanes, as illustrated by the example below:

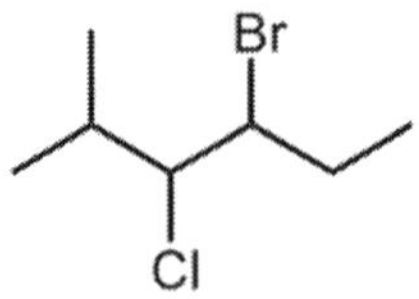

4-bromo-3-chloro-2-methylhexane

Lesson I.14.3. Naming Substituted Cycloalkanes/ cis- and trans- Isomers

Alkanes that have a cyclic structure, rather than linear, are called **cycloalkanes**. The structures and names for the most common unsubstituted cycloalkanes are provided below.

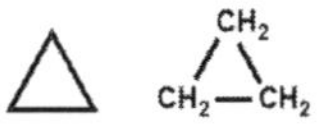

cyclopropane

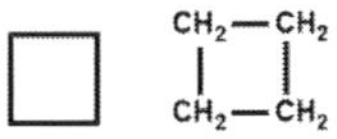

cyclobutane

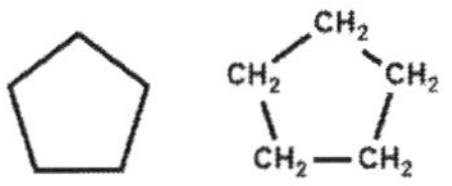

cyclopentane

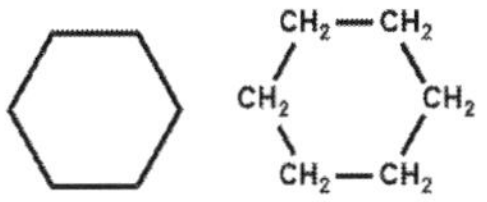

cyclohexane

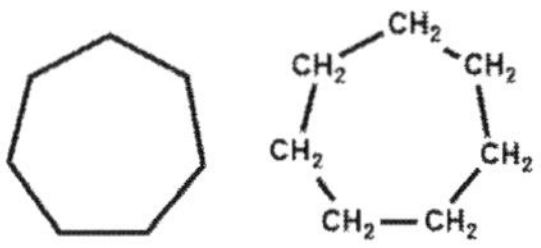

cycloheptane

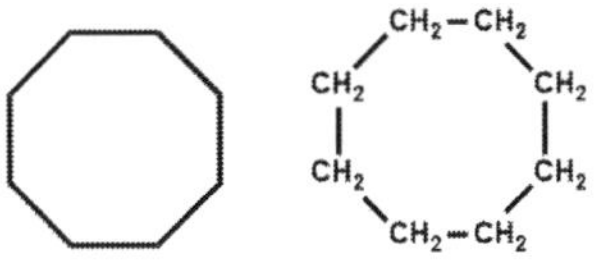

cyclooctane

A cycloalkane may be substituted just as a linear alkane may be substituted. The cycloalkane can also be the parent chain, if it is the longest contiguous chain of carbon atoms in the molecule. A parent chain may be composed *either* of C atoms in a cycloalkane *or* C atoms in a linear chain. One may not mix C atoms from both types of alkanes to form a parent chain, so:

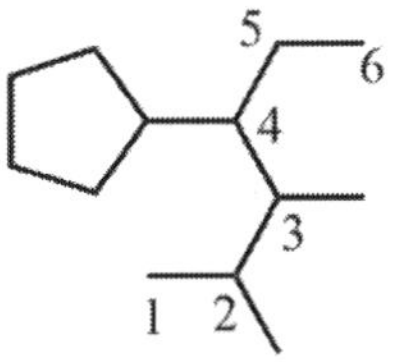

4-cyclopentyl-2,3-dimethylhexane

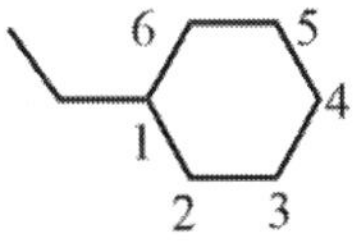

1-ethylcyclohexane

The rules for naming linear alkanes also apply to cycloalkanes. There is one additional consideration when naming cycloalkanes, however. A ring of atoms has a "top" face and a "bottom" face, as indicated in the picture below.

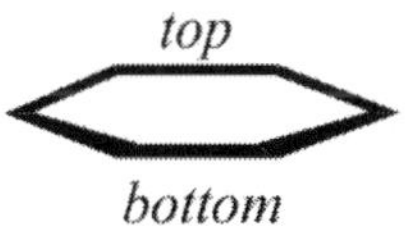

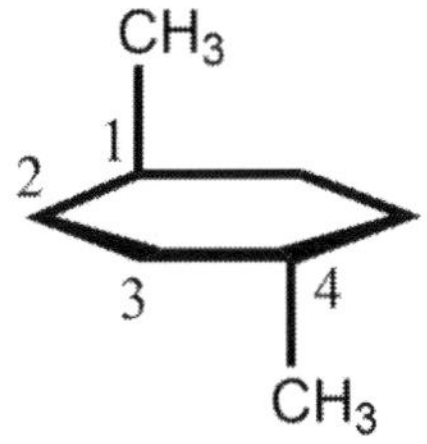

trans-1,4-dimethylcyclohexane

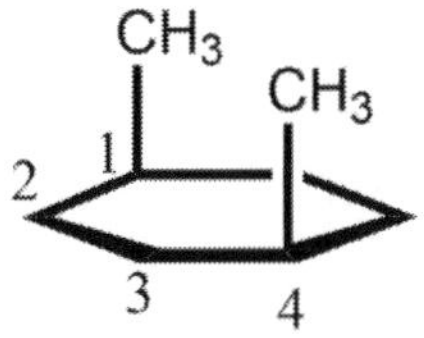

cis-1,4-dimethylcyclohexane

Because there are two faces of the ring, we will need to provide a prefix in front of the name of a disubstituted cycloalkane to indicate whether one group is oriented a face opposite of the other (we use the *trans*- prefix for this) or if the two groups are oriented on the same face (we use the *cis*- prefix in such cases). The two examples above illustrate this convention. The *trans*- and *cis*- forms of a disubstituted cycloalkane are examples of **configurational isomers**. Configurational isomers have the same constitution (all atoms are attached to the same atoms in both cases) and differ only in directions that the groups point off of the structure. Configurational isomers also **cannot interconvert without σ-bond breakage**: if we wanted to change the *cis*- into the *trans*- isomer, we would have to break the σ-bond to the Me substituent and switch its position, for example. We will cover other types of configurational isomers in more detail throughout the rest of this text.

Lesson I.14.4. Types of Carbon Atoms

In discussing organic molecules, it is often useful to discuss functional groups in terms of what type of carbon is attached to the functional group. Carbons can be referred to a primary (1°), secondary (2°) or tertiary (3°).

A "primary" carbon is bonded to **one carbon**. Hydrogens on this carbon are "primary".
A secondary carbon is bonded to **two carbons**. Hydrogens on this carbon are "secondary".
A tertiary carbon is bonded to **three carbons**. Hydrogens on this carbon are "tertiary".
A quaternary C is bonded to **four carbons**. These C have no hydrogens.

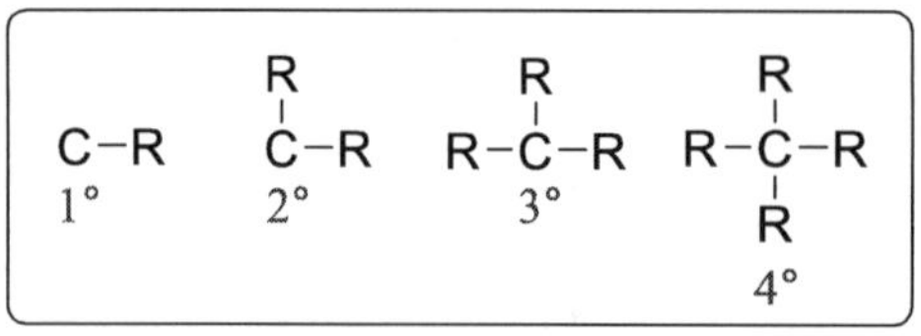

R = some group with a C bound to the C that is shown

Example I.14.1.

Label each carbon as 1°, 2°, or 3°

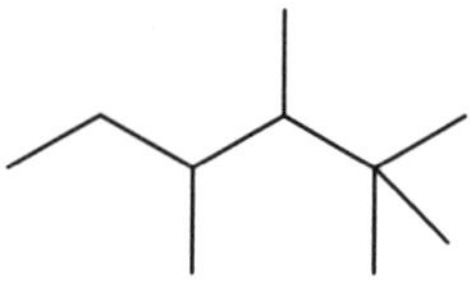

Solution I.14.1.

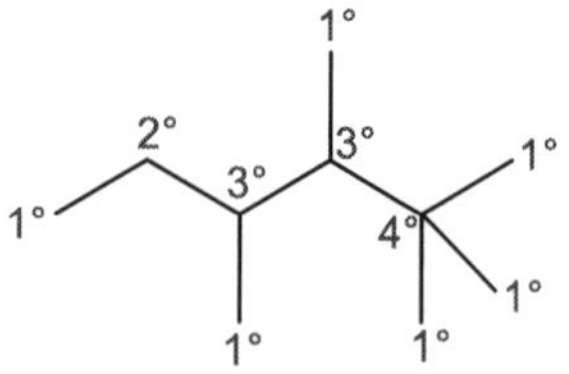

The types of Alkyl Halides (R—X) can be subdivided based on what type of carbon is attached to the halogen (X = F, Cl, Br or I). A primary alkyl halide as the halogen on a primary C, etc.:

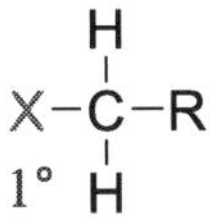

primary alkyl halide

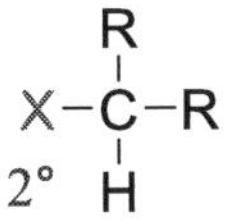

secondary alkyl halide

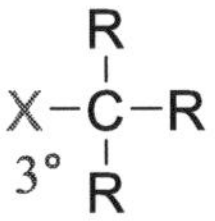

tertiary alkyl halide

Lesson I.14.4. Naming Alcohols

For alcohols, use the rules for naming alkanes and cycloalkanes as a starting point, with the following adjustments: Replace the "e" at the end of the alkane name with "ol".

1) The alcohol is always given the lowest possible number. Note that this means that the alcohol is always given the "1" position in cycloalkanes (so there is no need to add a number, because it is always 1).
2) Place the number indicating the position of the alcohol directly before the parent chain name (which now ends in "ol").

Again, we will illustrate how the rules are applied with examples:

CH_3OH methanol

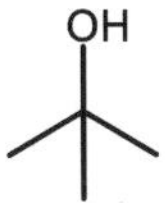

2-methyl-2-propanol (or *tert*-butanol)

CH_3CH_2OH ethanol

1-propanol

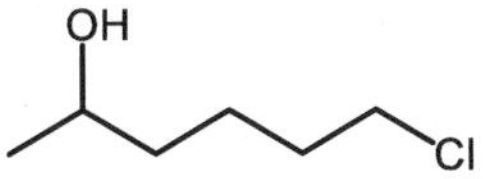

6-chloro-2-hexanol

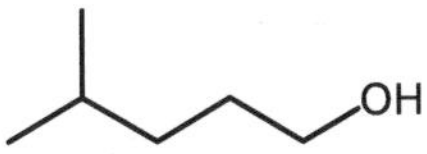

4-methyl-1-pentanol

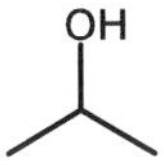

2-propanol

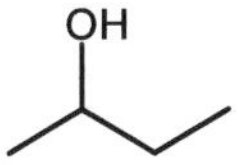

2-butanol (or *sec*-butanol)

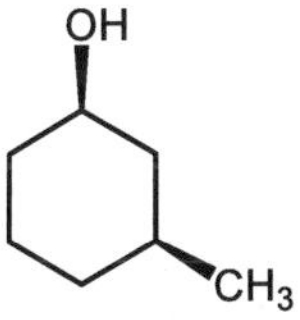

cis-3-methylcyclohexanol

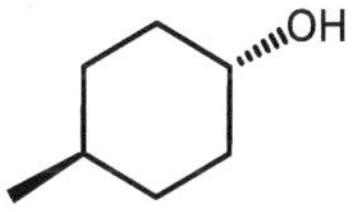

trans-4-methylcyclohexanol

Just as we saw for alkyl halides, alcohols can be referred to as primary (1°), secondary (2°) or tertiary (3°) depending on the carbon attached to the hydroxyl (–OH) group:

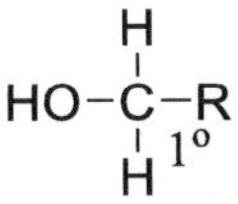

primary alcohol

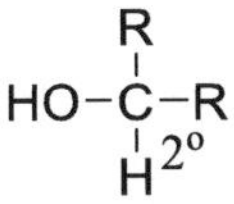

secondary alcohol

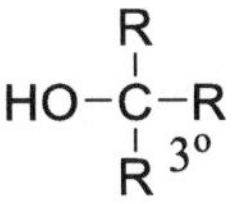

tertiary alcohol

Lesson I.15. Isomerism and Conformational Analysis I: Linear Alkanes and Newman Projections

Lesson I.15.1. Newman Projections and Conformational Isomers of Ethane

Now that we have a strong foundation in analyzing attractive and repulsive forces in molecules, we can begin to evaluate how these forces can influence the geometries of molecules and the relative spatial orientation of their bonds. Consider a simple molecule like ethane. We can draw ethane in several possible ways, two of which are illustrated below:

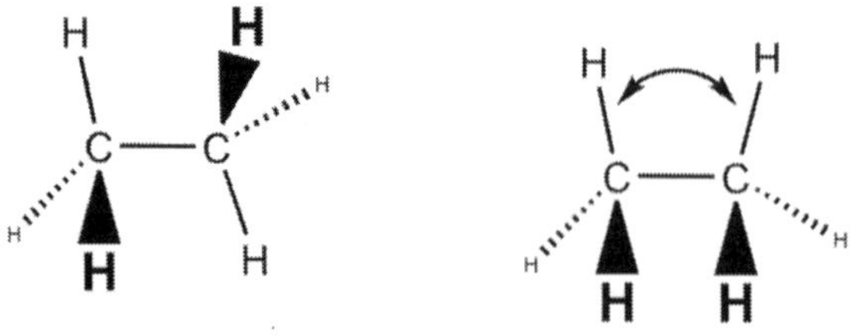

The only difference between these two representations of ethane is how the H atoms on one C are rotated relative to the H atoms on the other C. Structures that differ only by the angle of rotation about a σ-bond are called **conformational isomers**. Unlike the configurational isomers discussed in Lesson I.14, conformational isomers can be interconverted by simple bond rotation, so we do not need to add a prefix in front of the name to distinguish these separate species (because they interconvert easily). Both of the conformations shown in the picture are simple called "ethane" as a molecular name. Note, however, that the two conformations do not have the same stability. As indicated by the double headed arrow in the structure on the right, the C–H bonds are pointed in the same way on each of the two carbons. Remembering that a bond comprises two electrons and that electrons repel one another, placing the C–H bonds close together like this will be a less stable arrangement than the conformation on the left, in which the bonds are staggered and thus farther apart from one another. This is easier to see if we take a look at the molecule from a perspective looking down the C–C bond:

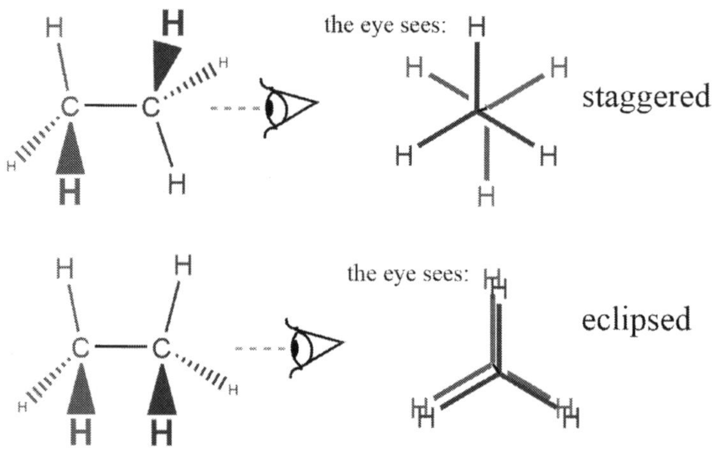

Note that it is somewhat difficult to discern which H atoms are attached to which C in the drawings on the right, so chemists use what are called **Newman Projections** for a clearer view of attachment:

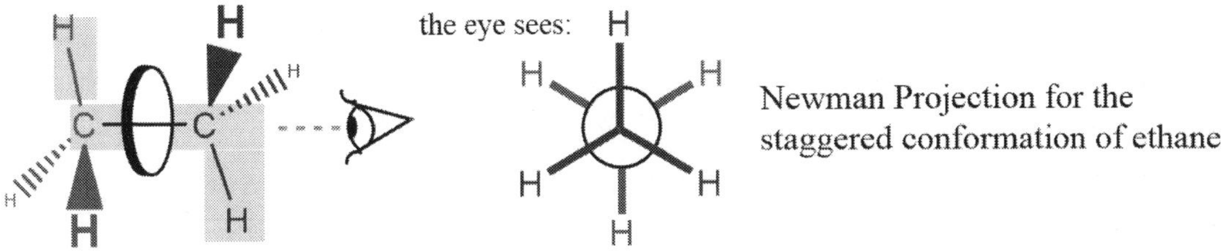

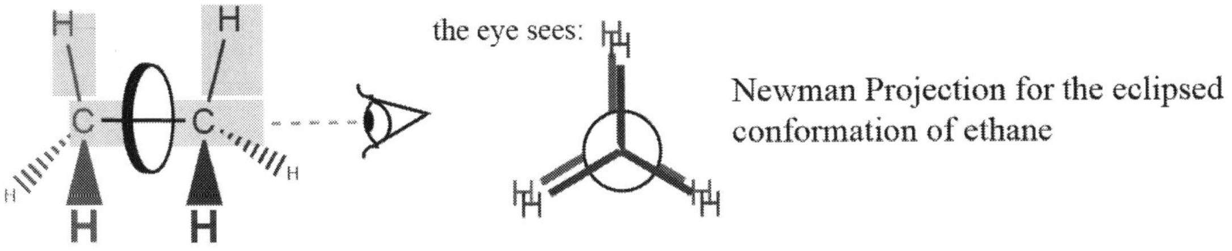

We can easily see from the Newman Projections on the right that the eclipsed conformation of ethane places the electrons in the C–H bonds on the front C closer to the C–H bonding electrons on the back carbon, whereas the staggered conformation minimizes this by placing these C–H bonds as far apart as possible. A staggered conformation along a bond is thus more stable than the eclipsed and is the conformation that most ethane molecules will have in a given sample.

Lesson I.15.2. Conformational Isomers of Butane

We can carry out the same type of analysis for more complex molecules as we did for ethane. When we undertake such conformational analysis of linear alkanes, we need to keep in mind two basic principles:

1) Eclipsed bonding pairs have more repulsion than do staggered bonding pairs
2) The larger the groups, the greater will be the repulsion

If we start off an analysis of butane, we will find that we can draw it in four unique staggered or eclipsed Newman Projections along the C(2)–C(3) bond:

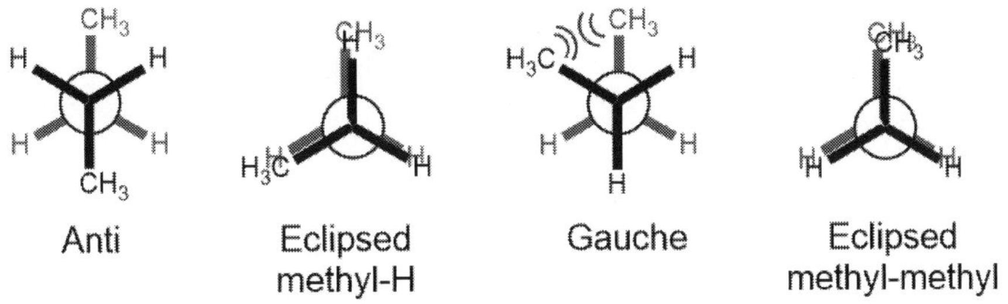

We can easily determine the relative stability of these structures by applying two principles. First, each of the two staggered conformations (Anti and Gauche) are more stable than either of the eclipsed conformations (Eclipsed methyl–H or Eclipsed methyl–methyl). Secondly, the Gauche conformation is less stable than the Anti conformation because the Gauche has the two methyl groups close enough to one another to induce some repulsion (called a **Gauche interaction**). Although not nearly as destabilizing as are eclipsing interactions, the Gauche interaction still makes the Gauche conformation less stable than the Anti by about 0.8 kcal/mol.

By a similar line of reasoning, it is evident that the Eclipsed methyl–methyl conformation is less stable than is the eclipsed methyl–H conformation because the eclipsed methyl–methyl places the two largest groups closest to one another. On the basis of the foregoing discussion, we can rationalize the diagram below, which is a plot of C(2)–C(3) σ-bond rotation angle versus energy:

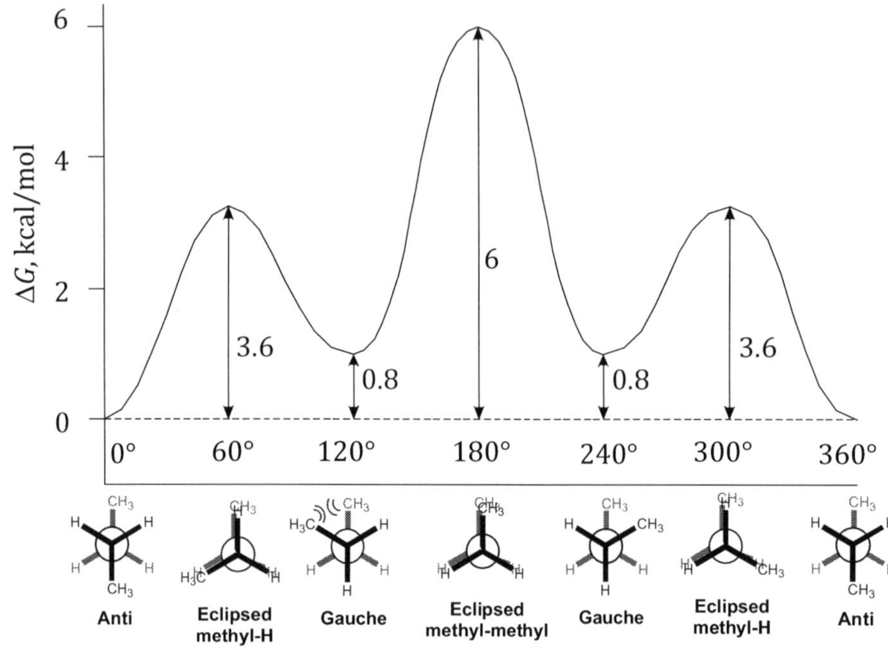

We can apply this type of analysis to any number of organic compounds to rationalize and predict stability trends, which we will do specifically for cycloalkanes in the next lesson.

Lesson I.16. Conformational Analysis II: Cycloalkanes and the Chair Conformation of Cyclohexane

Lesson I.16.1. Strain in Cycloalkanes

Cycloalkanes can be subject to constrained geometries that lead to deviations from ideal bond angles. For example, cyclopropane with equal C–C bond angles will be an equilateral triangle, with angles of 60°. This angle differs from the ideal angle of 109.5° for an sp^3-hybridized C atom. The 'ideal' angle is the angle that gives the strongest bond, thus holding the molecule together most tightly (i.e., making it most stable). Deviating from the ideal angle thus leads to weaker bonds and less stable molecules. The decreased stability that results from the bonds being strained from the ideal angle is called **angle strain**. The angles that would be present in the planar forms of some cycloalkanes are provided here:

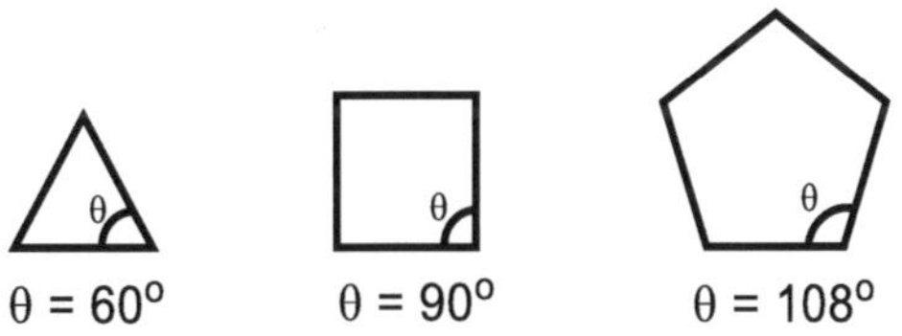

We saw in the previous lesson that certain conformations of molecules, specifically those in which there are eclipsing interactions, are less stable than are staggered conformations. When atoms are constrained in a ring, there is not free rotation about a C–C bond (we could not, for example, rotate 360°), so it is sometimes impossible to alleviate eclipsing interactions in the way we did with ethane and butane. For this reason, there can also be some steric strain induced in cycloalkanes. The sum of all steric and angle strains together is often referred to as **ring strain**.

An example of steric stain induced by molecular geometry is in cyclopropane. Cyclopropane is locked in a conformation in which every H is eclipsing two others (the Newman Projection below shows the perspective down one C–C bond in the right-hand image):

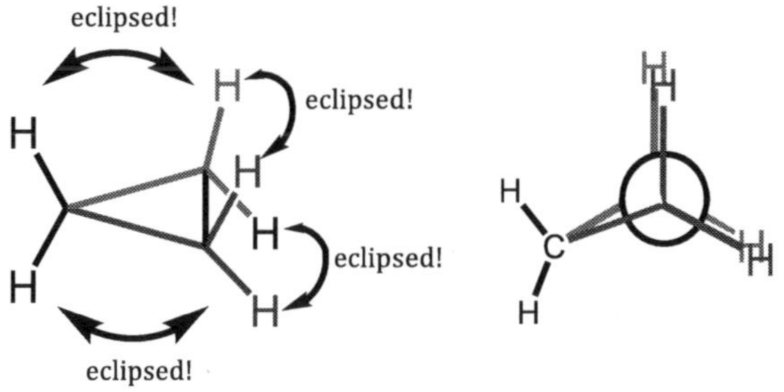

If cyclobutane were planar, it would suffer from the same destabilizing forces (left image, below). As ring size increases, however, the flexibility of the ring also increases. This flexibility allows the cyclobutane to "pucker" a bit (i.e., slight bond rotations occur) in response to eclipsing interactions to alleviate some of the steric strain (right image, below):

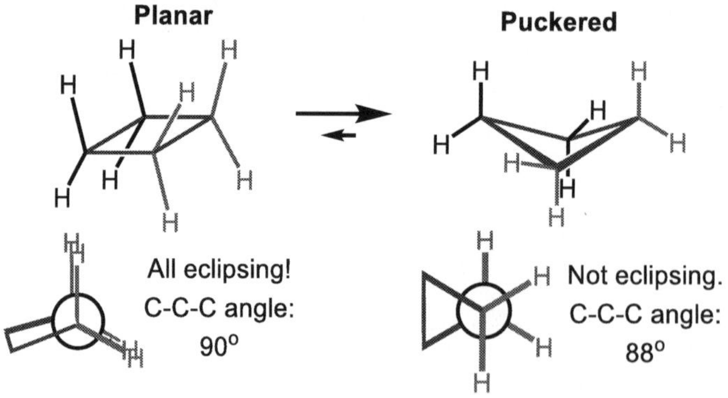

Similarly, cyclopentane distorts from a planar conformation to an envelope conformation, which is the actual form found in solution, via what is known as an **envelope distortion** (right image):

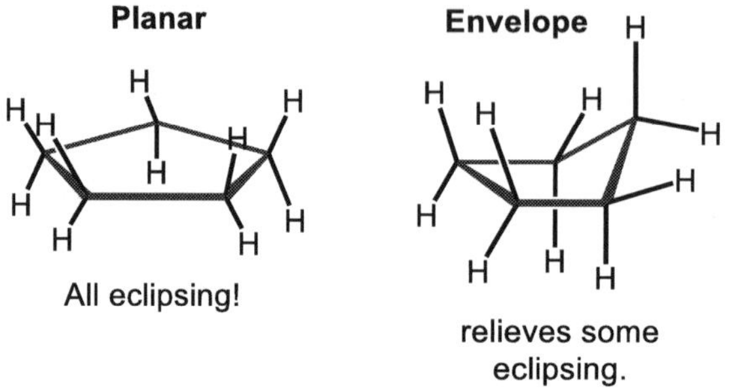

The relative strain per CH_2 unit in cycloalkanes trends in the order cyclopentane < cyclobutane < cyclopropane because flexibility increases as the ring size increases (which allows greater C–C bond rotation), thus allowing greater ability to alleviate steric strain. This effect becomes dramatic in the case of cyclohexane, as detailed in the next section.

Lesson I.16.2. The Chair Conformation of Cyclohexane

In cyclohexane, the flexibility of the ring allows two envelope distortions to occur (one up and one down) to produce what is called the **chair conformation** of cyclohexane:

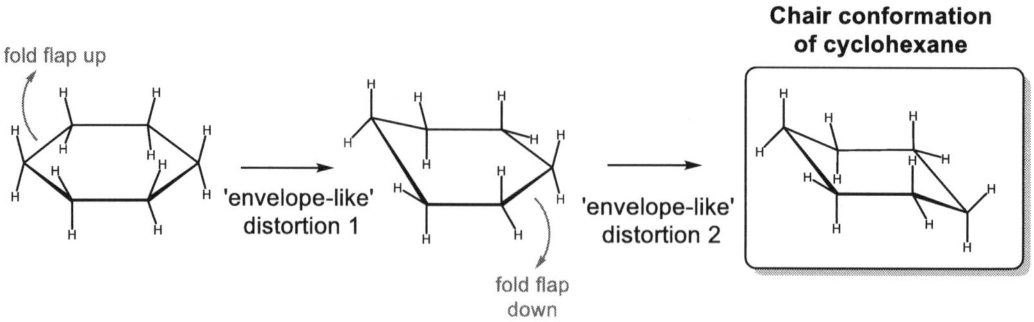

In the chair conformation, all bond angles are ideal, so there is no angle strain. Additionally, there are no eclipsing steric strains. Thus, **the chair conformation of cyclohexane thus has ~0 ring strain:**

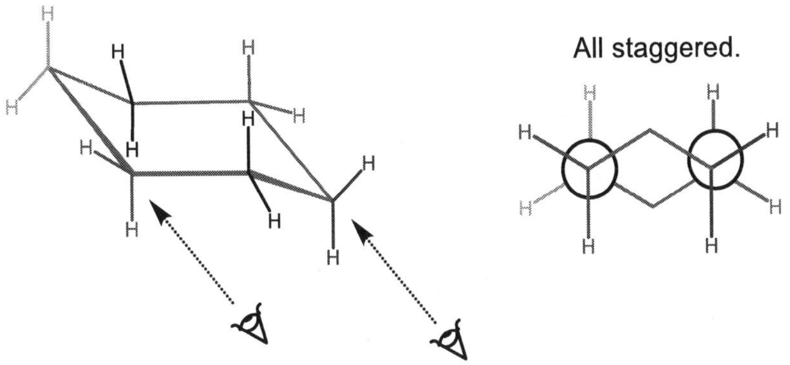

One notable feature of the chair conformation that becomes important when we start adding substituents is that there are two types of sites with H atoms: those that point straight up or straight down around are at **axial** sites, those that project horizontally from the ring are at **equatorial** sites:

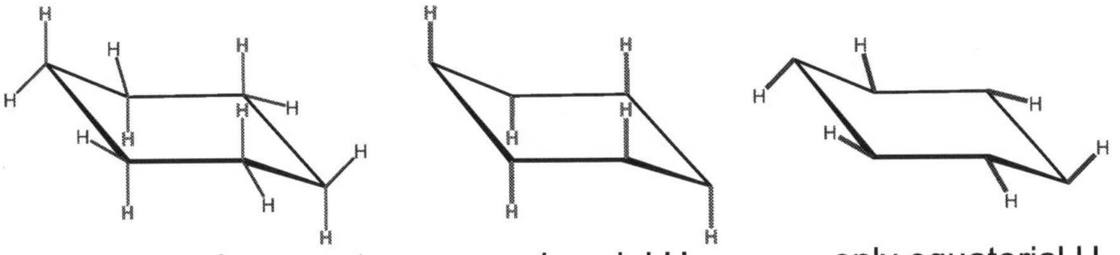

If there is enough thermal energy (i.e., at sufficiently high temperatures), the cyclohexane molecule can undergo a **ring flip**, where a conformation in which the "left tip" is up and the "right tip" points down interconverts with a conformation in which these "tips" are inverted:

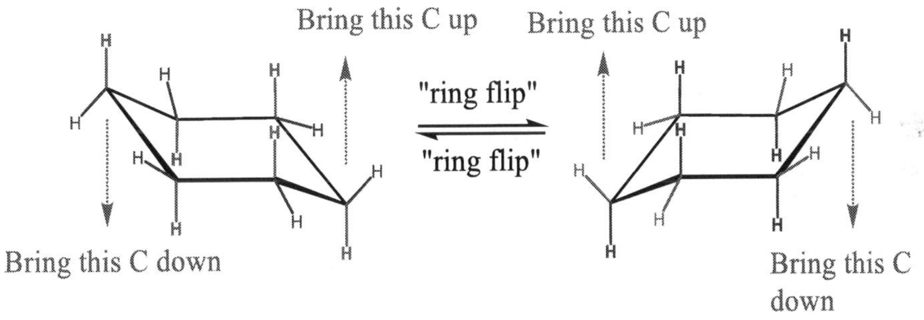

Ring flips are common at room temperature and at reaction temperatures, and later in this text we will encounter cases in which the flip is required to accommodate a reaction process. Note that each C has an axial and an equatorial site. Each C also has one 'up' and one 'down' site. When substituents are added, the ring flip may become more energetically unfavorable, as we will discuss in the next section.

Lesson I.16.3. Substituted Cyclohexane

Now that we have a good grasp of the parent chair cyclohexane, we can examine the extent to which substituents may influence stability of the ring. If we place anything larger than an H atom on one of the axial sites of cyclohexane, it will introduce a destabilizing Gauche interaction to the structure similar to what we saw in butane:

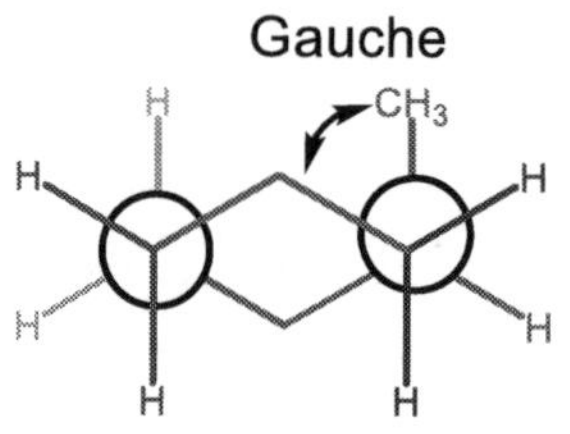

Methylcyclohexane

If two larger-than-H groups are placed at axial sites on the same face of the ring, the two may even be able to undergo some steric repulsion known as a 1,3-diaxial interaction:

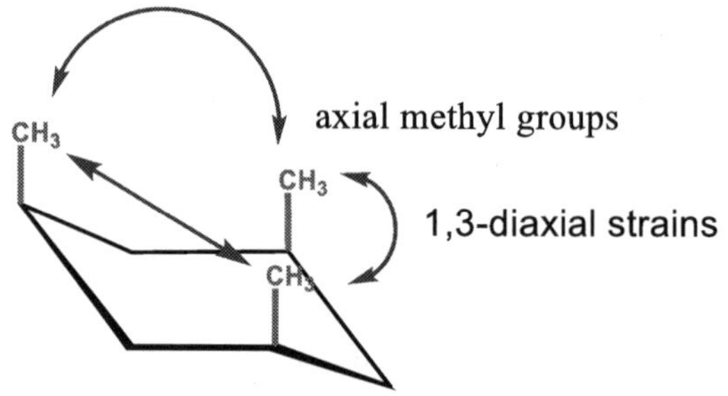

1,3,5-trimethylcyclohexane

If, however, the larger-than-H substituents are at equatorial sites, then a fully-staggered, Gauche-free conformation is maintained, because the equatorial substituent is *anti* to the ring on the adjacent C:

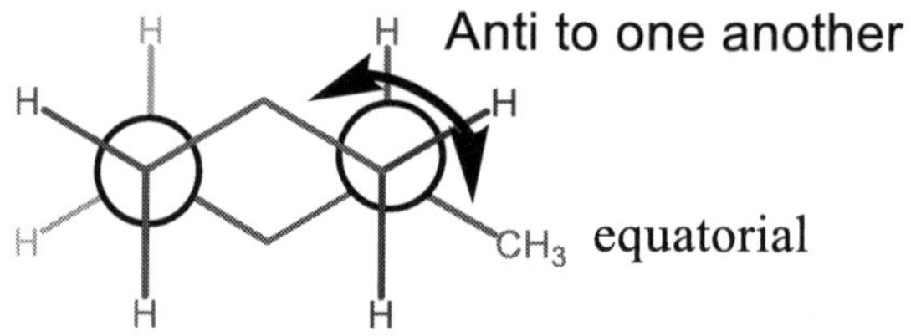

The overarching lesson we learn from these observations is that the more stable conformation is the one in which the larger substituent is in the equatorial position.

Example I.16.3

Draw the most stable conformation of *cis*-1-*t*-butyl-4-ethylcyclohexane.

Solution I.16.3

Start by placing the largest substituent in the equatorial site so that its steric strain is minimized, and fill in the equatorial and axial spots on carbon #4, where the next substituent needs to go:

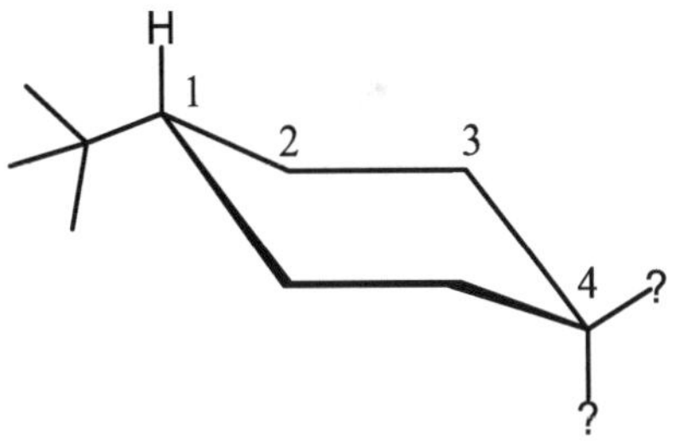

Next, use the information in the name of the molecule to place the ethyl group. The ethyl group is at position 4, and it is *cis*- to the *t*-butyl group. We placed the *t*-butyl group pointing down, and *cis*- indicates that the two groups must point the same direction, so the ethyl group must also point down. We have only one choice for its placement:

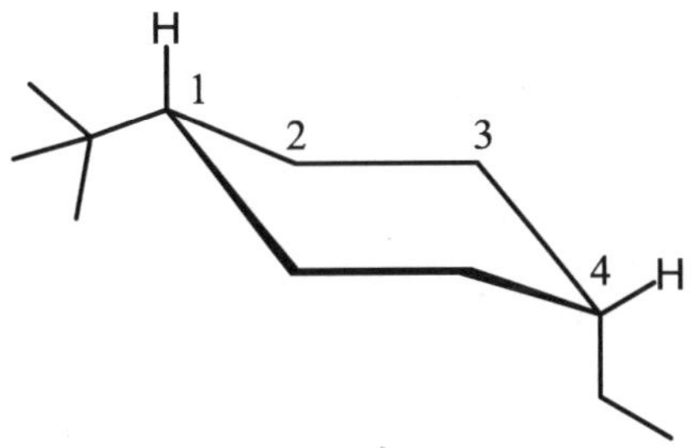

Lesson I.17. Stereochemistry I: Chirality and Optical Activity

Lesson I.17.1. Chirality and Configurational Isomerism

We saw in Lesson I.14 that isomers that have the same constitution but cannot be interconverted without bond breakage are called configurational isomers. The configurational isomers that we discussed were the *cis*- and *trans*-isomers of cycloalkanes. We will now discuss another type of configurational isomerism. A molecule that has "handedness" is said to be chiral. To identify what molecules have "handedness", it might be helpful to think about what makes a right hand and a left hand different. After all, both hands have four fingers and a thumb attached to a palm. No matter how we rotate a right hand, however, it is not superimposable (not identical) with a left hand. They are mirror images of each other. If we think of a hand has being four surfaces, it may help. A hand has 1) a palm, 2) a backhand, 3) a thumb side and 4) a little finger side. If the hand was the same on front and back or if the hand was symmetric on the front (like a cartoon hand with three fingers), then we could superimpose them. However, because all four 'sides' are different, the hands have what we have come to call "handedness". For molecules, we use the term "**chirality**" in place of "handedness". A molecule that possesses chirality is said to be **chiral**. A tetrahedral C atom has four substituents much as a hand has four "sides". If all four substituents coming off the tetrahedral C are different, we say the C atom is a **chiral center** (alternatively known as a **chirality center**, **stereocenter**, or **stereogenic center**, in various books). If any two substituents coming off of the C are identical, the C is not a chiral center. We sometimes indicate stereocenters in a molecule with an asterisk (*).

Example I.17.1

Which carbon atoms in these structures are stereocenters?

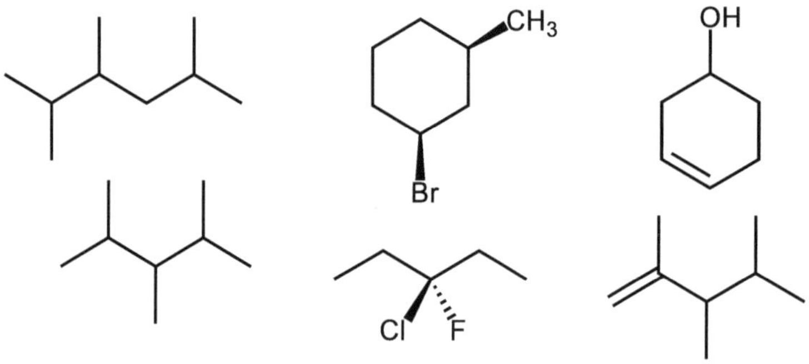

Solution I.17.1

Our goal here is to find each C atom that has four different groups coming off of it. It may help to draw out the H atoms to visualize this. The stereocenters are labelled with asterisks below:

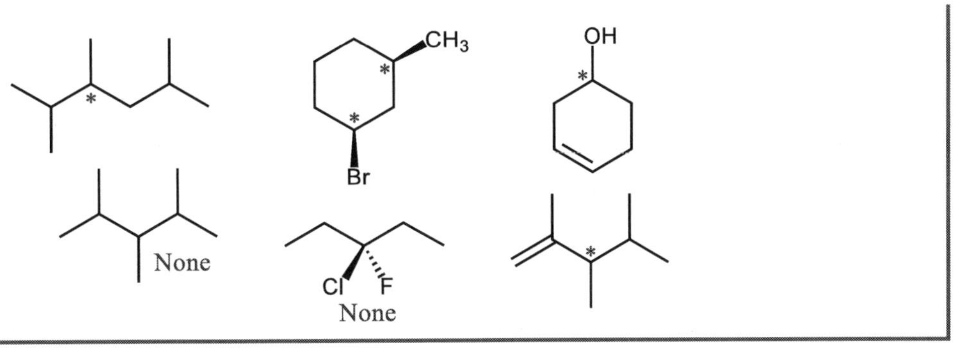

Lesson I.17.2. Properties of Stereoisomers

A chiral molecule and its mirror image share many of the same chemical properties: the two will have the same stability, solubility, boiling point, and melting point, for example. How, then, do we tell the two apart, and why does chirality even matter? One property of chiral molecules that differs between the two mirror images is the direction in which they rotate plane polarized light. For this reason, a chiral molecule is also said to be **optically active**. An instrument called a polarimeter is used to measure the angle by which plane polarized light is rotated by a given sample (units are degrees rotated from vertical; positive values for clockwise rotations, negative values for counter-clockwise rotations). A simplified drawing of a polarimeter is provided here:

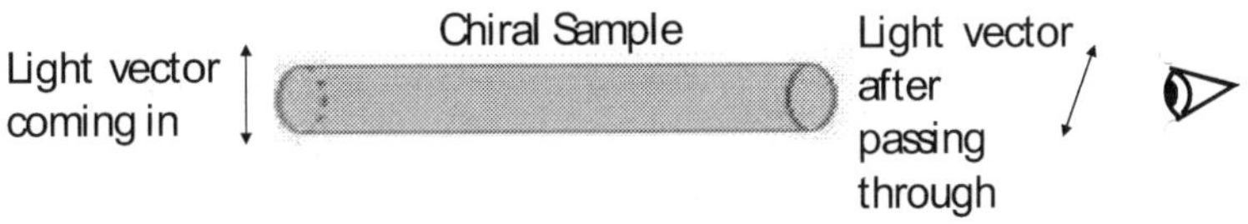

A chiral molecule that causes clockwise rotation of the plane polarized light vector is called a **dextrorotatory** molecule, and a "(+)-" symbol is placed in front of that molecule's name. If the mirror image of this dextrorotatory isomer is measured at the same concentration, we would instead observe a counterclockwise rotation of the light vector by the same angle. The molecule that is the mirror image of a dextrorotatory molecule is **levorotatory**. A "(–)-" symbol is placed in front of a molecule's name to indicate it is a levorotatory molecule. A sample of **achiral** (not chiral) molecules will not cause the plane polarized light vector to rotate at all, and such samples are said to be **optically inactive**. Because the dextrorotatory and levorotatory isomers rotate the vector by the same angle but with opposite signs (i.e., in opposite directions), a sample containing dextrorotatory and levorotatory isomers in equal amounts will exhibit no optical activity (the + and – cancel out). A 1.1 mix of dextrorotatory and levorotatory isomers is termed a **racemic mixture** (or **racemate**) to distinguish it from samples of achiral molecules.

Lesson I.18. Stereochemistry II: Cahn-Ingold-Prelog Rules and Assigning *R*- and *S*- Labels of Configuration

Lesson I.18.1. The Cahn-Ingold-Prelog Rules

In the previous lesson, we observed that the dextrorotatory and levorotatory isomers of a molecule rotate plane polarized light in different directions. The direction of light rotation, however, does not correlate with molecular structure in a predictable way. For this reason, scientists have come up with a system to identify chiral molecules that does not involve rotation of plane polarized light. The naming system is based on the relative spatial orientation of different substituents in order of priority about the chiral center. To name chiral molecules, then, we must first learn the convention for assigning priority to the substituents coming off a stereogenic atom, which is called the **Cahn-Ingold-Prelog** (CIP) rules. Consider this hypothetical structure to illustrate how these rules work:

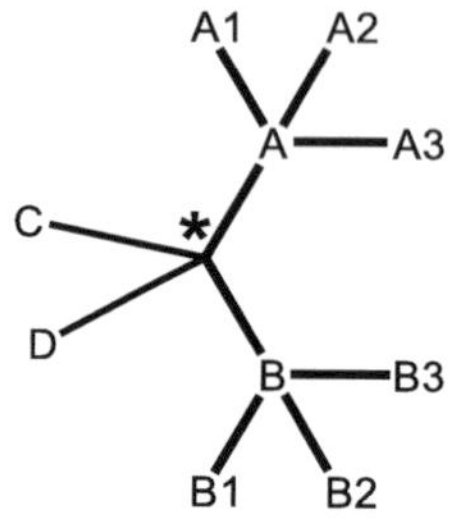

Cahn-Ingold-Prelog (CIP) rules:

1) First look at the atoms directly attached to the stereogenic atom. Higher atomic number = higher priority (A, B, C and D in the figure above).
2) If same atomic number, higher mass = higher priority (Deuterium > H, $^{13}C > ^{12}C$, etc.)
3) If atoms A and B are identical, move to highest priority atom attached to A and B until a break in the tie is found (first compare A1 to B1 priority. If tie, compare A2 to B2. If tie, compare A3 to B3, etc.)

Example I.18.1

Prioritize the substituents on the stereogenic atom in the structure below (1 for highest priority to 4 for lowest priority).

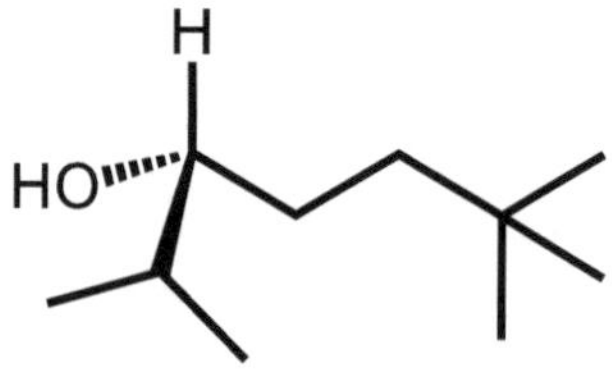

Solution I.18.1

The directly-attached atoms are O, H, C and C. The OH substituent is the highest priority (1), the H is the lowest (4), and the two C-attached substituents are tied, so we move out:

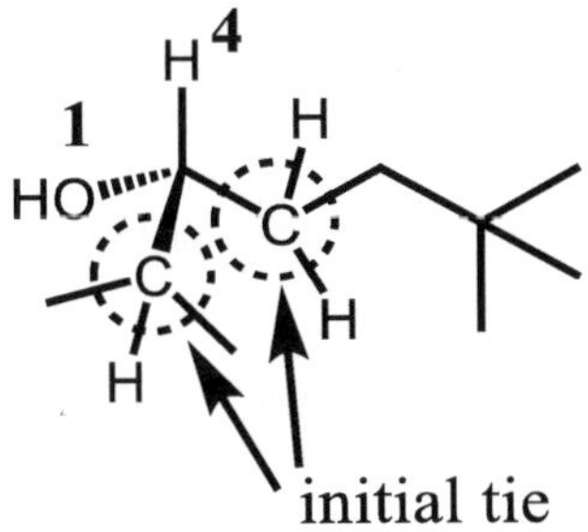

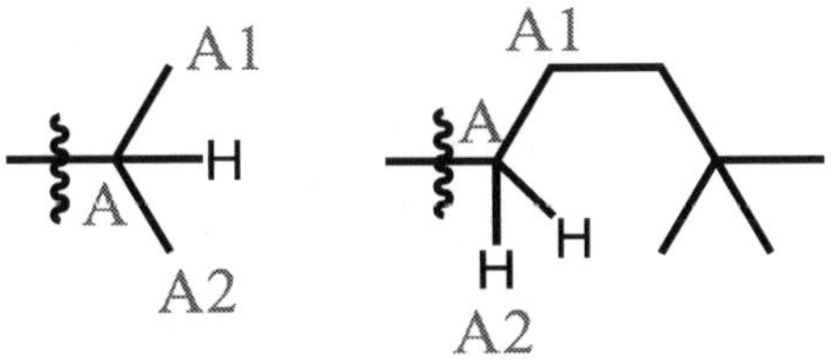

comparison of two initially-tied groups

We still have a tie upon comparing the "A1" atoms (both C), so we compare "A2" atoms. Finally, the tie is broken because the isopropyl substituent has a C at "A2", whereas the other substituent has an H. Our final prioritization is thus:

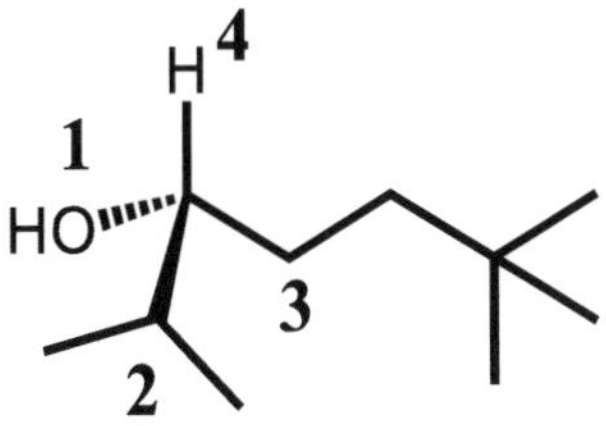

Cahn-Ingold-Prelog (CIP) rules (cont'd):

4) If a substituent is doubly or triply bonded to another atom, use the 'break and duplicate' strategy to create 'false atoms' as a visual aid to prioritize:

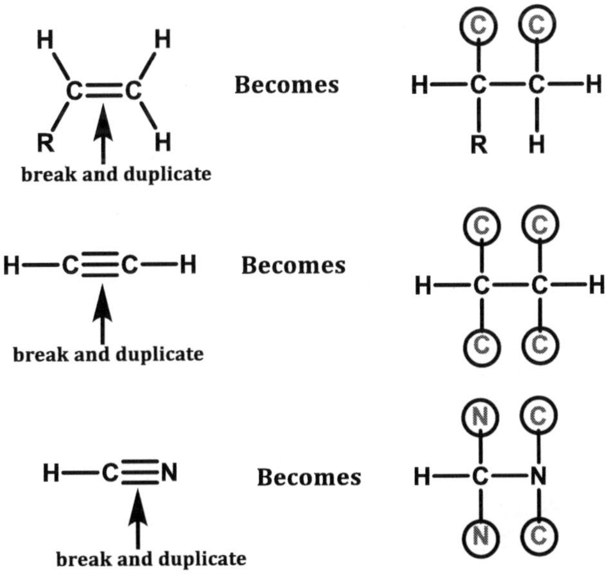

Lesson I.18.2. Assigning R- and S- Labels of Configuration

In the previous section, we learned the CIP rules to prioritize substituents as a first step to naming chiral molecules from their structures. We will now learn how to name the chiral molecules. When we label hands, we use the terms "right" and "left" as indicators of the handedness. When naming molecules, we use the Latin terms for right (*rectus*) and left (*sinister*) to indicate molecular handedness. We place a label of configuration in front of the molecule's name (i.e., a stereochemical prefix) so the reader knows to which isomer the name refers: "(*R*)-" for "rectus" and "(*S*)-" for sinister.

To determine whether an isomer is (*R*)- or (*S*)-, we follow these steps:

1) Assign priorities to the four groups (using CIP rules)
2) Point the **lowest** priority group (4th place) away from you
3) Totally ignore the 4th priority group now and determine the direction of procession from **1→2→3** priorities. If the procession is clockwise, that chiral center has an "(*R*)-" configuration. If the procession is in the counterclockwise direction, that chiral center has an "(*S*)-" configuration.

Example I.18.2

Assign a configurational label to this molecule:

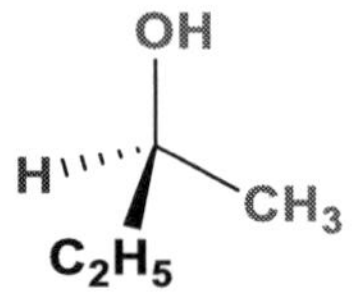

Solution I.18.2

First assign priorities for each of the substituents:

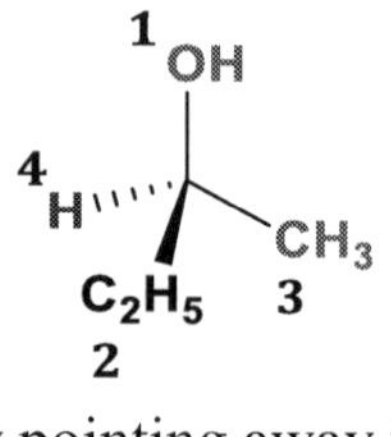

Here, the 4th place substituent is already pointing away from you, so simply draw an arc from 1 to 2 to 3, ignoring the 4th place substituent:

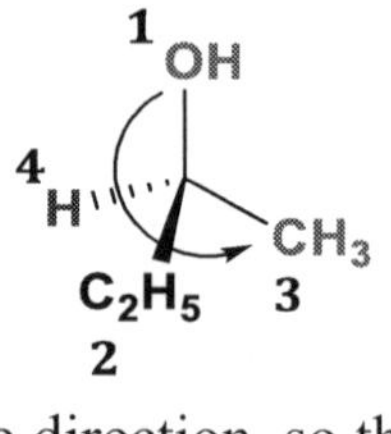

Here, the arc goes in a counterclockwise direction, so this molecule is in the (*S*)- configuration. Its full name would be written as (*S*)-2-butanol.

If the lowest priority substituent is not initially pointing back, you will need to rotate the molecule to point the 4th substituent back. You can do this by building a model and physically turning it or, as you become more proficient with these problems, mentally picturing the molecule and rotating it. One alternative strategy to consider is, if the lowest priority atom is pointing towards you (opposite of what you want), you can instead count 3→2→1 (opposite of normal counting), and you will still attain the correct configuration. If the lowest priority substituent is in the plane of the page, however, this alternate strategy cannot be used, and you will have to reorient the molecule or otherwise manipulate the structure. In the next lesson, for example, we will learn a convenient way to convert the wedge/hashed line structures into a representation of chiral molecules drawn such that all substituents are pointed either towards or away from the viewer. For now, though, a useful trick for assigning a configuration to a molecule presented with the 4th priority group in the plane involves knowing that **if you switch any two of the groups attached to the chiral center, you change its configuration**. Consider "Molecule A", in which the 4th priority H is in the plane:

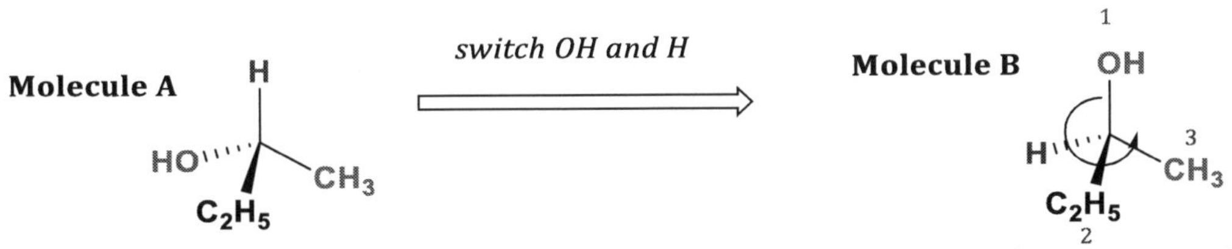

What is the configuration?

Has the opposite configuration compared to Molecule A!

We can see that once we switch the OH and the H so that H is in the back, generating Molecule B, it is relatively easy to assign an *S*- configuration to Molecule B by counting 1→2→3 in a counterclockwise fashion. Because we know that switching the two groups switches the configuration, we know that Molecule A must be in the *R*- configuration if Molecule B is in the *S*- configuration.

Helpful video examples for manipulating 3D structures to determine configuration are provided on the Proton Guru YouTube channel (search "Proton Guru Channel" on youtube.com to find the channel).

Lesson I.19. Stereochemistry III: Fischer Projections

Lesson I.19.1. Fischer Projections

Our study of stereochemistry and chirality has made the importance of three-dimensional shapes of molecules evident. The **Fischer Projection** is another way to represent the three-dimensional structure of chiral molecules. In this convention, each stereogenic atom is placed at the intersection of orthogonal lines. The groups on the horizontal lines are understood to represent substituents pointing towards the viewer from the chiral center, with groups on the vertical lines representing substituents pointing away from the viewer from the chiral center.

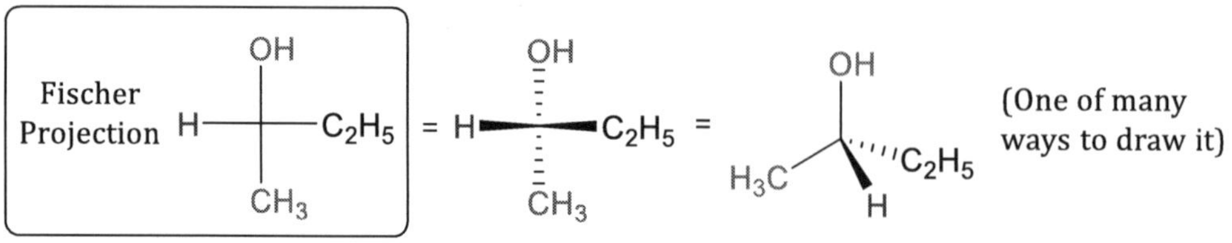

Example I.19.1

Assign the configuration of the chiral center in the Fischer Projection shown below:

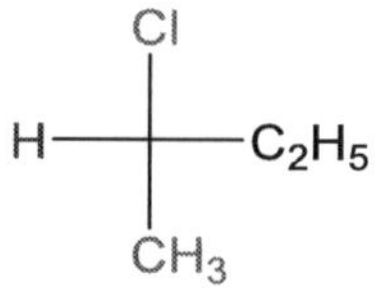

Solution I.19.1

First, prioritize the substituents:

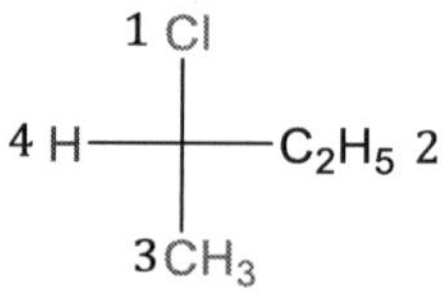

We see that the 4th place substituent (H) is towards us in the Fischer Projection convention. This is a "backwards" molecule, so we count 3→2→1, instead of 1→2→3 (see Lesson I.18.2):

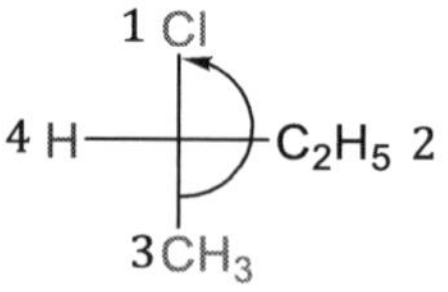

We see a counterclockwise precession from 3→2→1, corresponding to the (*S*)-configuration.

Lesson I.20. Stereochemistry IV: Enantiomers, Diastereomers and Meso Compounds

Lesson I.20.1. Enantiomers

A molecule may have only one chiral center or it may have many stereocenters. Each stereocenter in a molecule may be in the (*R*)- or (*S*)- configuration, so a molecule with "*n*" stereocenters can have up to 2^n stereoisomers. As an example, a molecule with two stereocenters at carbons 1 and 2 can have up to 2^2, or 4, stereoisomers: the "(1*R*,2*R*)-" isomer, the "(1*S*,2*S*)-" isomer, the "(1*R*,2*S*)-" isomer, and the "(1*S*,2*R*)-" isomer. Some vocabulary is needed to delineate the relationship between these isomers. Two chiral molecules that are non-superimposable mirror images are referred to as **enantiomers** and form an **enantiomeric pair**. A molecule and its enantiomer will have opposite configurations at *every* stereocenter in the molecule. If a molecule has a "(1*R*,2*R*)-" configuration, its enantiomer will have a "(1*S*,2*S*)-" configuration. Likewise, the enantiomer of a "(1*R*,2*S*)-" isomer is a "(1*S*,2*R*)-" isomer.

Lesson I.20.2. Diastereomers

Stereoisomers that are not mirror images are known as **diastereomers**. In contrast to enantiomers, two diastereomers do not necessarily have the same stability, solubility, melting point, boiling point, etc. Consequently, it is possible to separate diastereomers by distillation, recrystallization, etc., much easier than for enantiomers. Referring back to the example in Lesson I.20.1, a "(1*R*,2*R*)-" isomer will be a diastereomer of both the "(1*R*,2*S*)-" and the "(1*S*,2*R*)-" isomers.

Example I.20.1

Provide the stereochemical relationships between the compounds shown below:

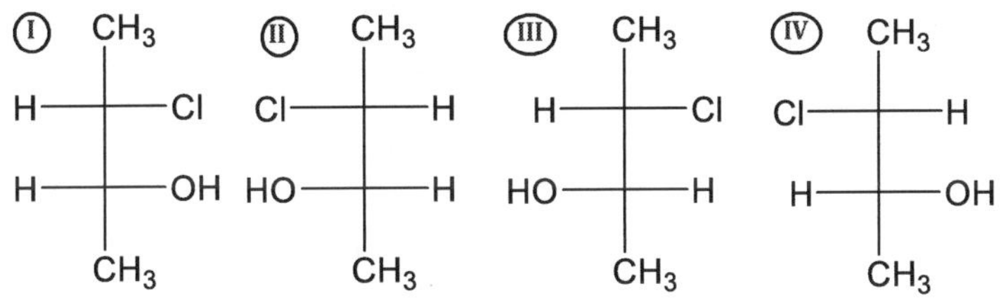

Solution I.20.1

We can see that I and II are mirror images of one another and are non-superimposable, so I and II form an enantiomeric pair. Likewise, III and IV are enantiomers. All other pairs we compare are diastereomers (I+III, I+IV, II+III, II+IV).

Lesson I.20.3. Meso Compounds

It is possible for a molecule to contain stereogenic *atoms* in its structure, but that the *molecule as a whole* is achiral. This condition occurs when one half of a molecule is the mirror image of the other half (i.e., has a plane of symmetry). Remember, a symmetric object cannot be chiral. Molecules that have stereogenic atoms but which are achiral molecules are called **meso compounds**.

Example I.20.2

Which of the following are meso compounds?

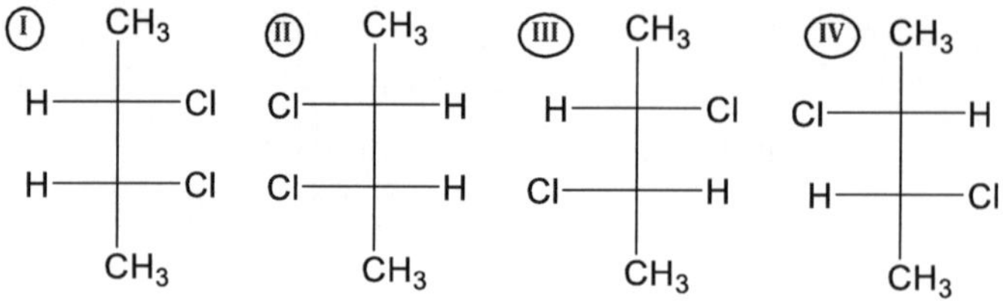

Solution I.20.3

Compounds I and II each possess a plane of symmetry (represented by the dashed lines) and are thus achiral, despite the fact that carbons 2 and 3 are stereocenters. Compounds I and II are therefore meso compounds. In fact, they are identical.

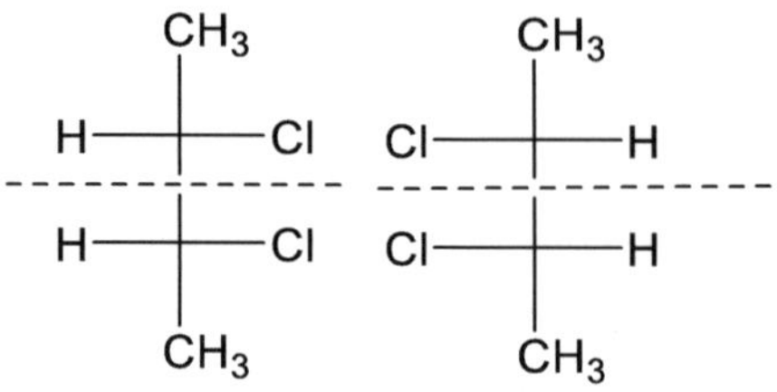

PART II. Substitution, Elimination and Oxidation

Lesson II.1. Tracking Stereocenters in Reactions

Lesson II.2. Nucleophilicity versus Basicity and Solvent Effects

Lesson II.3. The SN1 Reaction of Alkyl Halides

Lesson II.4. The SN2 Reaction of Alkyl Halides

Lesson II.5. Stereochemistry of Substitution Reactions

Lesson II.6. The E1 Reaction of Alkyl Halides

Lesson II.7. The E2 Reaction of Alkyl Halides

Lesson II.8. Stereoelectronic Effects in the E2 reaction

Lesson II.9. Factors Leading to Non-Zaitsev Products in the E2 Reaction

Lesson II.10. Competition Among Substitution and Elimination Reactions of RX: Predicting Pathways

Lesson II.11. Substitution and Elimination of Alcohols

Lesson II.12. Oxidation and Reduction: Definitions

Lesson II.13. Oxidation Reactions of Alcohols

Lesson II.14. Nomenclature and Acid Cleavage of Ethers

Lesson II.15. Ring-Opening of Epoxides: Steric/Electronic Effects on Major Product

Lesson II.1. Tracking Stereocenters in Reactions

Lesson II.1.1. Stereocenters Formed in the Course of Reaction

In Lesson I.9, we learned that more stable products often form more rapidly because they often have lower activation barriers. As a result, the most stable product possible is often the product formed in highest yield. In Lesson I.20, we learned that two enantiomers have the same stability as one another. It follows that, *if an achiral starting material undergoes a reaction leading to two enantiomers as products, the two enantiomers will be produced in equal amounts.*

Example II.1.1

A reaction between an alkene and H_2O in the presence of catalytic acid is observed to produce 2-butanol as shown:

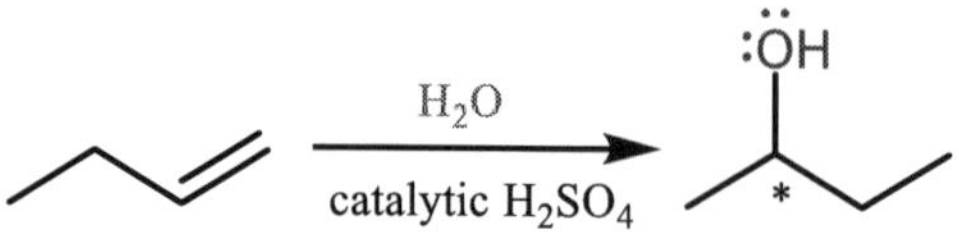

Draw the major products of this reaction, indicating stereochemistry where relevant.

Solution II.1.1

We recognize that a stereocenter has been generated and we have labeled it with an asterisk. Two enantiomers of equal stability are possible products. We thus draw both structures and label them with the labels of configuration "(*R*)-" or "(*S*)-". The product will be a racemic mixture of the two enantiomers.

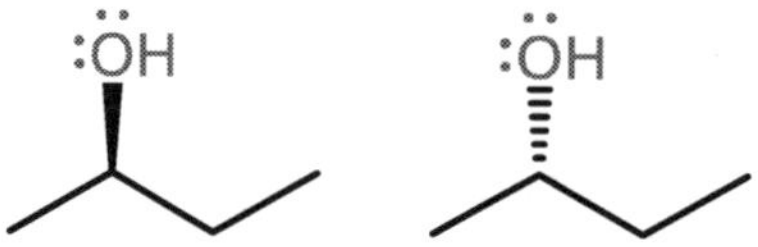

(*R*)-2-butanol (*S*)-2-butanol

What if a product having *two* stereocenters is generated from an achiral starting material? For example, consider the reaction of HBr/H_2O_2 with an alkene to form an alkyl bromide. We have not covered this reaction yet, but suppose we are told that the reaction produces the product shown, with no special requirements for which way the added Br and H atoms have to point:

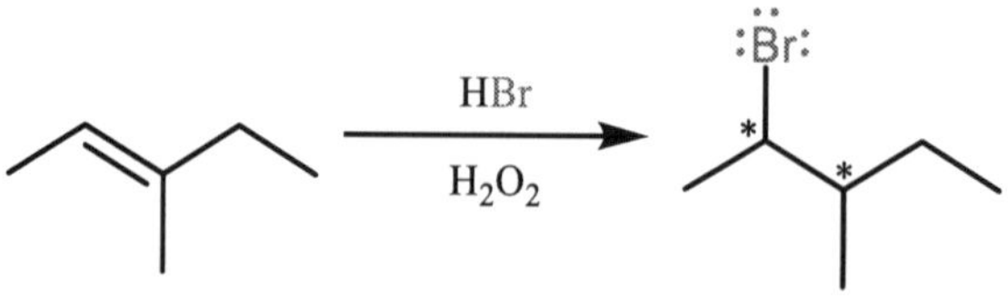

Because we have generated two chiral centers and there is no specificity for what direction the groups add, each of the two chiral centers generated could be in either the *R*- or the *S*- configuration. In this case, the product will consist of four stereoisomers:

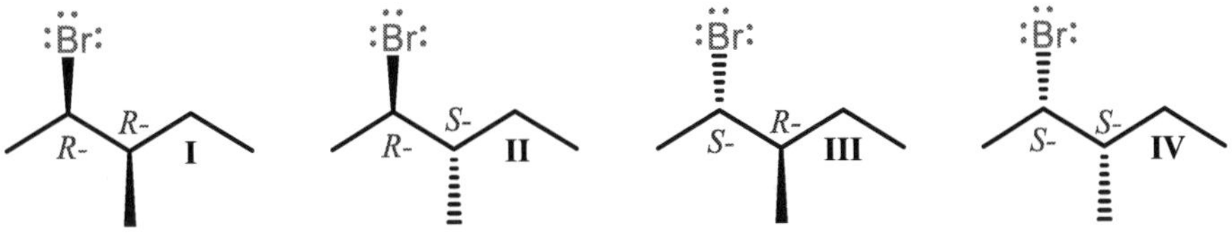

Some reactions are **stereospecific**, meaning that the reactants have to come together in a specific geometry to form the product. Consider the two reactions below, each of which adds two –OH units to a C=C in place of the π-bond, but with different stereospecificity:

<u>Reaction A Stereospecificity:</u> The two OH groups added point in opposite directions when added (one towards and one away from us)

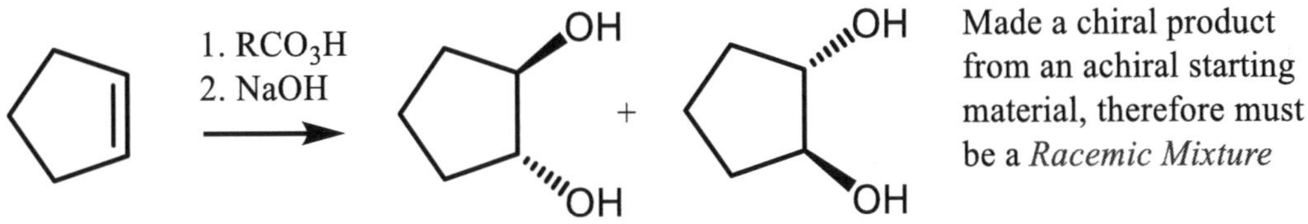

<u>Reaction B Stereospecificity:</u> The two OH groups added point in the same direction when added (both towards or both away from us)

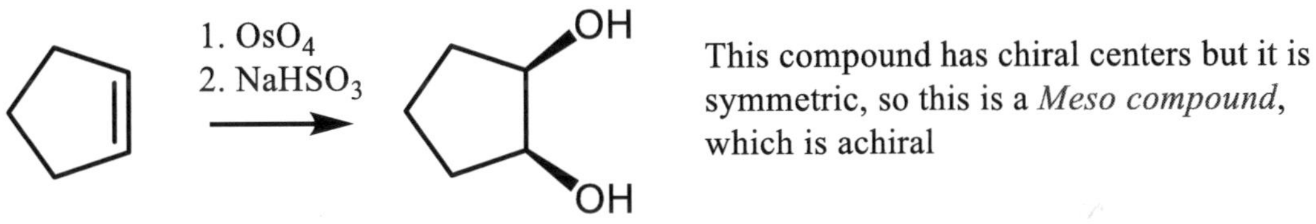

Reaction A reflects our general principle that if a chiral product is made from achiral reagents it must be formed as a racemic mixture. The stereospecificity of the reaction limits us to only isomers having the two OH groups pointing in opposite directions for reaction A. Reaction B leads to a product that has two chiral atoms, but the *molecule* is achiral because it is symmetric; it is a meso compound. This meso compound is the only product of the reaction because it the only stereoisomer that meets the stereospecificity requirement of the reaction. We will discuss stereospecificity requirements of each new reaction as we learn their mechanisms, and we must keep these requirements in mind when we try to predict major products of reactions.

Lesson II.1.2. Stereocenters Present Before a Reaction Occurs

If a starting material has stereocenters in it prior to a reaction, there are several considerations one must take into account. The first question is whether this reaction affects the bonds to the stereogenic atom in question. If not, then the configuration of that stereogenic atom does not change. If bonds to the chiral atom are affected during a set of reactions, its configuration may be inverted, it may be retained, or it may be scrambled to a mixture of *R*- and *S*- configurations. Which of these occurs depends on what arrow-pushing steps occur between the starting material and product. You will learn how to predict which steps occur and how those steps impact stereocenters as you continue your study of organic chemistry. For now, consider these three illustrative examples (only substitution products are shown):

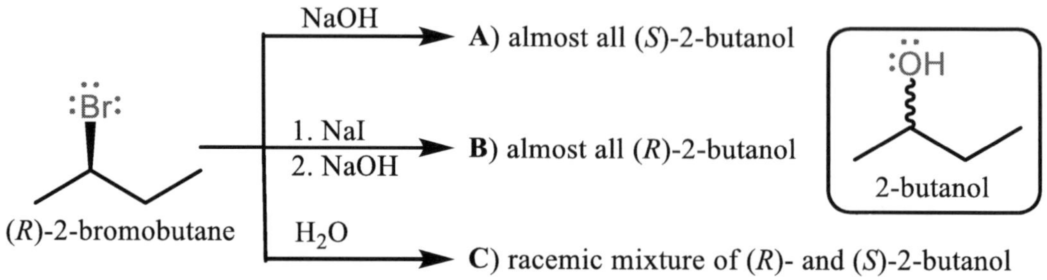

In pathway A, the arrow-pushing mechanism involved to change the starting material into the product involves one S_N2 step. Later in the book, you will learn that an S_N2 step inverts configuration, so by doing one S_N2 step the configuration is inverted from *R*- to *S*- in this case.

Pathway B requires two S_N2 steps in succession. The first step inverts *R*- to *S*-, then the second step inverts that *S*- back to *R*-, and you end up with net retention of stereochemistry.

Pathway C involves heterolysis followed by coordination (a sequence known as the S_N1 reaction, which we will see later on):

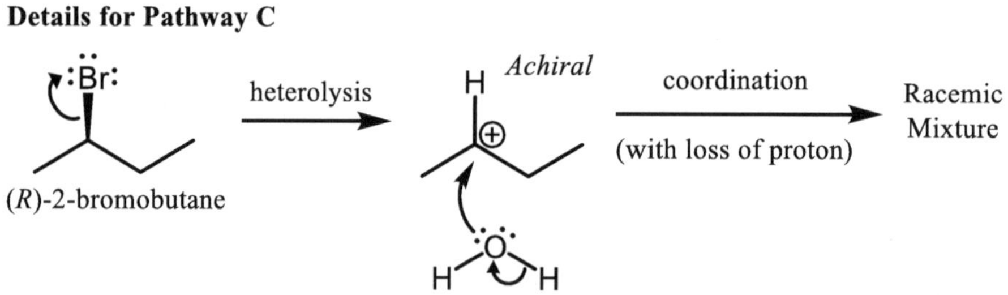

Note that the carbocation formed by heterolysis of the C–Br bond is planar, and therefore achiral. The final product has a chiral center, so we apply the concepts discussed in the prior section. The two enantiomers are of equal stability and are thus formed in equal amounts (a racemic mixture).

Lesson II.2. Nucleophilicity versus Basicity and Solvent Effects

Lesson II.2.1. Basicity is a Measure of Thermodynamic Stability

We know that acid strength is measured by pK_a values, which is calculated from the equilibrium constant (K_a) of the acid dissociation reaction (Lesson I.9). Equilibrium constants are thermodynamic parameters, with $\Delta G = -RT\ln K$. As a result, acid and base strengths do not tell us whether the reaction to add/remove H^+ is fast or slow, just whether that reaction is favorable/unfavorable. In the current lesson, we will focus on base strength: *A less stable anion is a stronger base*. The first strong base typically taught in chemistry is hydroxide (^-OH). Knowing that **$^-$OH is a strong base**, and therefore an anion less stable than $^-$OH will also be a strong base, we can make a generalization to help us quickly identify strong bases: **an anion with negative charge that is (1) located on H or *sp*^{3}-hybridized O/N/C and (2) is not stabilized by resonance delocalization will be a strong base**.

To identify which reagents could produce basic **anions** in solution, we must recall (Lesson I.3) that only ionic compounds dissociate into cations and anions in solution. Covalent compounds do not dissociate into ions and will thus never produce anions that could be strong bases. An ionic compound must have a counter-cation associated with the anion, most commonly a proton, a group I/II metal (Li, Na, K, Mg, Ca) or an ammonium ($[NH_4]^+$) cation. Here are some examples of covalent and ionic reagents that are commonly used in introductory organic chemistry:

Covalent	Ionic	
CH_3OH	NaOH	$[NH_4][O_2CCH_3]$
H_2O	$KOCH_3$	HBr
NH_3	$LiNH_2$	NaH

No covalent reagents dissociate into ions in solution, so none can function as strong bases. All the ionic compounds will dissociate into ions upon dissolution. Among those compounds in the list, HBr (negative charge on Br) and $[NH_4][O_2CCH_3]$ (negative charge stabilized by resonance; you must draw its Lewis structure to see that) do dissociate to anions, but the anions are not strong bases.

Lesson II.2.2. Nucleophilicity is a Measure of Reaction Rate (Kinetics)

It is easy to confuse nucleophilicity with basicity because bases and nucleophiles both function as electron donors. The key difference is that basicity is a measure of *stability*, whereas nucleophilicity is a measure of *reaction rate*. **The faster a species donates its electron pair to an electrophile, the better a nucleophile it is**. Note the difference in terminology as well: bases are referred to as being 'strong' or 'weak', whereas nucleophiles are classified as 'good' or 'poor'. A given anion may be a "strong base and good nucleophile", a "strong base and poor nucleophile", a "weak base and good nucleophile", or a "weak base and poor nucleophile".

We know what anions are strong bases, so now let us identify anions that are good nucleophiles. Because nucleophilicity is not tied to stability in the way that basicity is, many more anions are good nucleophiles than are strong bases. What factors might make an anion a poor nucleophile? One factor is electronegativity. Fluorine is the most electronegative element, so it is slow to give up its electrons, making fluoride a poor nucleophile. The steric bulk of a nucleophile also influences its ability to rapidly give electrons to an electrophile. Consider the difference between the small HO^- vs. the bulky *t*-BuO^- trying to approach the partial positive charge of ethyl chloride:

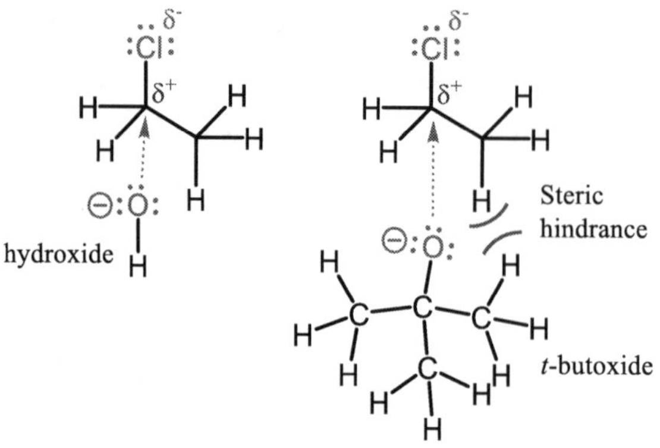

It is not difficult to see that the bulkier *t*-BuO^- (also written as $^-OC(CH_3)_3$ or $^-O^tBu$) must overcome a greater steric repulsive force (a higher energy of activation) than the smaller HO^-. In general, **bulky anions are poor nucleophiles**. "Bulky" in this context means that the sites adjacent to the anionic atom have three or more non-H branches (*t*-BuO^- has three methyl branches, for example). In fact, the *t*-BuO^- anion can only act as a good nucleophile if it attacks a methyl group, which has only small H atoms in the way of the nucleophile's approach:

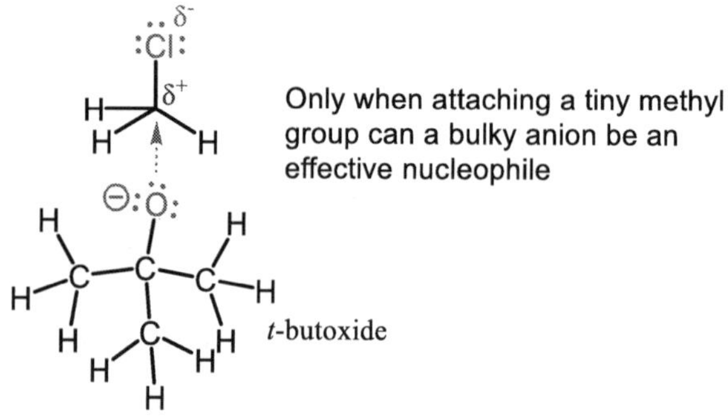

To summarize: **non-bulky anions other than fluoride are mostly good nucleophiles**. Now consider neutral compounds. Most common neutral compounds are significantly more stable than the hydroxide anion (our prototypical strong base), so there are few *strong* neutral bases. Nucleophilicity, however, only requires rapid donation of electrons. Large atoms (3rd row of periodic table or lower, e.g., P and S) with lone pairs can thus serve as good nucleophiles. These atoms are not as electronegative as are the second-row atoms like O and N, so there is a lower activation barrier for pulling electrons from such species.

To summarize our evaluation of nucleophiles:

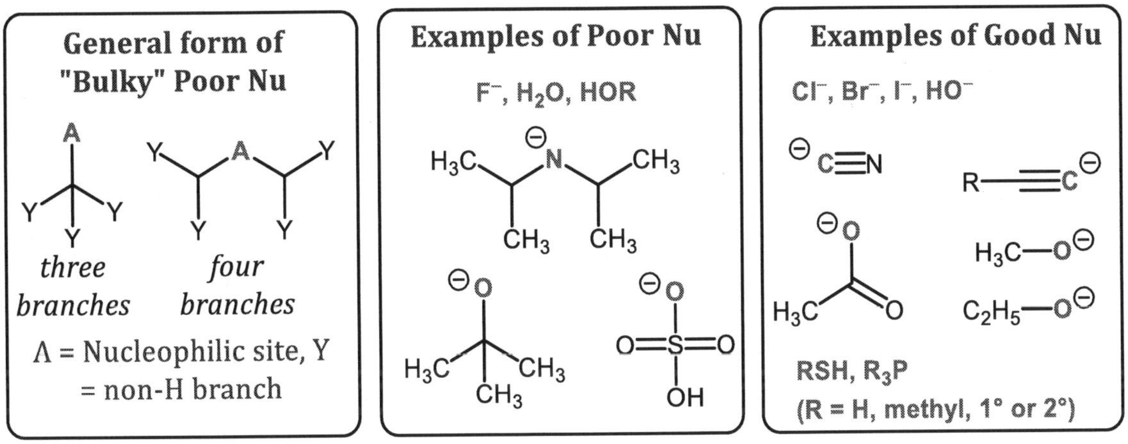

Example II.2.1

Label each species as covalent or ionic. Give each a second label as a good nucleophile (GN) or poor nucleophile (PN). Finally, give each compound a third label as a strong base (SB) or weak base (WB).

NaOH	NaCN
NaH	$KOCH_3$
H_2O	$NaOOCCH_3$
KI	$LiOC_2H_5$
C_2H_5OH	$KOC(CH_3)_3$
$HSCH_3$	$NaNH_2$
$HOC(CH_3)_3$	

Solution II.2.1

We need to first recognize which are ionic and which are covalent compounds. We have not studied any neutral compounds that are strong bases. The only neutral compound we are considering a good nucleophile is HSR. With this in mind, we can label all of the covalent compounds:

NaOH	NaCN
NaH	$KOCH_3$
covalent, WB, PN H_2O	$NaOOCCH_3$
KI	$LiOC_2H_5$
covalent, WB, PN: C_2H_5OH	$KOC(CH_3)_3$
covalent, WB, **GN:** $HSCH_3$	$NaNH_2$
covalent, WB, PN: $HOC(CH_3)_3$	

Next, we look at the ionic compounds and see what anion they will make in solution. deduce the anion that is produced. Remember that an anion with negative charge that is (1) located on H or *sp*^{3}-hybridized O/N/ C and (2) is not stabilized by resonance delocalization will be a strong base. We also know that non-bulky anions other than fluoride are mostly good nucleophiles. For many of the cases it is easy to determine the labels:

ionic, SB, GN NaOH	**NaCN**
ionic, SB, GN NaH	$KOCH_3$ ionic, SB, GN
covalent, WB, PN H_2O	**$NaOOCCH_3$**
ionic, WB, GN KI	$LiOC_2H_5$ ionic, SB, GN
covalent, WB, PN: C_2H_5OH	**$KOC(CH_3)_3$**
covalent, WB, **GN:** $HSCH_3$	$NaNH_2$ ionic, SB, GN
covalent, WB, PN: $HOC(CH_3)_3$	

This leaves us with the three ionic compounds in bold, above. These need to be considered more carefully by drawing out the structures of the anions and deducing whether they are bulky, have resonance, etc.:

$KOC(CH_3)_3$ anion:

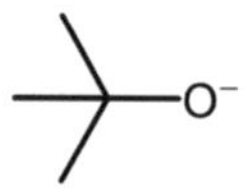

- Bulky anion, so PN.
- Negative on *sp*^{3} O, so SB.

$NaOOCCH_3$ anion:

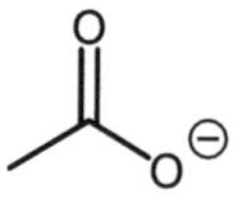

- nonbulky anion, not F, so GN.
- resonance-stabilized anion; WB

NaCN anion:

$^{\ominus}C\equiv N$

- nonbulky anion, not F, so GN.
- Negative on *sp* C, so WB.

Bases and nucleophiles must both donate electron pairs in the course of their usual reactivity, so any factors that hinder electron pair donation will diminish their basicity and nucleophilicity. One way that the ability to donate an electron pair is diminished is by strong attraction of solvent molecules for the lone pair or negative charge. The strongest type of interactions that can occur between a neutral solvent molecule and an anion (whether it is a base or a nucleophile) is hydrogen bonding (see Lesson I.3 for a review of intermolecular forces). It is easy to see how the **solvent cage** surrounding an anion could influence its basicity and/or nucleophilicity. Consider A^- solvated by water

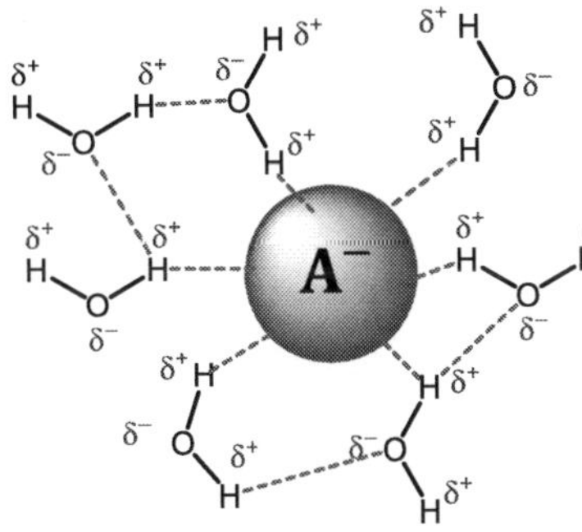

The pull of the $H^{\delta+}$ in the H_2O molecules makes the anion less basic and less nucleophilic, as can be visualized in the simplified reactions below, in which B^- is a base and Nu^- is a nucleophile:

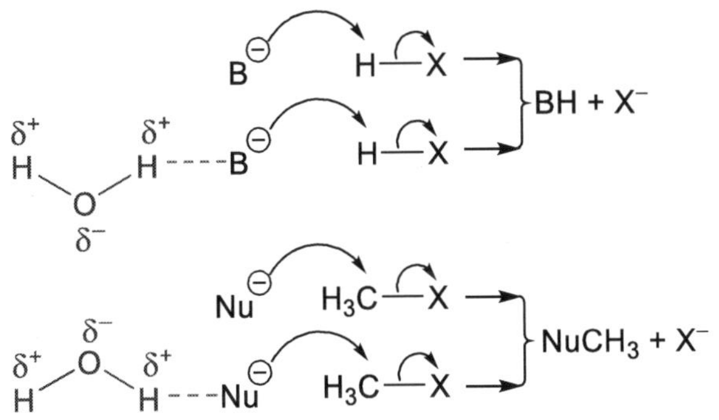

The weakening of basicity/nucleophilicity is most pronounced in H-bonding solvents, which are often referred to as **polar protic solvents**. Common polar protic solvents include water and alcohols.

Polar solvents that cannot engage in H-bonding of bases/nucleophiles are called **polar aprotic solvents**. Common polar aprotic solvents include acetone ($CH_3C(O)CH_3$), acetonitrile (CH_3CN), *N,N*-dimethylformamide ($HC(O)N(CH_3)_2$, DMF) and dimethylsulfoxide ($CH_3S(O)CH_3$, DMSO). Although polar aprotic solvents are incapable of H-bonding, they can engage in dipole–ion interactions with the anion. Dipole–ion interactions are weaker than H-bonding, so attenuation of basicity/nucleophilicity is not as pronounced in polar aprotic solvents as it is in polar protic solvents.

Nonpolar solvents (like alkanes) do not engage in strong intermolecular interactions with anions, so basicity and nucleophilicity are highest in these solvents. The general trend for how strongly a type of solvent interacts with an anion is:

nonpolar solvent (most basic/nucleophilic)	$<$	polar aprotic solvent	$<$	polar protic solvent (least basic/nucleophilic)

From a practical standpoint, very polar species that typically dissociate into ions upon dissolution (i.e., ionic salts) are usually insoluble in nonpolar solvents. For this reason, chemists most often do reactions requiring a strong base or a good nucleophile in a polar aprotic solvent to optimize reactivity while maintaining the solubility of their reagents.

Lesson II.3. The S_N1 Reaction of Alkyl Halides

Lesson II.3.1. S_N1 is Heterolysis then Coordination

In Lesson I.8, we saw some of the elementary steps of organic reactions. We are now going to learn our first reaction mechanism that involves more than one of these steps: the S_N1 reaction. The S_N1 reaction involves: 1) heterolysis to form a carbocation (sometimes referred to informally as "leaving group leaves"), then 2) coordination of a nucleophile to the carbocation:

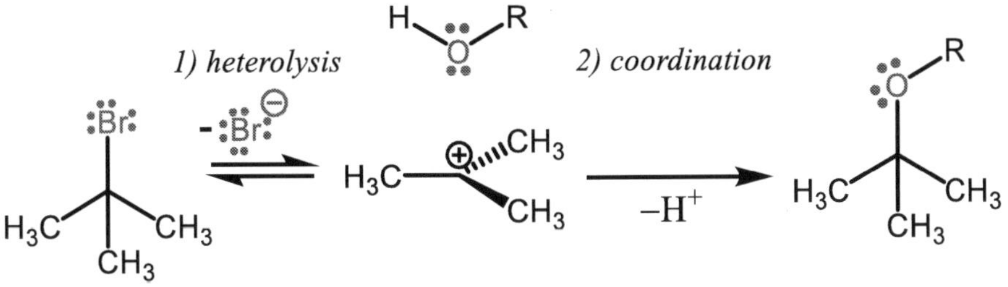

Note that the carbocation to which the nucleophile coordinates is very reactive, so that even a weak nucleophile – like the neutral alcohol in the example above – will suffice for this reaction.

Here are some examples of net S_N1 reactions, without showing the mechanistic steps:

General:

$$R{-}LG + H{-}Nu \longrightarrow R{-}Nu + H^+ + LG^-$$

$$R{-}LG + Nu^- \longrightarrow R{-}Nu + LG^-$$

Br^- as LG
H-OH as H-Nu

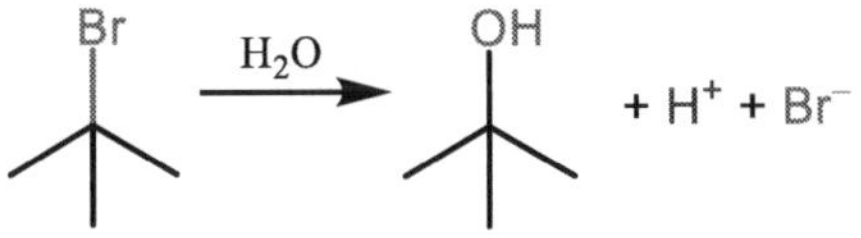

**note that H^+ comes off of the H_2O after H_2O adds so that the O will be neutral*

H_2O as LG
Br^- as Nu^-

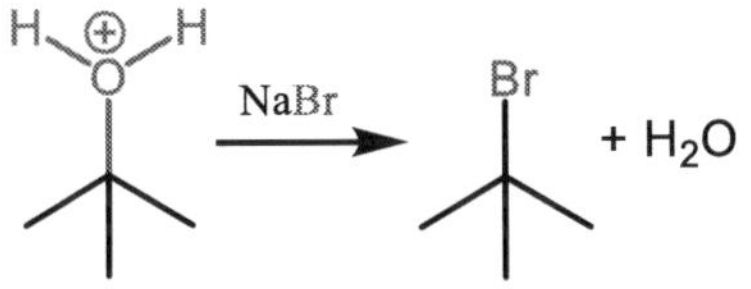

**note that NaBr dissociates in solution to form Na^+ and Br^- and that the LG in this case is a neutral molecule*

The net result is that the *leaving group* (LG) – the group removed from C in the heterolysis step – is substituted by the nucleophile. The "S" in S_N1 stands for substitution and the subscript "N" stands for nucleophilic. The "1" indicates that the kinetics of the reaction are *unimolecular*: the reaction rate only depends on the concentration of *one* species. In this case, we look at the mechanism and see that the carbocation is the least stable species formed on the way to the product, so the step in which it forms is *rate-limiting*. A qualitative reaction coordinate diagram for a thermodynamically-favorable S_N1 reaction might look like this:

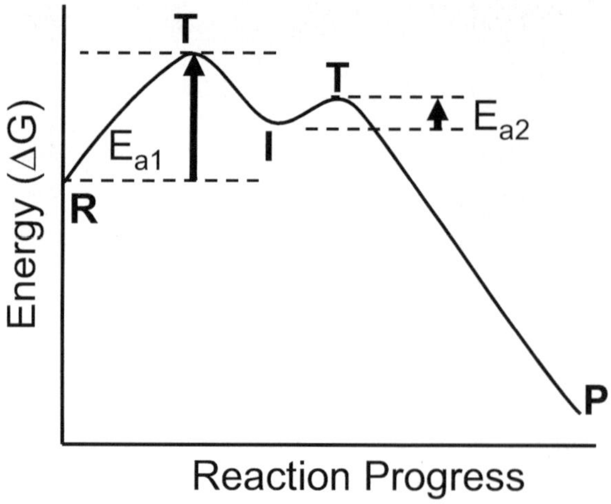

Where **R** is the reactant, **I** is the carbocation intermediate, the peaks **T** are transition states, and **P** is the product. The first 'hill' represents the activation barrier for heterolysis and the second 'hill' represents the activation barrier for coordination.

The rate of the overall two-step process is governed by the high E_a of the heterolysis step, so the rate is only dependent on the concentration of the starting material (the substrate, R–LG), so

$$\text{Rate} = k[\text{R–LG}]$$

Note that, since the nucleophile does not show up in the rate law expression, its concentration does not influence the rate. If we double the nucleophile concentration, the reaction rate is unchanged. If we double the concentration of the substrate (R–LG), which *is* a term in the rate law, the reaction rate will double.

The rate of the S_N1 reaction can also be influenced by the leaving group identity, substrate substitution degree, and reaction solvent. The better the leaving group (the more stable the species displaced from the substrate), the faster the reaction will be because there is a lower activation barrier to leaving group departure.

In the S_N1 mechanism, carbocation formation is rate-limiting, and more stable carbocations form faster. So, in terms of S_N1 reaction rate: methyl halide < 1° alkyl halide < 2° alkyl halide < 3° alkyl halide (fastest). Indeed, **the carbocations that would form from methyl and 1° alkyl halides would be so unstable that methyl and 1° halides will never undergo S_N1 reactions.**

Because the solvent influences the stability of ions, we must choose a solvent that best stabilizes the carbocation and the leaving group that are formed in the rate-limiting step. As discussed in Lesson II.2, polar protic solvents have the strongest intermolecular interactions with ions. For this reason, the **S_N1 reaction is fastest in polar protic solvents**.

Lesson II.3.2. Carbocation Rearrangement

In Lesson I.12, we learned that we can predict reaction spontaneity by comparing reactant stability and product stability. A carbocation is capable of rearrangement (elementary step 2 in Lesson I.8), and **if this rearrangement affords a more stable carbocation, then carbocation rearrangement will happen spontaneously and rapidly** (see carbocation stability in Lesson I.11). In fact, this rearrangement happens so quickly that the major product of a reaction which proceeds through a carbocation will always be derived from the rearranged, most stable carbocation if a spontaneous carbocation rearrangement is possible.

Below are two examples of spontaneous carbocation rearrangement reactions, in which a hydride (H with a pair of electrons) or an alkyl group shift from one carbon to an adjacent carbocationic C:

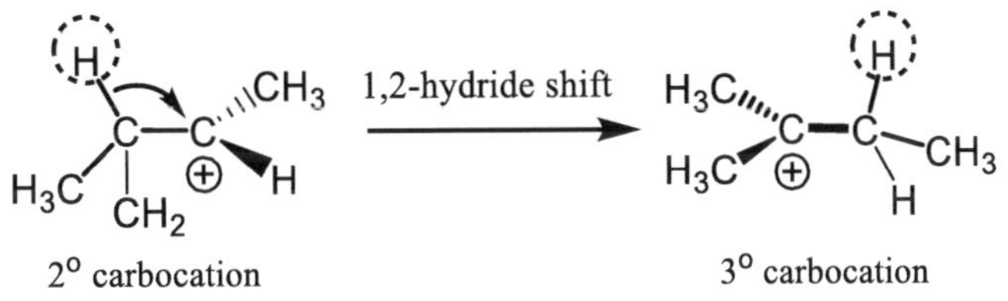

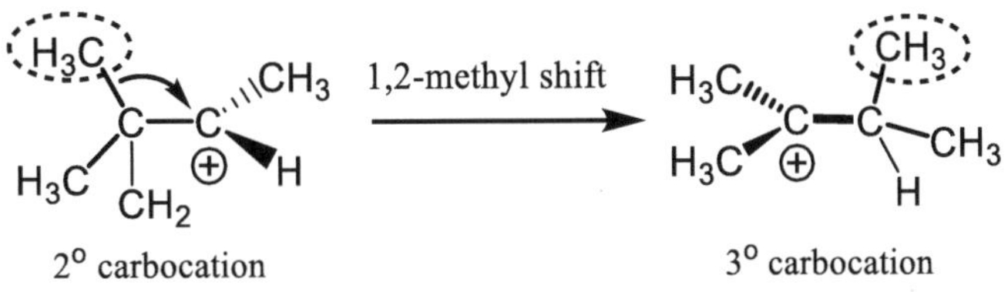

Generally, a hydride shift occurs selectively when it is possible, and an alkyl group will only shift if a hydride is not present to shift, or under other special circumstances.

Example II.3.1

Provide the major products and intermediates needed to form them via S_N1 reaction:

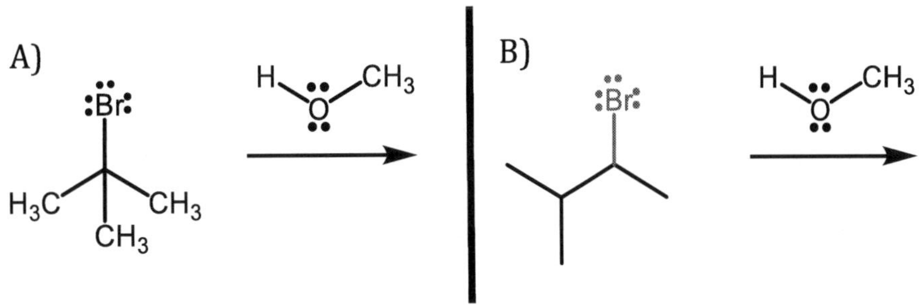

Solution II.3.1

First, consider reaction A). As with all S_N1 reactions, the sequence it 1) heterolysis of LG then 2) coordination of nucleophile. Look out for the possibility of carbocation rearrangement anytime a carbocation forms! Here, the carbocation is tertiary and will not rearrange:

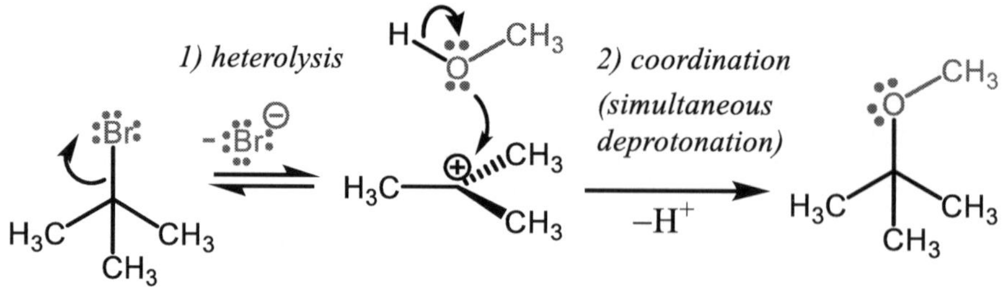

Next, consider reaction B). Here, the carbocation that initially forms is secondary, but it will spontaneously rearrange to become tertiary on the path to the major product:

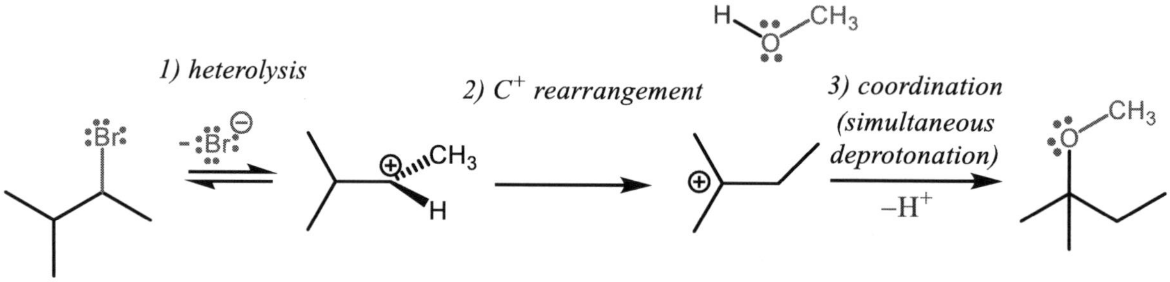

Lesson II.4. The SN2 Reaction of Alkyl Halides

Lesson II.4.1. SN2 is a Concerted Elementary Step

In Lesson I.8, we saw that the SN2 reaction is an important elementary step in many organic reaction mechanisms:

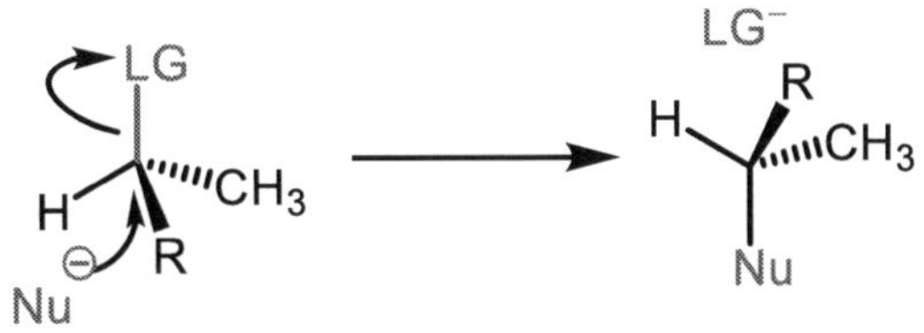

In contrast to the SN1 reaction (Lesson II.3), in which a nucleophile coordinates to a carbocation, a nucleophile in the SN2 reaction is attracted to the partial positive charge on an sp^3-hybridized carbon atom. Because the partial positively-charged C is a poorer electrophile than a carbocation, **a good nucleophile is required for the SN2 reaction** (see Lesson II.2 for how to identify good nucleophiles).

In SN2, everything happens in a single concerted step, so the rate law is easy to determine. Two species come together to facilitate this reaction, so it is a **bimolecular reaction**, with the rate law:

$$\text{Rate} = k[\text{R–LG}][\text{Nu}^-]$$

Both the substrate and the nucleophile concentrations appear in the rate expression. If we double the concentration of the nucleophile, the rate will double. Likewise, if we double the concentration of the nucleophile, the rate will double. If we double the concentration of *both* the substrate and the nucleophile, the reaction is four times faster.

The rate of the SN2 reaction can also be influenced by the leaving group identity, substrate substitution degree, nucleophile strength, and reaction solvent. The better the leaving group (the more stable the species displaced from the substrate), the faster the reaction will be because there is a lower activation barrier to leaving group departure.

In the SN2 mechanism, the nucleophile must pass between any branches at the attached C before it can form a bond to it. The more non-H branches on the electrophilic C, the slower the reaction rate will be, due to the greater activation barrier caused by steric repulsion during nucleophile approach. So, in terms of rate: methyl halide > 1° alkyl halide > 2° alkyl halide > 3° alkyl halide (slowest). Indeed, **the 3° alkyl halides are so sterically hindered that they will never undergo SN2 reactions**. Even the presence of a β-branch (a branch off of the parent chain at a site *beside* the C with the leaving-group) will slow down the reaction:

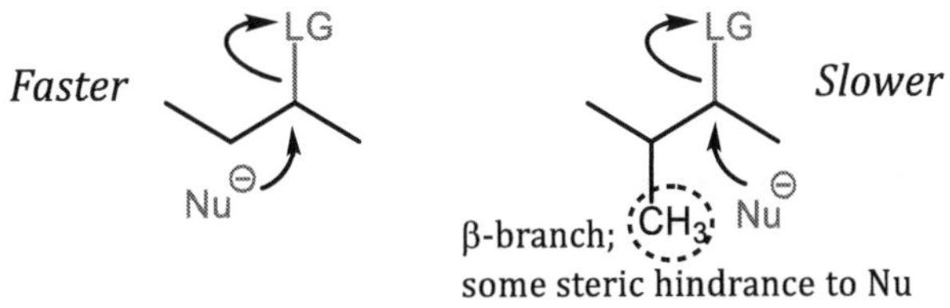

Recall that nucleophilicity is by definition a measure of how quickly a species donates electrons to an electrophile. This definition tells us that **the SN2 reaction is faster with better nucleophiles.** Because the solvent influences nucleophile strength (Lesson II.2), we must choose a solvent that does not diminishing the nucleophile strength. As discussed in Lesson II.2, this means that **the SN2 rate is fastest in polar aprotic solvents.**

Example II.4.1

Provide the major substitution products for these reactions:

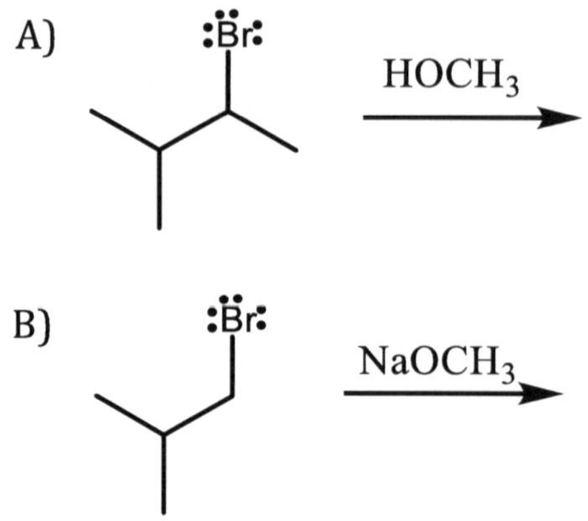

Solution II.4.1

For reaction A, there is no good nucleophile (methanol is a covalent compound and will not make ions in solution), so this will have to undergo substitution via SN1.

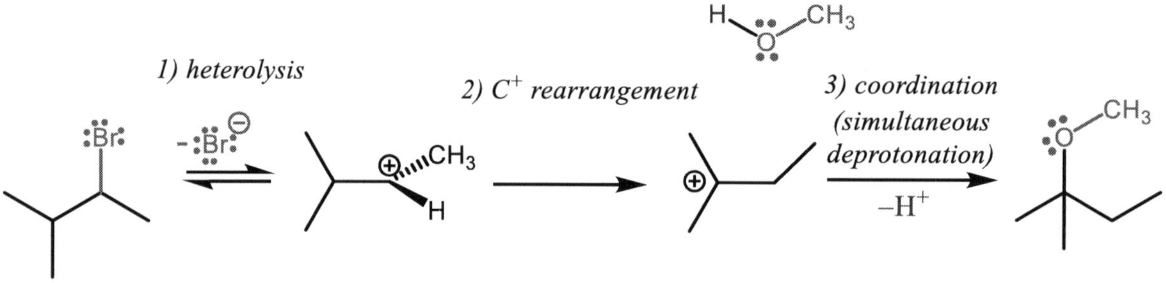

For reaction B, there is a good nucleophile ($NaOCH_3$ will dissociate to make methoxide anion) and the primary substrate is appropriate for SN2 reaction:

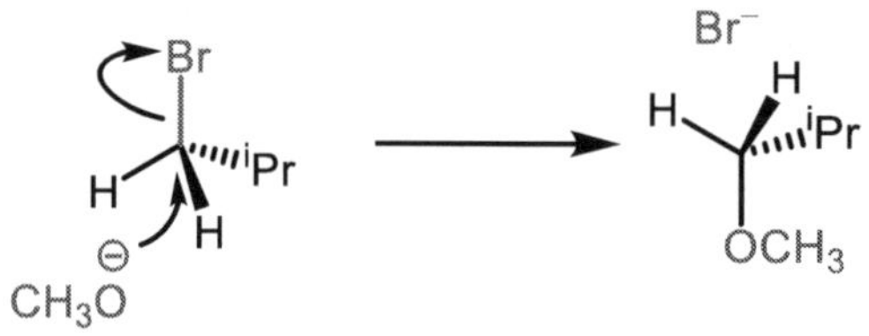

Lesson II.5. Stereochemistry of Substitution Reactions

Lesson II.5.1. The S_N1 Reaction is Stereorandom

Recall from Lesson II.1 that a single chiral product cannot be obtained from an achiral starting material for any step of a reaction sequence. Consider a chiral alkyl halide reacting via an S_N1 pathway:

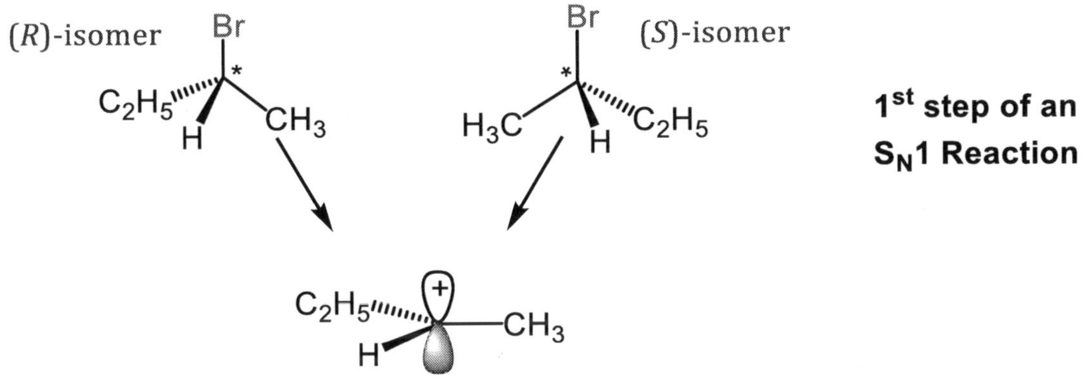

After heterolysis, chirality is lost: the carbocation has trigonal planar geometry and is therefore achiral. We cannot even tell from which of the two isomers this cation was formed. The nucleophile (H_2O in this case) may attack either face of the carbocation plane with equal probability to afford two alcohol product stereoisomers:

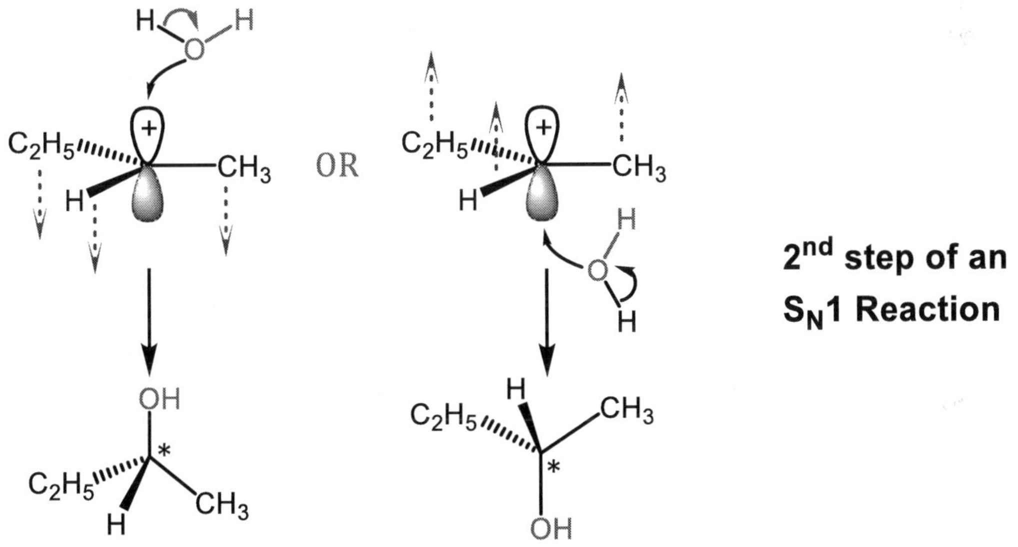

Each of the two enantiomers has the same stability, so we would expect them to be formed in equal amounts in this reaction. We can now see that, regardless of whether we start with the *R*- or *S*-configuration of the alkyl bromide, we will always end with a 50:50 mixture of the two enantiomeric products (a racemic mixture). For this reason, the S_N1 reaction is a stereorandom reaction.

Lesson II.5.2. Backside Attack and Walden Inversion in the SN2 Reaction

The SN2 reaction is concerted, and the nucleophile must attack to form a bond to the carbon 180° apart from where the leaving group is attached. This is sometimes referred to as **backside attack**. One consequence of this concerted, backside attack by the nucleophile is that, if a nucleophile attacks a stereogenic site, a single chiral compound will be produced:

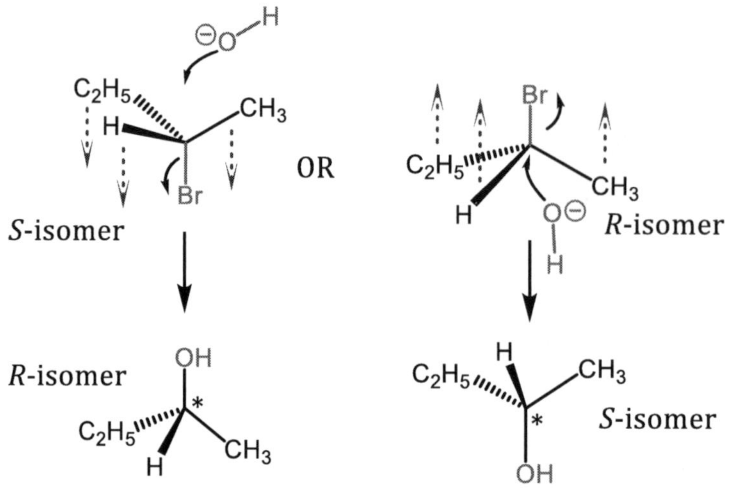

Note that the bonds from the stereogenic C to the three non-leaving groups will be repelled by the electrons on the hydroxide nucleophile in this example, pushing them in the opposite direction to reestablish the tetrahedron. The resulting change of configuration is referred to as **Walden inversion**. Walden inversion means that **the SN2 reaction is a stereospecific reaction**.

Example II.5.1

Provide the major substitution products for these reactions:

A) 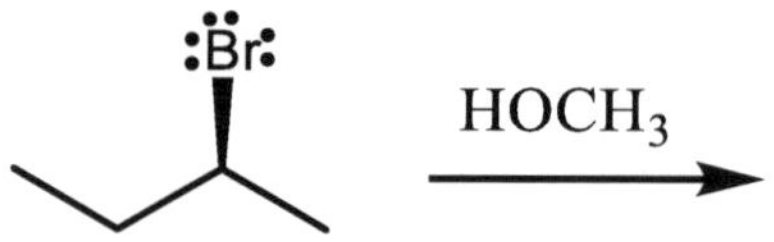

B) 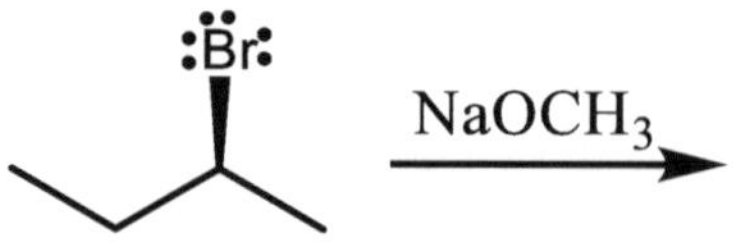

Solution II.5.1

To determine whether substitution will proceed by an S_N1 (fails on primary or methyl alkyl halides) or an S_N2 (fails on tertiary alkyl halides) pathway, first check the degree of substitution. The substrate is secondary in both cases, so both pathways will work. Next, remember that the S_N2 reaction requires a good nucleophile. If a good nucleophile is in the reaction solution, S_N2 will be favored (no need to make the unstable carbocation with S_N2!). If you do not recall how to figure out whether a good nucleophile will form in solution, consider reviewing Lesson II.2 or you will struggle with some of the material that is to come later on in the book.

For reaction A, there is no good nucleophile (methanol is covalent), so this has to undergo S_N1. The coordination to the achiral carbocation leads to a stereorandom product.

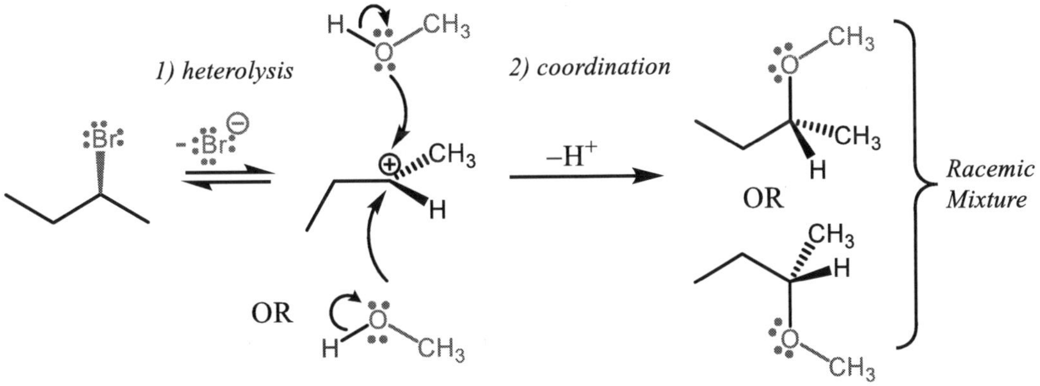

For reaction B, there is a good nucleophile ($NaOCH_3$ will dissociate to make methoxide anion) and the primary substrate is appropriate for S_N2 reaction, leading to inversion of configuration:

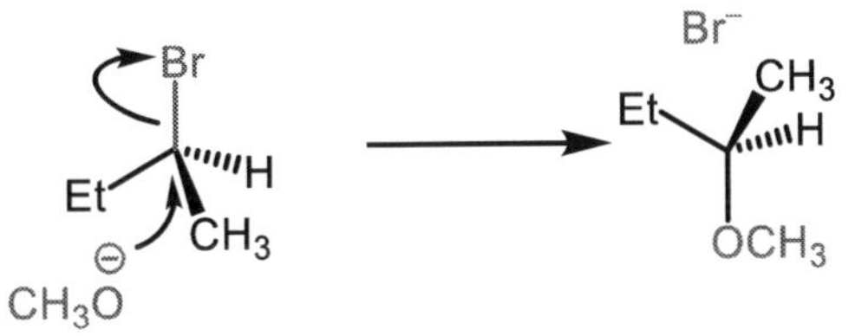

Lesson II.6. The E1 Reaction of Alkyl Halides

Lesson II.6.1. E1 is Heterolysis Followed by Electrophilic Elimination

Like the S_N1 reaction, the E1 reaction is a name given to a sequence of two specific elementary steps: 1) heterolysis to form a carbocation, then 2) electrophilic elimination of a proton:

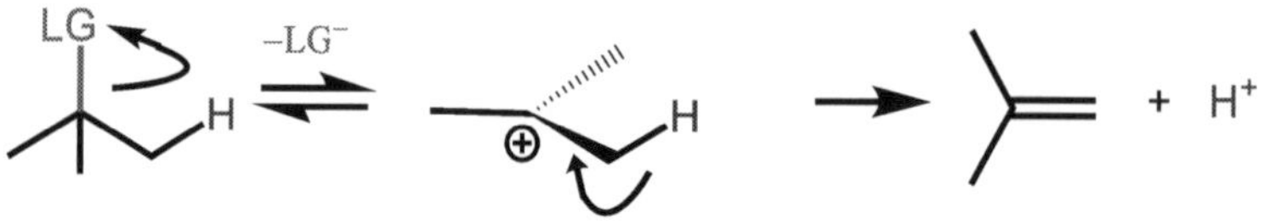

The net result is that the leaving group and an H on the adjacent C are eliminated to give a π-bond between the two C atoms from which the LG and H eliminated. The "E" in E1 stands for elimination and the subscript "N" stands for nucleophile. The "1" indicates that the kinetics of the reaction are *unimolecular*. Only the concentration of *one* species influences the reaction rate. In this case, we look at the mechanism and see that thc carbocation is the least stable species formed on the way to the product, so this step is *rate-limiting*. Shown visually, the qualitative reaction coordinate diagram for a thermodynamically-favorable E1 reaction might look like this (the same as for S_N1!):

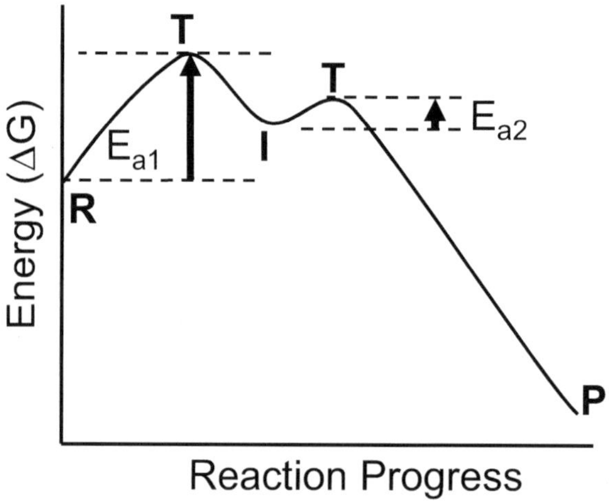

Where **R** is the reactant, **I** is the carbocation intermediate, the peaks **T** are transition states, and **P** is the product. The first 'hill' represents the activation barrier to heterolysis and the second 'hill' represents the activation barrier to electrophilic elimination.

The rate of the overall two-step process is governed by the high E_a of the heterolysis step, so the rate is only dependent on the concentration of the starting material (the substrate, R–LG), so

$$\text{Rate} = k[\text{R–LG}]$$

If we double the concentration of the substrate (R–LG), the rate is doubled.

The rate of the E1 reaction can also be influenced by the leaving group identity, substrate substitution degree, and the reaction solvent. The better the leaving group (the more stable the species displaced from the substrate), the faster the reaction will be because there is a lower activation barrier to leaving group departure.

In the E1 mechanism, carbocation formation is rate-limiting. More stable carbocations form faster. So, in terms of rate: methyl halide < 1° alkyl halide < 2° alkyl halide < 3° alkyl halide (fastest). Indeed, **the carbocations that would form from methyl and 1° alkyl halides would be so unstable that methyl and 1° halides will never undergo E1 reactions.**

Because the solvent influences the stability of ions, we must choose a solvent that stabilizes both the carbocation and leaving group formed in the rate-limiting step. As discussed in Lesson II.2, polar protic solvents provide the strongest intermolecular forces with ions. For this reason, the **E1 reaction is fastest in polar protic solvents**.

Lesson II.6.2. Zaitsev's Rule for Determining the Major Products of Elimination

Early in the history of elimination reactions, a chemist named Zaitsev noticed that the major product formed by elimination reactions was almost always the most substituted alkene, that is to say, the alkene in which the carbons in a C=C bond had the most non-H substituents. This empirical rule for predicting major products of elimination reactions became known as **Zaitsev's Rule** (sometimes spelled Saytseff's Rule). As we saw in Lesson I.11, this is because the most substituted alkene is typically the most stable. **Zaitsev's Rule, then, is essentially telling us that the major product is the most stable product**, which we have seen is generally true of many reaction types.

Example II.6.1

Predict the major elimination product of this reaction:

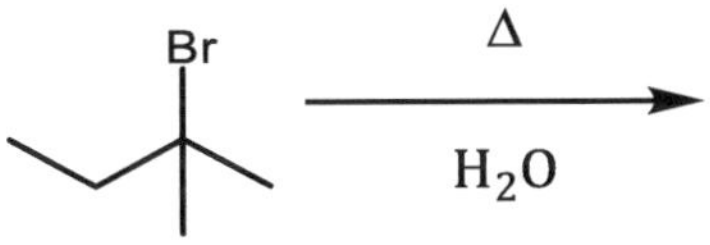

Solution II.6.1

Heterolysis to form the carbocation is the first step:

For the second step, electrophilic elimination, there are two possible sites from which to eliminate the proton:

Choice A 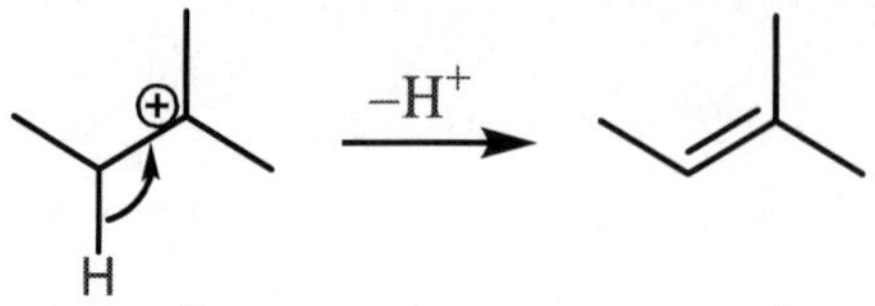

Choice B 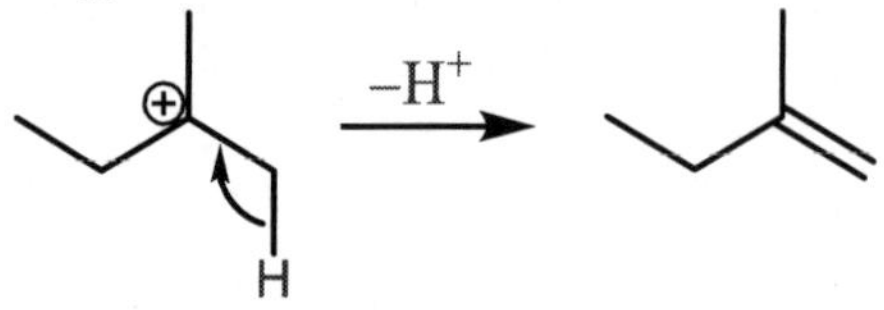

Choice A leads to the more substituted, more stable alkene and is the Zaitsev Product. This will be the major elimination product of the reaction.

Lesson II.7. The E2 Reaction of Alkyl Halides

Lesson II.7.1. E2 is a Concerted Elementary Step

Like the S_N2 reaction, the E2 reaction is one of the elementary steps that we learned in Lesson I.8.

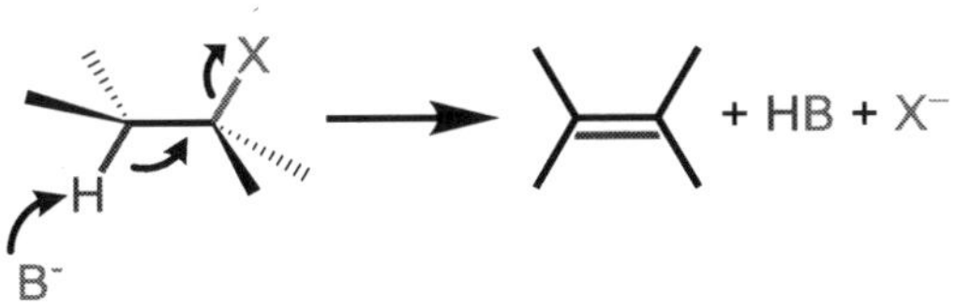

In contrast to the E1 reaction (Lesson II.6), in which H^+ eliminates from a carbocation to yield a more stable neutral alkene, the H^+ in the E2 reaction must be removed from a neutral species. For this reason, **a strong base is necessary for the E2 reaction, to remove this proton** (see Lesson II.2 for help on identifying strong bases).

Everything happens in a single, concerted step, so the rate law is easy to determine. There are two species coming together to facilitate this reaction, so it is a **bimolecular reaction**, with the rate law:

$$\text{Rate} = k[\text{R–LG}][\text{B}^-]$$

Both substrate and base appear in the rate expression. If we double the substrate concentration, the rate will double. Likewise, if we double the concentration of the base, the rate will double. If we double the concentration of *both* the substrate and the base, the reaction is four times faster.

The rate of the E2 reaction can also be influenced by leaving group identity, substrate substitution degree, base strength, and reaction solvent. The better the leaving group (the more stable the species displaced from the substrate), the faster the reaction will be because there is a lower activation barrier to leaving group departure.

Recall that more non-H substituents on the carbons of a C=C bond makes the alkene more stable. Because the C with the halide leaving group on it ends up in the C=C bond, more non-H substituents at that site leads to more stable products. So, **in terms of E2 rate: methyl halide < 1° alkyl halide < 2° alkyl halide < 3° alkyl halide (fastest)**.

A strong base is required for the E2 reaction, and the stronger the base, the faster the E2 reaction. Because the solvent influences base strength (Lesson II.2), we must choose a solvent that solvates the reagents without diminishing the base strength. As discussed in Lesson II.2, this means that **the E2 rate is fastest in polar aprotic solvents**.

Lesson II.7.2. Zaitsev's Rule Applies to Most E2 Reactions

Similar to the E1 reaction (Lesson II.6), the major products of an E2 reaction can generally be predicted by Zaitsev's rule. We will see cases where some mechanistic details can lead to non-Zaitsev products in Lessons II.8 and II.9, but for the examples in this Lesson, the Zaitsev rule works well:

Example II.7.2

Predict the major elimination product of this reaction:

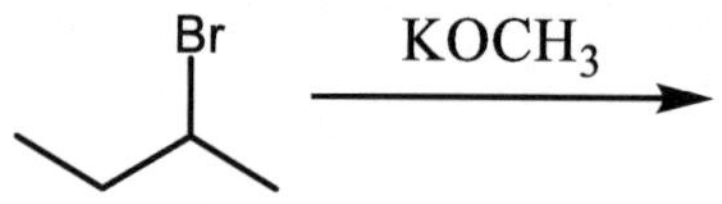

Solution II.7.2

In this concerted reaction, there are two possible sites for deprotonation by $KOCH_3$ (a strong base):

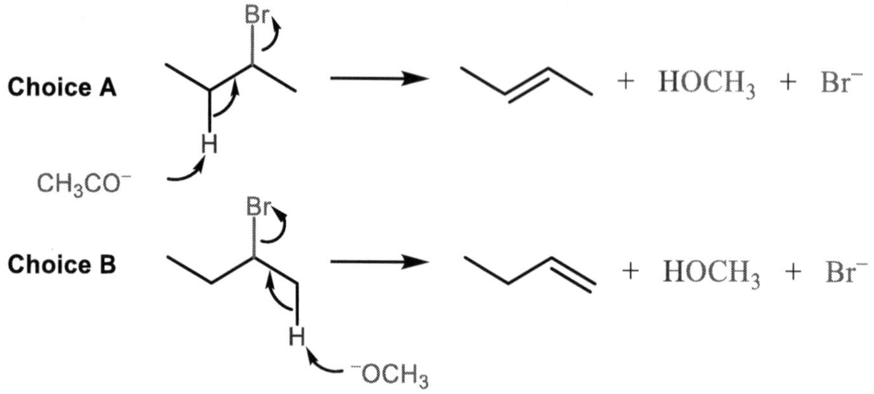

Choice A leads to the more substituted, more stable alkene and is the Zaitsev Product. This will be the major elimination product of the reaction.

Lesson II.8. Stereoelectronic Effects in the E2 reaction

Lesson II.8.1. Eliminated Groups must be Antiperiplanar

For an E2 reaction to occur, the H and leaving group (X) that will undergo elimination must be **antiperiplanar**. Antiperiplanar means the two groups are coplanar and in an *anti-* conformation:

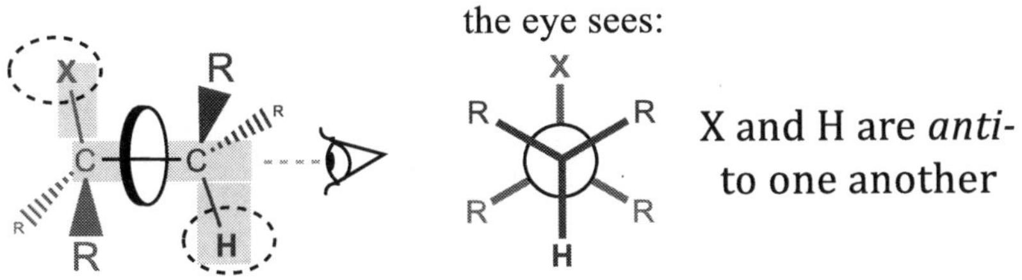

This is what is known as a **stereoelectronic effect**, an effect of the spatial orientation of orbitals (and consequently of the substituents connected via those orbitals). The stereoelectronic requirements of the E2 reaction must be considered when we attempt to predict products of a reaction.

Example II.8.1

Predict the major elimination product of each reaction:

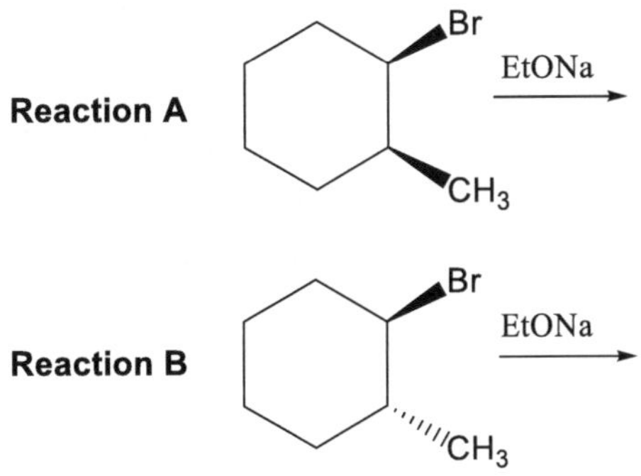

Solution II.8.1

EtONa is ionic, so it will dissociate to form ethoxide anion (EtO^-) and sodium cation. The ethoxide anion is a strong base, so the E2 reaction is possible. Now we know that, for an E2 reaction, the leaving group (here Br) and the H to be eliminated must be *anti-* to one another, we must be more cautious in selecting the H to be removed by the base. In Lesson I.16, we saw that each C in a cyclohexane ring has an "up" and a "down" position, or in the perspective of this problem a "towards" (wedge) and "away" position. We can fill these H atoms in on the sites adjacent to the leaving group:

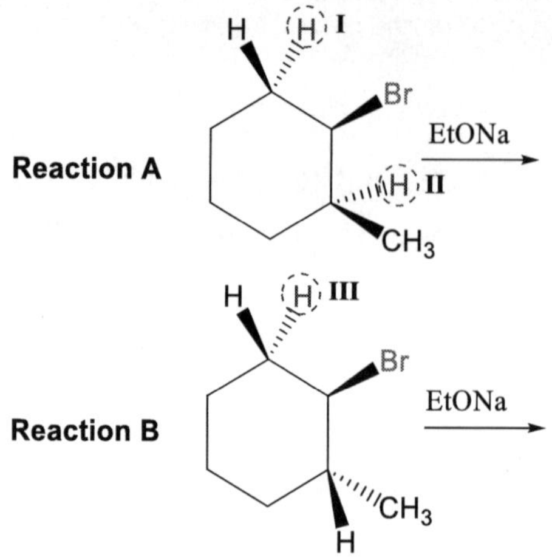

Only H atoms pointing in the opposite direction of the Br (i.e., *anti-* to Br) are mechanistically viable reaction sites. These are H atoms I, II and III (circled) in the substrates above.

For reaction A, removal of hydrogen II gives a more substituted product, so the major E2 product is:

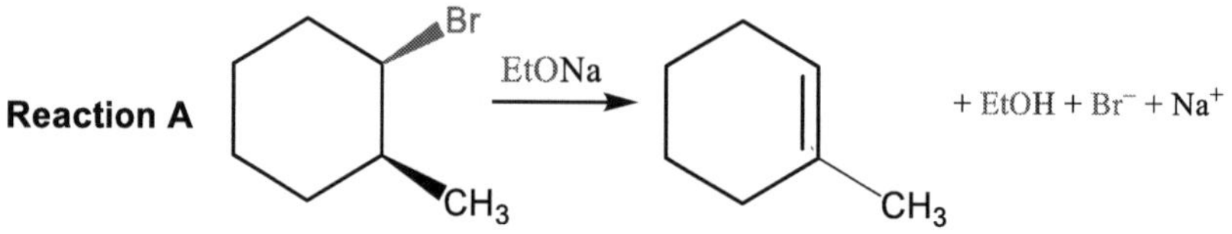

For Reaction B, there is only one choice (H atom III). Removal of H atom III then gives the major product:

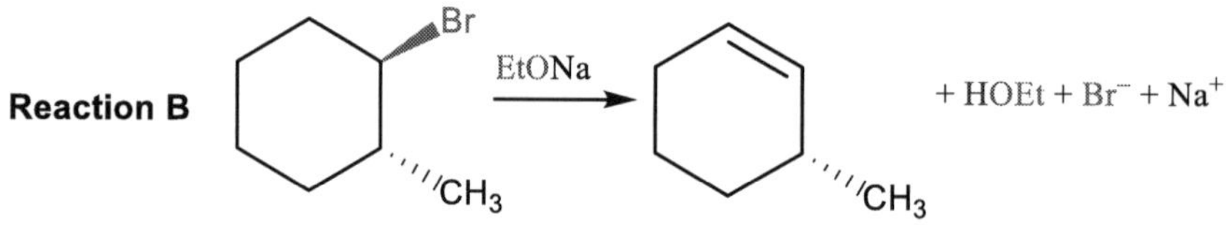

Note that none of the bonds to the chiral center with the Me substituent changed in Reaction B, so the stereochemical configuration of that Me substituent is retained in the product.

Lesson II.9. Factors Leading to Non-Zaitsev Products in the E2 Reaction

Lesson II.9.1. Steric Hindrance Between Base and Substrate

The more stable/substituted alkene (i.e., the **Zaitsev product**) is the major product in many E2 reactions. There are situations, however, that raise the activation energy of the pathway leading to the Zaitsev product and thus increase the amounts of less substituted, non-Zaitsev products. The non-Zaitsev product is sometimes called the Hofmann product.

Steric hindrance between a bulky base and substrate will generally be greater for deprotonation of a more substituted site (below left) than the less substituted site on the substrate (below right):

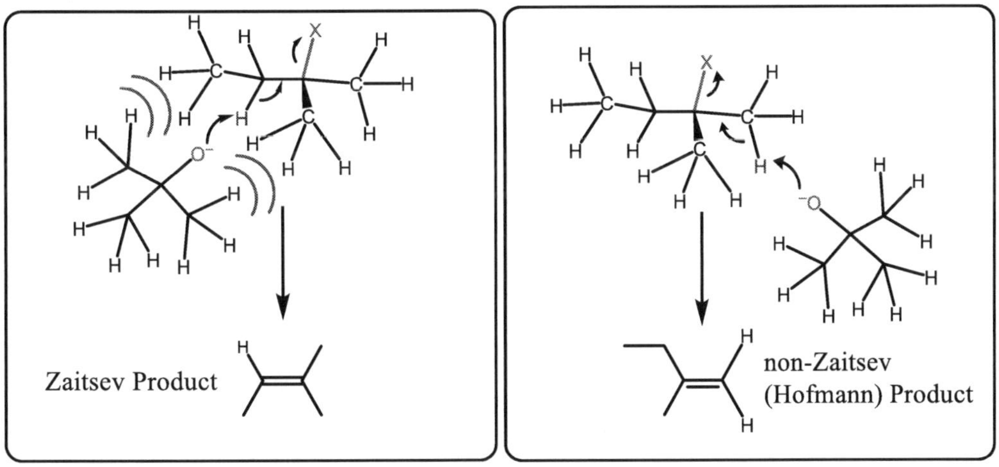

The take home lesson from this is that **as the steric encumbrance of the base increases, we will get a larger percentage of the Hofmann product** in an E2 reaction. In fact, if *t*-butoxide is the base, the Hofmann product is the major product!

Example II.9.1

Which base will lead to a higher amount of 1-hexene upon reaction with 2-bromohexane: NaOH or $NaOCH(CH_3)_2$ (sodium isopropoxide)?

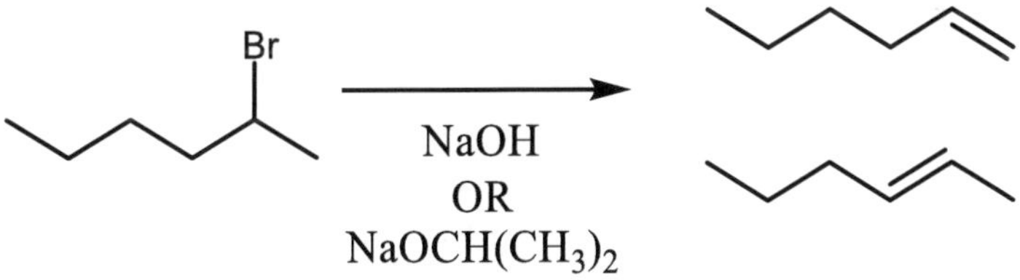

Solution II.9.1

The isopropoxide anion is a much bulkier base than is hydroxide, so it will have a greater steric encumbrance to deprotonate the more substituted site needed to access 2-hexene. A greater yield of the non-Zaitsev product 1-hexene will thus be formed when sodium isopropoxide is added as the base.

Lesson II.9.2. Steric Hindrance in the Substrate can lead to Hofmann Products

There may be cases where the Hofmann product would be our desired product, so we would want this to be the major product. A variation of the E2 reaction called the **Hofmann elimination reaction** was developed and has proven quite useful in this regard. In the Hofmann elimination reaction, the leaving group is an amine, which requires the reactant to contain an ammonium substituent:

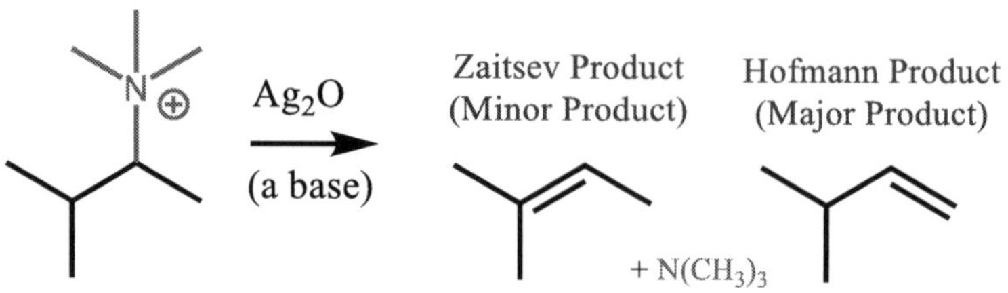

How can we explain the observed products? As with all E2 reactions, Hofmann elimination requires the H and the leaving group to be *antiperiplanar*. It is therefore helpful to compare the relative stabilities of the *anti*-conformation leading to the Zaitsev product vs. the Hofmann product:

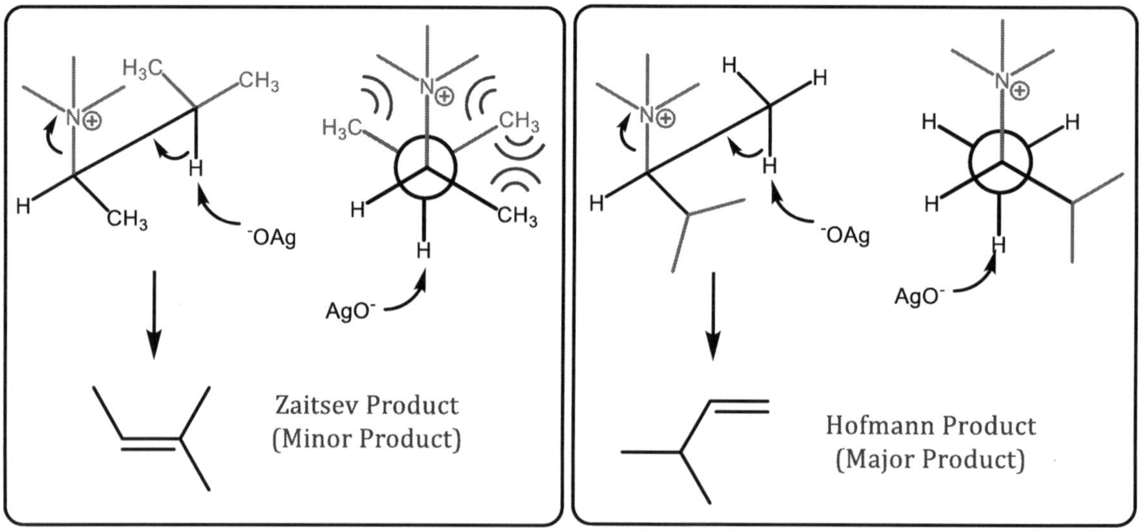

It is evident in the Newman representations that there is significant steric repulsion between the bulky trimethylammonium ($-N(CH_3)_3^+$) group and the adjacent $-CH_3$ substituents in the conformation required to produce the Zaitsev product. The conformation required to access the Hofmann product, however, places two small H atoms adjacent to the leaving group. The energy difference between these two conformations is sufficiently great that the Hofmann product forms significantly faster than the Zaitsev product, so it is the major product observed. The take home lesson from this observation is that **if the leaving group is bulky (has three or more non-H branches), then the Hofmann product is the major product.**

Example II.9.2

Provide the major product of this reaction:

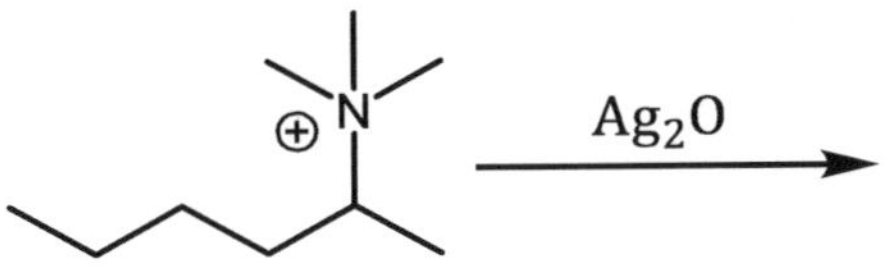

Solution II.9.2

The leaving group hear is a very bulky, 3-branched trimethylammonium unit. The very bulky leaving group favors the conformation analogous to that shown on the previous page, in which the bulky leaving group and the less-substituted (less crowded) carbon are involved in the E2 reaction. Therefore, we will observe the less substituted alkene (non-Zaitsev or Hofmann product):

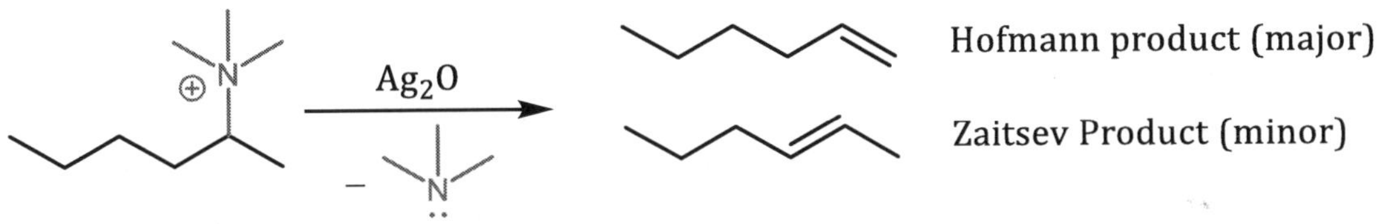

Lesson II.10. Competition Among Substitution and Elimination Reactions of RX: Predicting Pathways

Lesson II.10.1. Recapping Factors Influencing Rate of Substitution and Elimination Reactions

You may have noticed that the substitution and elimination reactions can all use alkyl halides as the starting material and that the conditions for the reactions also can look rather similar. These facts mean that we must be careful and methodical when we encounter a reaction and try to predict which reaction mechanism will predominate and therefore what major products we will actually get. If we look back at Lessons II.3-7 and consider all of the factors that influence the rate of the S_N1, S_N2, E1 and E2 reactions, we can summarize these data as follows:

	S_N1 rate	E1 rate	S_N2 rate	E2 Rate
Reagent strength	No effect	No effect	> for better Nu	> for stronger B
Substrate	3° > 2° > (1° without resonance) > CH_3	3° > 2° > (1° no resonance) > CH_3	CH_3 > 1° > 2° > (3°)	3° > 2° > 1° > CH_3
Solvent	> in polar protic	> in polar protic	> in polar aprotic	> in polar aprotic

Nu = nucleophile, B = base. **Substrates in parentheses will not work for the indicated reaction. For E1 and S_N1, secondary reacts at a rate similar to that of RX having primary sites with resonance stabilization of the carbocation formed by hydrolysis.**

We can summarize this data more simply if our task is to simply determine which of the pathways will predominate for a given substrate/reagent combination:

Substrate	**Reagent**			
	Poor Nu/Weak Base	Good Nu/Weak B	Good Nu/Strong B	Poor Nu/Strong B
CH_3–X	No reaction	S_N2	S_N2	S_N2; there's only one C, so cannot make a C=C!
1° R–X	No reaction!*	Mostly S_N2	Mostly S_N2	Mostly E2
2° R–X	(Slow) S_N1/E1 mix	Mostly S_N2	Mostly E2	Mostly E2
3° R–X	S_N1/E1 mix	Mostly S_N1	Mostly E2	Mostly E2

Nu = nucleophile, B = base.

***If the primary carbocation formed by hydrolysis of RX is resonance stabilized, that RX will undergo a mix of E1 and S_N1.**

With these facts in mind, we can design a flowchart (on the next page) to determine which reaction(s) will predominate under a given set of circumstances.

Below is a flowchart that incorporates all of the information from the above tables:

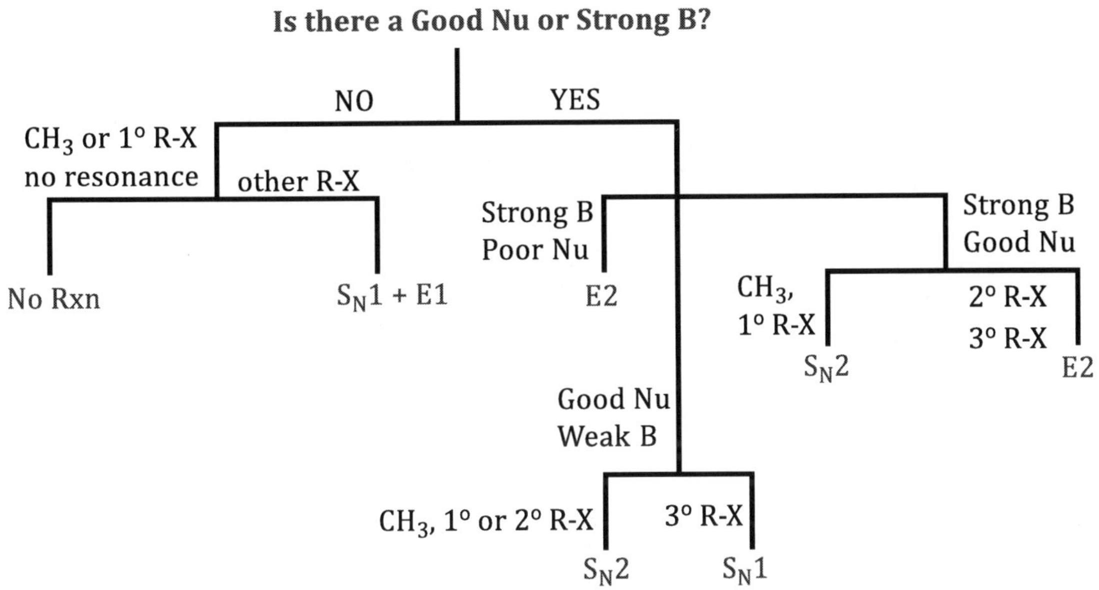

Coupled with our knowledge of what constitutes a good nucleophile or a strong base in this context, the flowchart can be an extremely powerful tool for determining which pathway predominates, allowing us to accurately predict the products under a given set of circumstances. Note that this chart loses its effectiveness completely if you are unable to recognize what is a good nucleophile and what is a strong base. Reviewing Lesson II.2 is a good idea at this point to help identify such reagents.

Example II.10.1

Determine the mechanism(s) that predominate for each of these reactions and show the major product(s). Be sure to indicate the proper stereoisomer(s) as necessary.

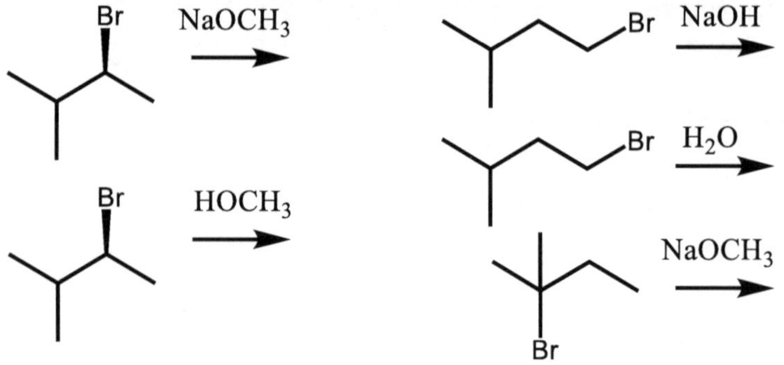

Solution II.10.1

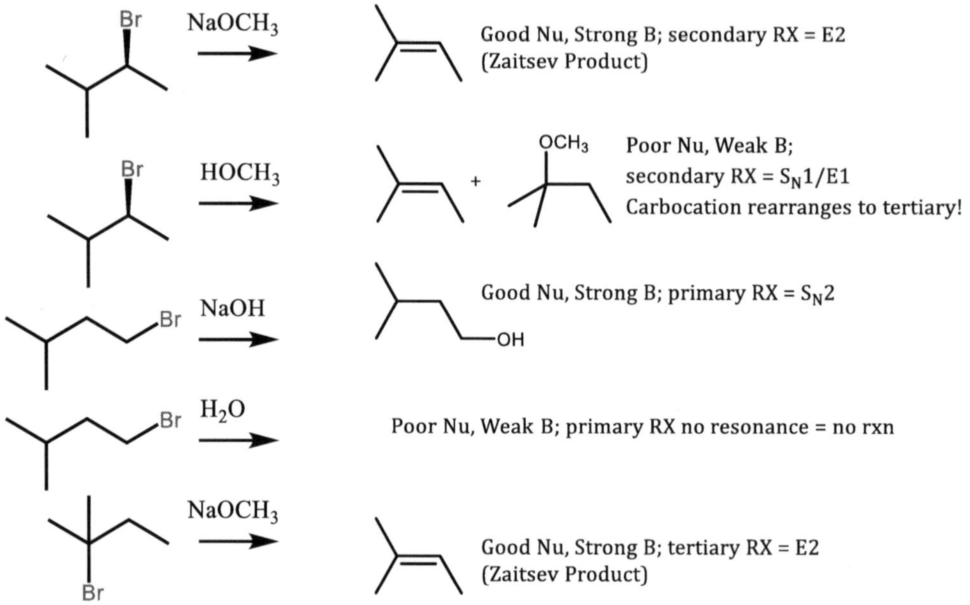

Lesson II.11. Substitution and Elimination of Alcohols

Lesson II.11.1. Activating a Poor Leaving Group with an Acid

In our previous studies on substitution and elimination reactions, we noted that the presence of a good leaving group on the substrate was required for reaction to occur. If we attempt any of these reactions on an unmodified alcohol, our efforts will fail because the leaving group (HO^-) would be a strong base, and therefore a very poor leaving group. However, there are several ways to activate the OH unit of an alcohol to convert it into a good leaving group. Once this activation has taken place, the substrate is viable for both substitution and elimination reactions.

The first way to activate an alcohol is to protonate the OH group with a strong acid:

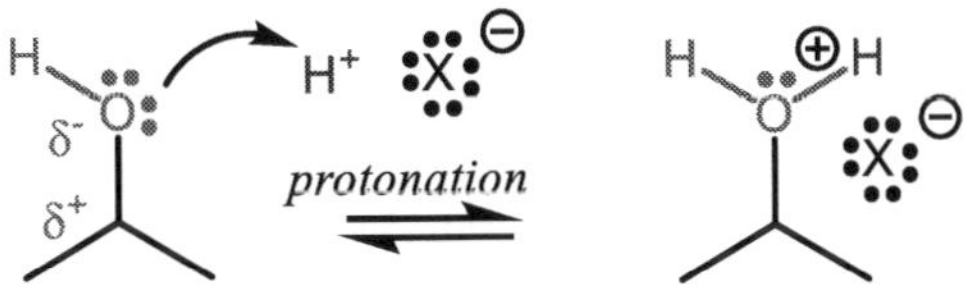

(X = Cl, Br, I or HSO_4) **Oxonium Intermediate**

Once protonated, the resultant **oxonium intermediate** has a good leaving group (H_2O). This makes it a viable substrate for both substitution and elimination reactions.

If a reasonably strong nucleophile is present (Cl^-, Br^- or I^-), a substitution pathway will dominate, with H_2O as the leaving group from the oxonium intermediate. **If an SN1 pathway is viable (2° or 3° alcohols), the SN1 product is the major product**. If the SN1 route is not viable (methyl or 1° alcohols), then the SN2 product will be the major product. All of the guiding principles for SN2 and SN1 reactions still hold, as they were described in Lessons II.3–II.5, for protonated alcohols.

<u>Example II.11.1</u>

Provide a reasonable mechanistic pathway leading to the major product of the reaction shown:

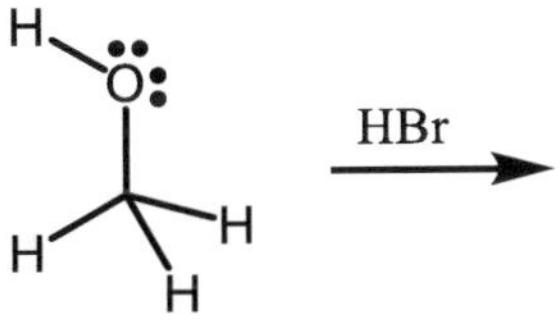

<u>Solution II.11.1</u>

The first step is protonation of the OH group to form an oxonium intermediate. This is a methyl alcohol, so the 2nd step, substitution, is an SN2 reaction on the activated substrate:

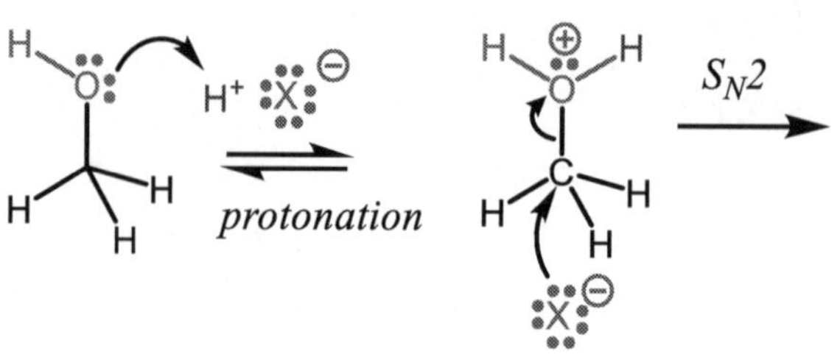

If the conjugate base of the acid used to protonate the alcohol is bulky (as is HSO_3^-, the conjugate base of H_2SO_4), then elimination will dominate. These reactions proceed by an E1 mechanism because E2 requires a strong base, which, of course, could not be present when we have added a strong acid! However, we know that the E1 reaction involves a carbocation intermediate. So, for primary substrates the E1 pathway will not be very effective. However, if we convert the alcohol to a good leaving group by some other route (as in Lesson II.11.3), it can then be reacted with a base to form an alkene via the E2 mechanism.

Example II.11.2

Provide a reasonable mechanistic pathway leading to the major product of the reaction shown:

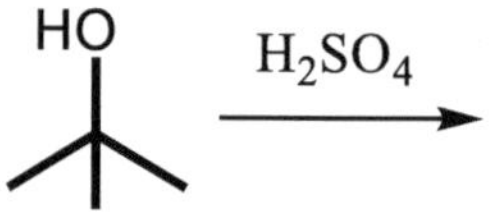

Solution II.11.2

The first step is protonation of the OH group to form the oxonium intermediate. There is no reasonable nucleophile, so an elimination reaction via the E1 pathway follows:

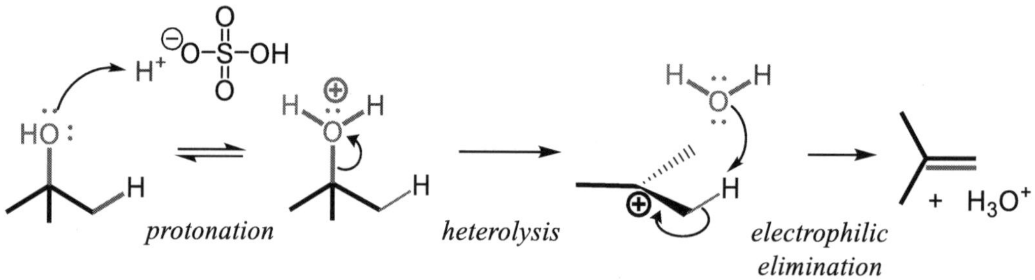

Lesson II.11.2. Activating a Poor Leaving Group with PBr_3 or $SOCl_2$

There are ways to activate the OH group of an alcohol for substitution/elimination reactions other than using a strong acid to protonate it. The two alternatives we will cover involve using thionyl chloride ($SOCl_2$) or phosphorus tribromide (PBr_3). The reaction of an alcohol with PBr_3 proceeds as follows:

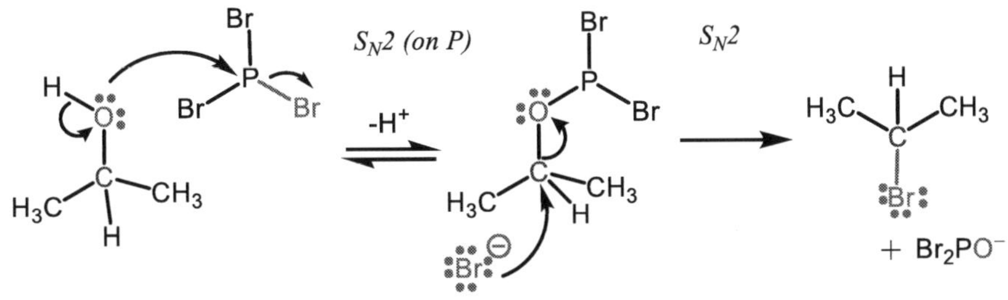

The reaction of an alcohol with $SOCl_2$ proceeds as follows:

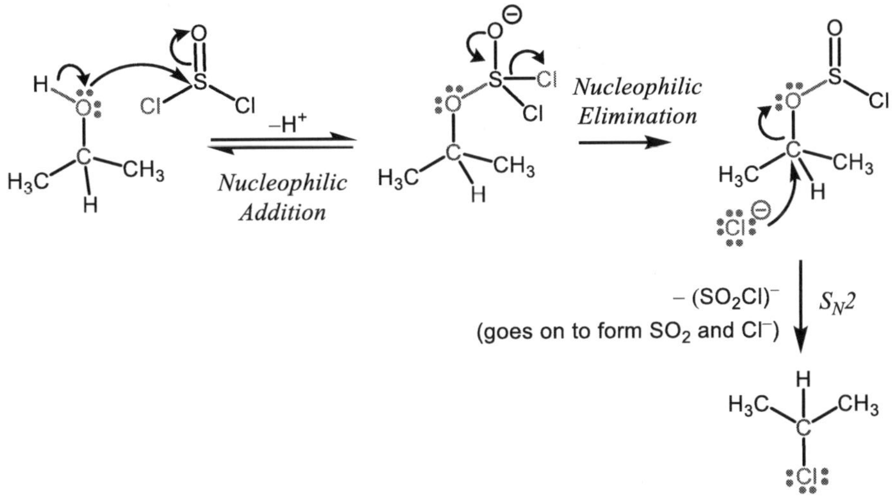

After the first activation step, the displacement of the $[OPBr_2]^-$ or $[OS(O)Cl]^-$ leaving group is an S_N2 reaction. Everything we learned about the S_N2 reaction in Lessons II.3 and II.5 still apply here, so a 3° alcohol still will not undergo an S_N2 reaction, even if we activate it with PBr_3 or $SOCl_2$, for example and if the site that is attacked s chiral, its configuration will be inverted.

Lesson II.11.3. Changing an Alcohol into a Sulfonate Ester

The substitution reactions of alcohols we have examined all result in net substitution of an OH group for a halide. If we want to substitute an OH group with something else, a different activation pathway will be needed. One convenient approach is to convert the OH into a **sulfonate ester**, which proceeds via the following mechanism:

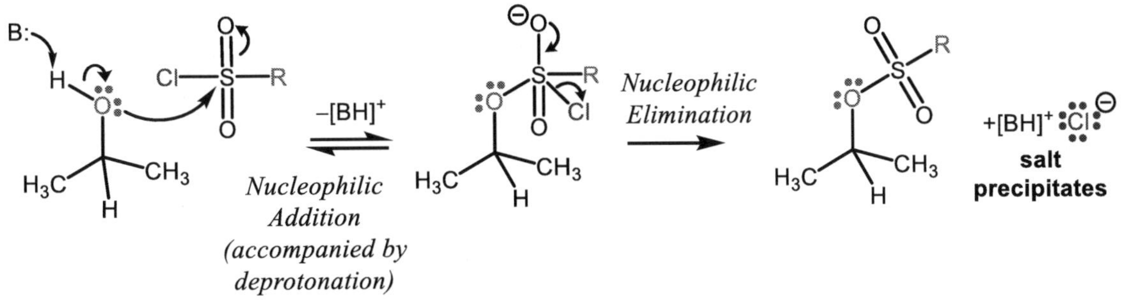

After the sulfonate ester unit has been formed, the compound is isolated. The sulfonate anion is an excellent leaving group due to its high resonance stabilization. Sulfonate anions are the best leaving groups that we will learn in this book, and they are so important that a few of them have been given common names and abbreviations (which you, the student, must memorize):

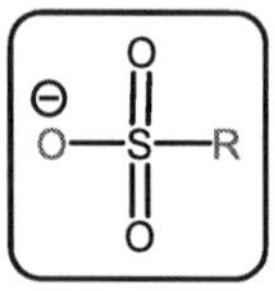

When R =	*Anion Name*	*Abbreviation*
	Tosylate	TsO^-
$-CH_3$	Mesylate	MsO^-
$-CF_3$	Triflate	TfO^-

Example II.11.3

Provide the product for each step leading to the major product of the reaction shown:

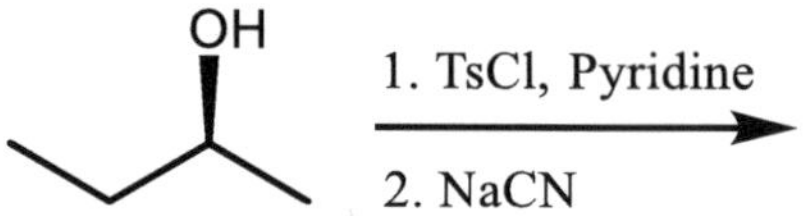

Solution II.11.3

The first step simply replaces the H on O with Ts, to form the tosyl group (OTs). Note that the stereochemistry of the C is unchanged because the reaction takes place at the O:

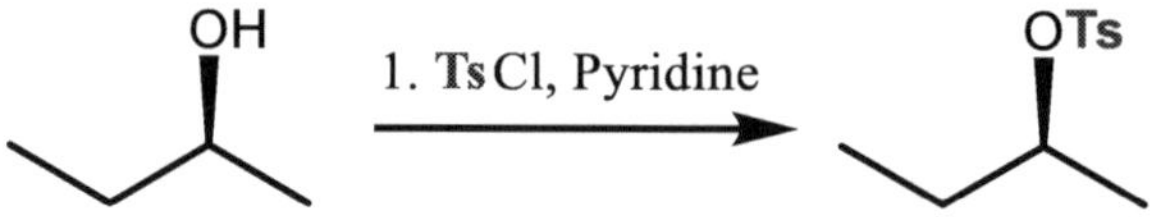

The tosyl group is a very good leaving group and will be readily displaced by the good nucleophile NC^- via an S_N2 pathway with inversion of configuration:

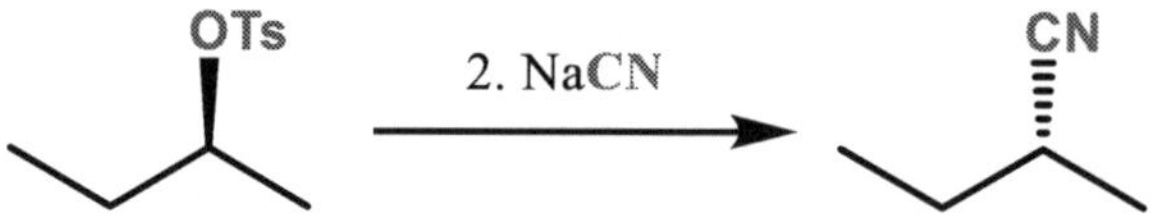

Lesson II.12. Oxidation and Reduction: Definitions

In organic chemistry, we use a simplified definition of oxidation and reduction, with a focus on the carbon atoms in a structure. In the simplest definition, oxidation is defined as a reaction leading to more C–O bonds and/or fewer C–H bonds, whereas reduction is defined as a reaction leading to fewer C–O bonds and/or more C–H bonds. For a broader definition, we would substitute "C–O bonds" with "C–EN" bonds, where "EN" is any element more electronegative than C.

Example II.12.1

Label each of the following reactions as being an oxidation, a reduction or neither:

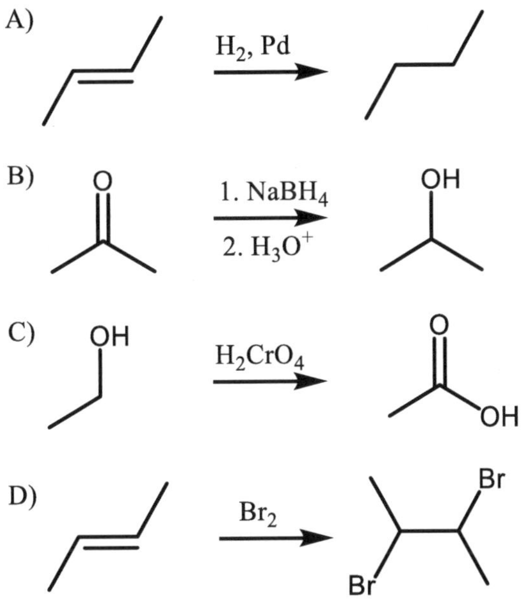

Solution II.12.1

Reaction A: has more C–H bonds in the product than in the reactant, so this is a reduction. Reaction B has fewer C–O bonds in the product than in the reactant, so this is a reduction. Reaction C has more C–O bonds in the product than in the reactant, so this is an oxidation. Reaction D has more C–EN bonds in the product than in the reactant, so this is an oxidation.

Lesson II.13. Oxidation Reactions of Alcohols

Lesson II.13.1. Oxidation of Alcohols Using Chromium Reagents

The carbon in an alcohol to which the OH group is attached (called the "carbinol" carbon) can be oxidized by several chromium reagents, typically with added acid. Those that we will cover in this book are: H^+/CrO_4^{2-}, $H^+/Cr_2O_7^{2-}$, CrO_3/H_2SO_4 (reagents for what is called the "Jones Oxidation"), and pyridinium chlorochromate (PCC). For PCC the proper solvent to assure the reactivity described here is CH_2Cl_2, so you will often see this written in the reaction conditions around the arrow as well.

PCC is a reagent that can replace one C–H bond on the carbinol with another bond to the O of the OH group (which must be accompanied by loss of H from the OH group). This leads to the formation of an aldehyde (from 1° alcohol) or ketone (from 2° alcohol):

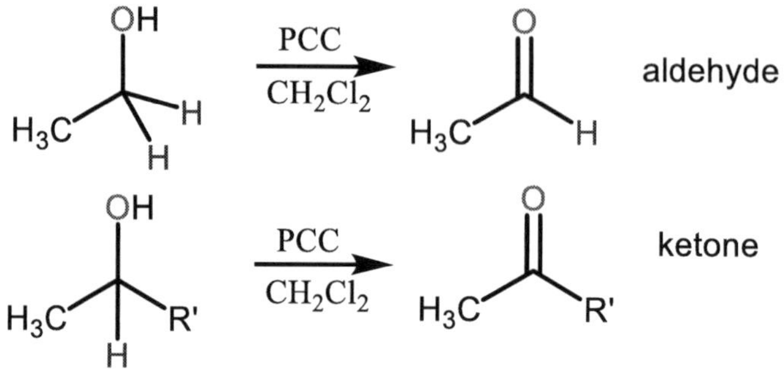

All of the chromium reagents listed above, other than PCC, are more powerful oxidants. They are sufficiently strong oxidants to replace *all* of the C–H bonds on the carbinol with bonds to O. This will lead to the formation of ketones (from 2° alcohol) or carboxylic acids (from 1° alcohol):

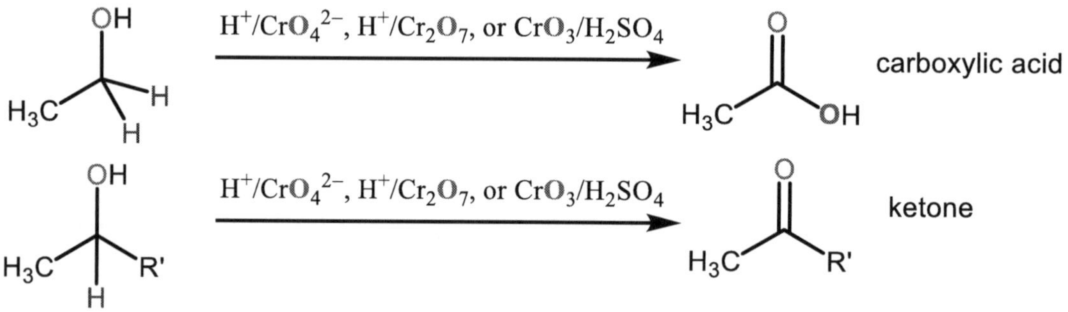

Note that, in the case of 2° alcohols, reaction with PCC or the more powerful oxidizing agents both afford ketones because there is only one H on the carbinol that can be changed to a C–O bond. However, when the reactant is a 1° alcohol, PCC will yield different products than Jones Oxidation. Note that these oxidation reactions can only remove H atoms on the same carbon as the –OH group; the adjacent groups, like the $-CH_3$ groups in the examples above, are not changed at all by these oxidation reactions.

Example II.13.1

Provide the major product for each of the following reactions:

A) 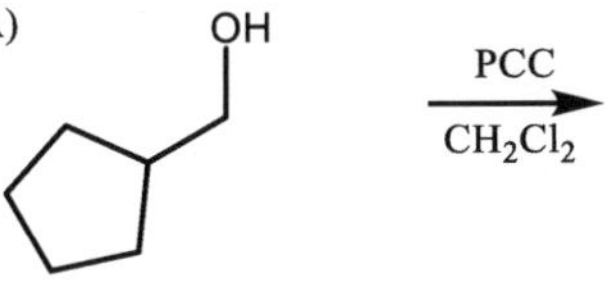

B) 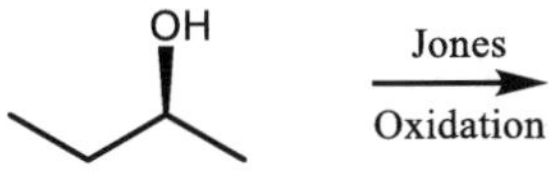

C) 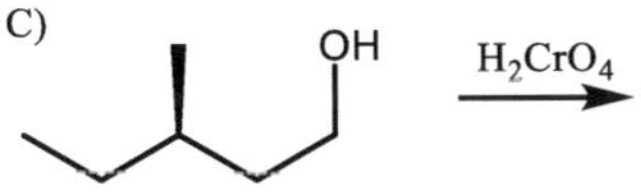

D) 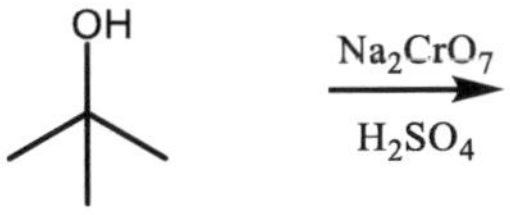

Solution II.13.1

A) 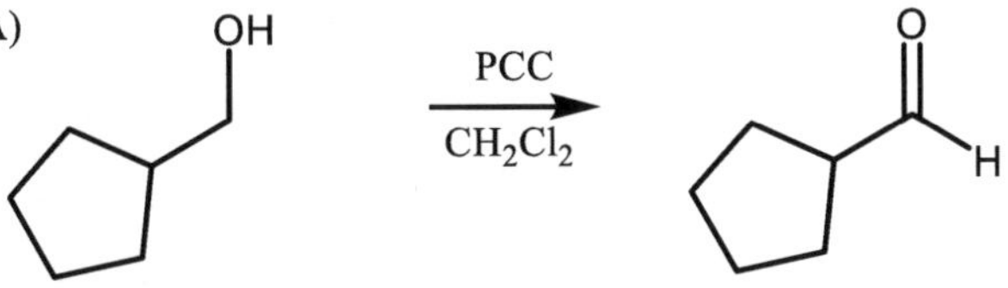

B) 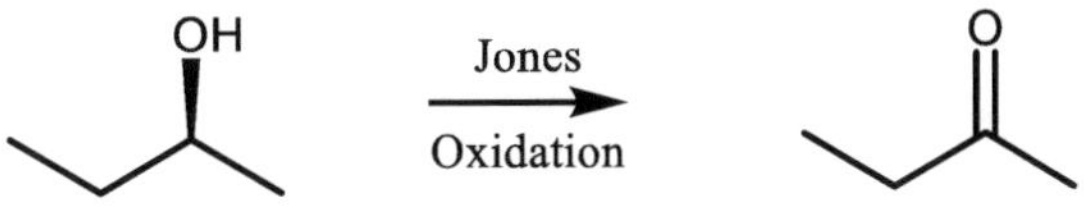

C) 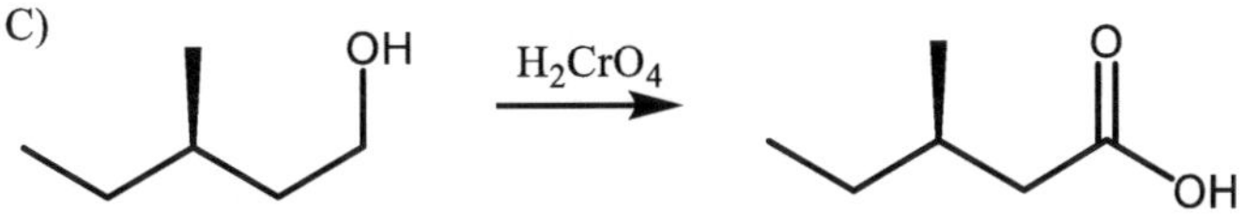

D) 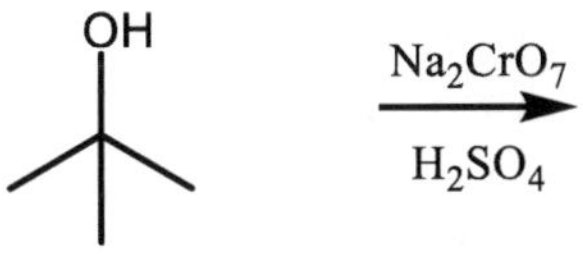

No reaction; no H on the alcohol C!

Lesson II.14. Nomenclature and Acid Cleavage of Ethers

Lesson II.14.1. Naming Ethers

When we name ethers, we divide them into two types: **symmetrical** (the two R groups are the same, as in $CH_3CH_2OCH_2CH_3$), and **unsymmetrical** (the two R groups are different, as in $CH_3OC_2H_5$).

We can effectively name ethers with either **common names** or **systematic names**. Common names are formulated by first naming each R group (the R and R^1 groups in the diagram below):

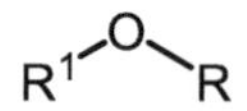

Alkyl (R^1) alkyl ether

The two alkyl groups are then listed alphabetically, followed by "ether". If the ether is symmetrical, it is simply named "di*alkyl* ether." These examples illustrate the common naming method:

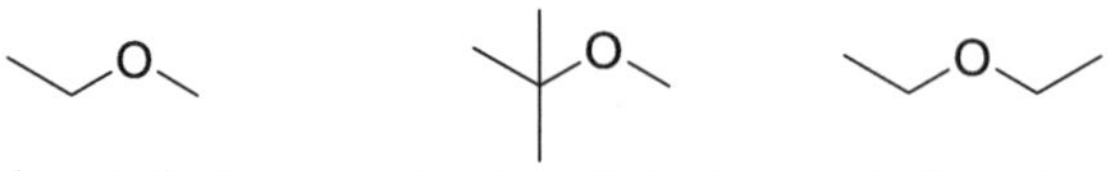

ethyl methyl ether *t-butyl methyl ether* *diethyl ether*

IUPAC systematic names are often used for ethers as well. In this approach, the longer alkyl group is named as the parent chain and the O with the shorter alkyl group is named as an **alkoxy** substituent (methoxy, ethoxy, or alkyloxy for longer C chains). These two examples illustrate the systematic naming method:

1-ethoxybutane *methoxycyclohexane*

Lesson II.14.1. Substitution Reactions of Ethers

In Lesson II.11, we saw that HX (X = Cl, Br or I) simultaneously provides H^+, which activates an alcohol OH group, and X^-, the nucleophile which displaces water from activated alcohols. These same HX acids can undergo similar reactions with ethers. Acid cleavage of ethers is an important reaction because ethers are relatively inert under most of the reaction conditions we have studied. As for alcohols **if S_N1 is possible in the ether, S_N1 will occur faster than S_N2**. This means that the major product will come from an S_N1 pathway for nucleophilic attack on 2° or 3° ether carbons, and only by S_N2 on methyl or 1° ether carbons.

The fact that the S_N1 pathway outpaces the S_N2 pathway for substitution has important ramifications, especially if the ether is not symmetrically substituted. Consider the reaction of ethyl isopropylether with HI. We have to choose which of the two carbon atoms on the O will lose the leaving group. In this case, the secondary side will be faster for the SN1 pathway (SN1 fails on primary sites), so the secondary C loses the leaving group to produce the most stable carbocation available:

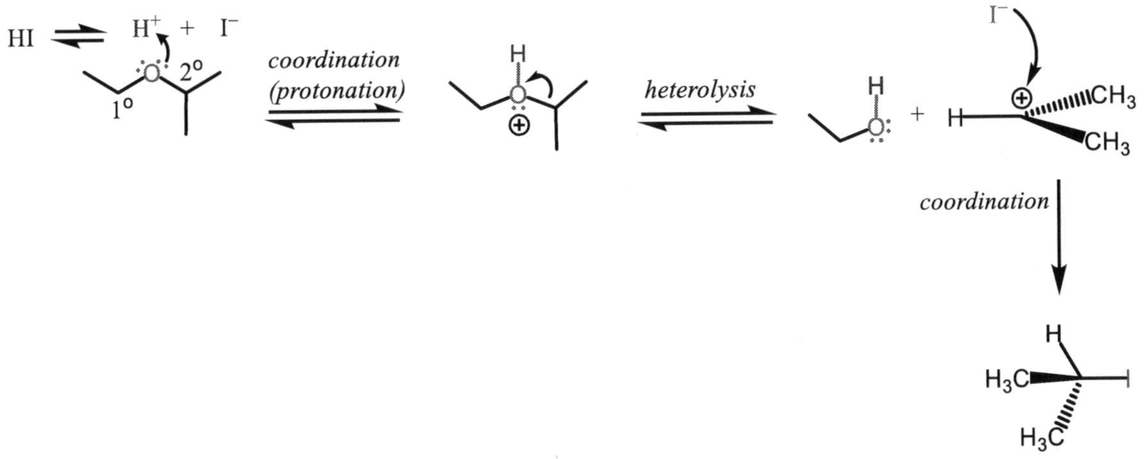

The net result is production of one equivalent of ethanol and one equivalent of 2-iodopropane as the major products. Once we have identified whether an SN1 or SN2 pathway is going to take place, we can use what we learned in Lessons II.3–II.5 to determine the major products of the reaction.

Example II.14.1

Provide the major product for each of the following reactions:

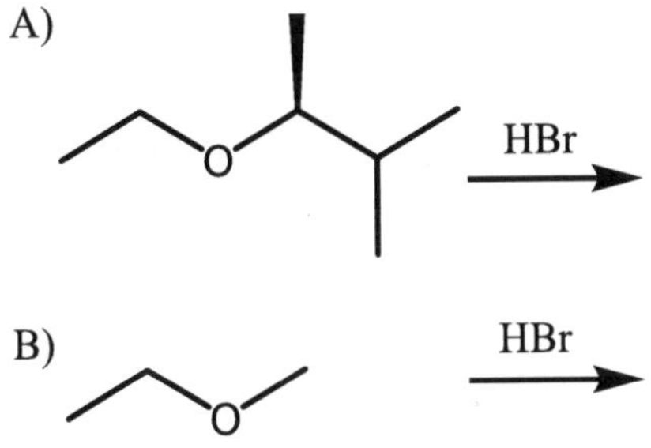

Solution II.14.1

In reaction A, the left-hand ether carbon is primary and the right-hand ether carbon is secondary. The secondary site can undergo SN1 reaction, so this will be the major pathway. The secondary carbocation will form (after protonation and heterolysis):

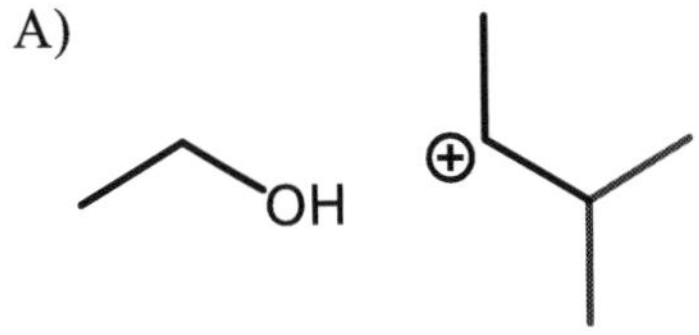

The secondary carbocation will rearrange to form the tertiary carbocation:

A)

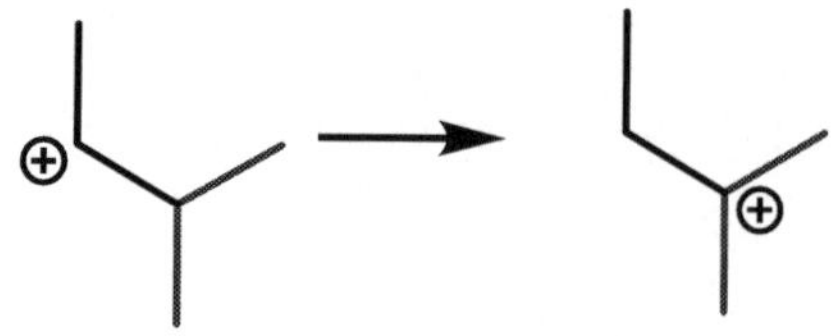

The tertiary carbocation will then undergo coordination by the bromide. The final net reaction for reaction A is thus:

A)

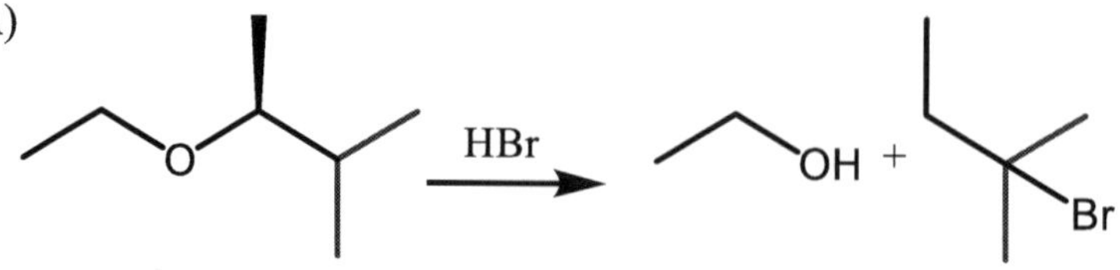

In reaction B, the left-hand ether carbon is primary and the right-hand ether carbon is methyl. Neither site can undergo S_N1 reaction, so S_N2 will be the major pathway. For S_N2, we know nucleophilic attack will occur more rapidly at the less sterically hindered site, so the major product comes from nucleophilic attack on the methyl site:

B)

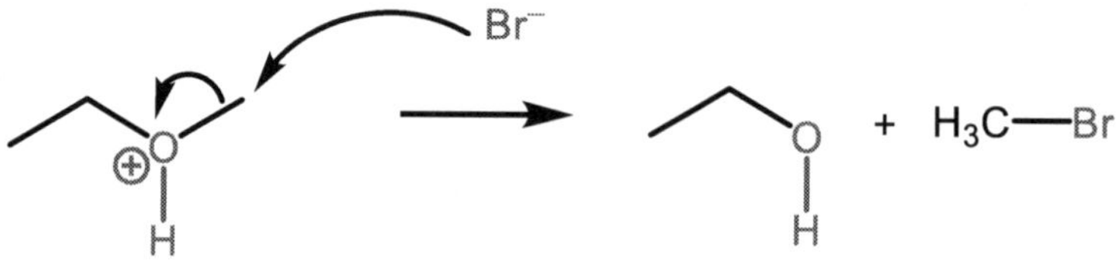

(shown after protonation)

One additional note about this reaction is that we can use more than one equivalent of the acid and drive the reaction further. We learned in Lesson II.11 that ROH reacts with HX to form RX and water. If we perform the reaction in Example II.14.1A with 2 or more equivalents of HBr, the first equivalent of HBr will react as shown in Solution II.14.1A, but the second equivalent of HBr will convert EtOH to ethyl bromide, so that the overall products will be as shown:

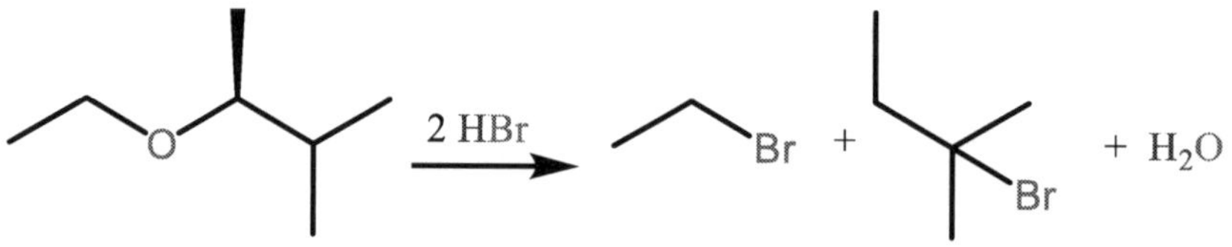

Lesson II.15. Ring-Opening of Epoxides: Steric/Electronic Effects on Major Product

Lesson II.15.1. Epoxides are More Reactive than Other Ethers

Epoxides are ethers in which the O belongs to a three-membered ring: 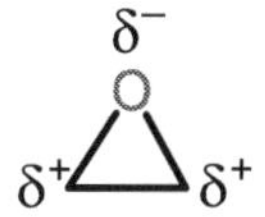

In Lesson I.16, we learned that a three-membered ring has the most ring strain of any ring size. The significant ring strain in the epoxide makes it much less stable (more reactive) than acyclic ethers. The enhanced reactivity of epoxides means that the epoxide can react with good nucleophiles to open the epoxide ring, even if the epoxide O is not protonated:

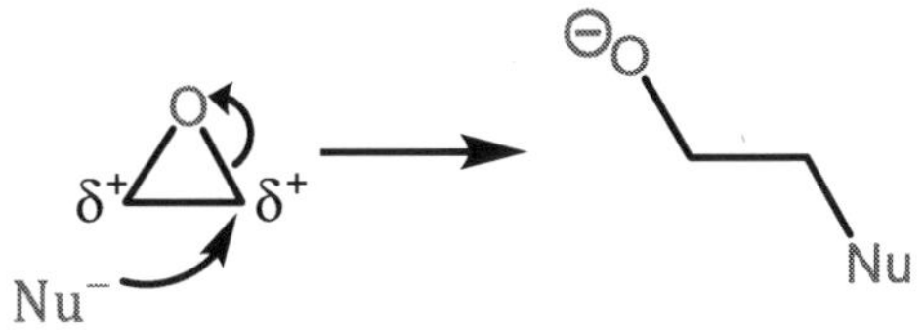

Note that, although ring strain is alleviated by ring opening, the anion produced is still a strong base and is thus relatively unstable. Because the anion produced by ring opening is a strong base, not every nucleophile will react with the neutral epoxide. **The nucleophile must be a relatively unstable anion (a strong base)**, such that the reactant anion has similar stability to the product anion. The net reaction is spontaneous due to relief of ring strain. It is also important to remember that, after nucleophilic attack under basic conditions, the resulting anion must be protonated to obtain the neutral alcohol product.

If the epoxide is reacted with a nucleophile under acidic conditions, however, even a poor nucleophile will react with the epoxide-derived oxonium intermediate, because the leaving group is a neutral OH group and quite stable:

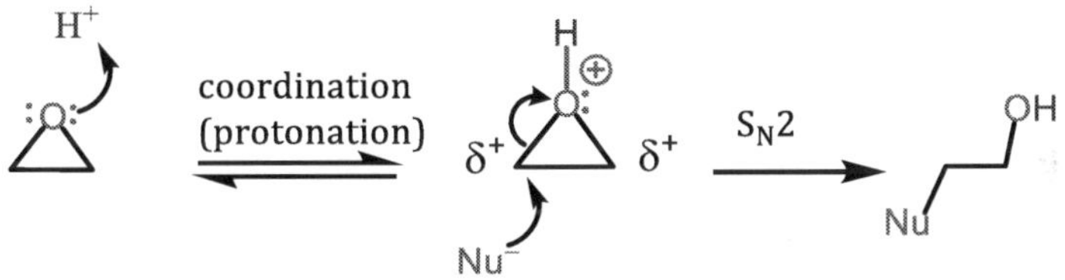

Lesson II.15.2. Regioselectivity in Ring-Opening of Epoxides

We must determine which side of an asymmetrically-substituted epoxide will be attacked by the nucleophile in a ring-opening reaction. Regardless of conditions, if one side of the epoxide is primary and one is secondary, the less-substituted side is attacked because it is less hindered; the sterics win out in this case, as seen for the following example under acidic conditions:

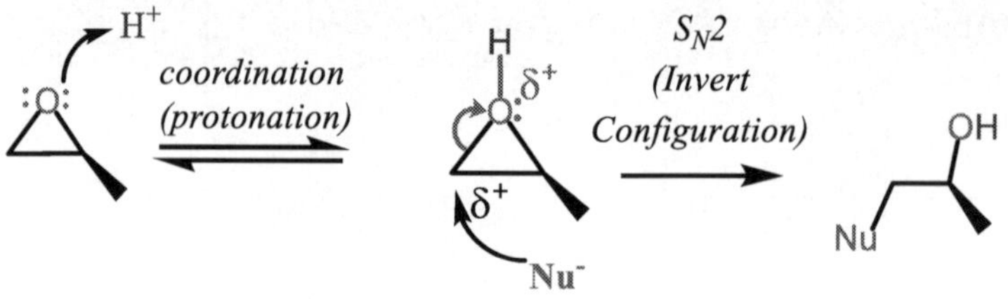

… and for this example, under basic conditions (followed by protonation):

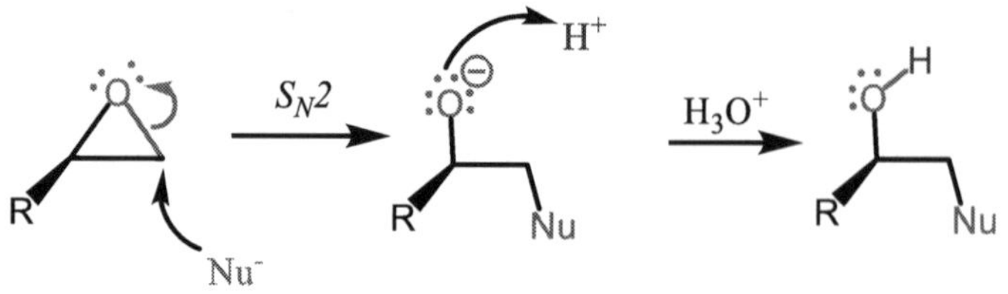

The only time that the more substituted side is attacked is if one side is tertiary and the reaction is under acidic conditions. The observed attack by the nucleophile on the tertiary site may seem to contradict the general trend that S_N2 reactions are faster at less sterically-hindered sites. In the case of an oxonium cation, however, there is so much more partial positive charge on the more substituted carbon that the attraction of the nucleophile's electrons to this site is much stronger than it would be if the nucleophile were attaching a neutral species. All epoxide ring opening reactions undergo an S_N2-like ring opening, with inversion of configuration:

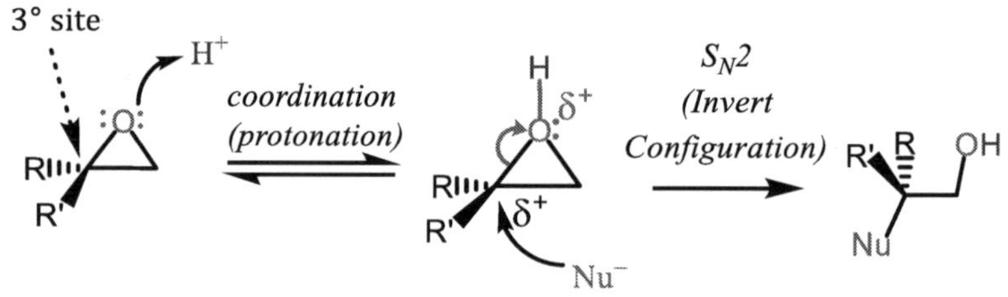

oxonium cation

Part III. Reactions Involving Alkenes, Alkynes and Radicals

Lesson III.1. Alkene Nomenclature and the E-/Z- Convention

Lesson III.1.1. The "-ene" Suffix

In Lesson I.14, we learned how to name alkanes. We will use the core set of nomenclature rules from that lesson as the basis to name alkenes as well. When a hydrocarbon contains a C=C bond, we replace the "-ane" suffix with "-ene". So, if we have a 6-carbon chain with one C=C bond, it is a "hex**ene**"; an 8-carbon chain with one C=C bond is an "octene". If two or three C=C bonds are present, we use the suffix "-adiene" or "-atriene". A 6-carbon chain with two C=C bonds is a "hex**adiene**", and an 8-carbon chain with three C=C bonds is an "oct**atriene**".

Lesson III.1.2. Alkene Priority in Hydrocarbons

We first pick as parent the longest chain that contains the C=C bond. If multiple C=C bonds are present, we select as parent the chain that contains the greatest number of C=C bonds (these parent chain rules supersede the ones for alkanes). We then number the parent chain to give the C=C bond(s) the lowest possible substituent numbers.

Example III.1.1

Provide the unambiguous systematic name for the molecule shown below:

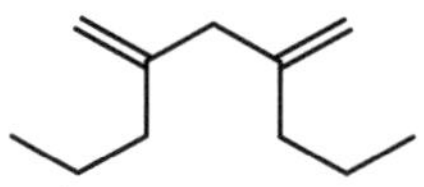

Solution III.1.1

Even though the longest chain present in this molecule has 9 carbons, we have to select as parent the chain that has the greatest number of C=C bonds. For this molecule, it is a 5-carbon chain with 2 C=C bonds. It does not matter if we start numbering left to right or right to left in this example, because the molecule is symmetric. The C=C bonds begin at carbons 1 and 4, so the parent is a 1,4-pentadiene. Two substituents are present at carbons 2 and 4. We would need to use the complex substituent rules to assign them names, and they are both propyl substituents. The complete systematic name is thus 2,4-dipropyl-1,4-pentadiene.

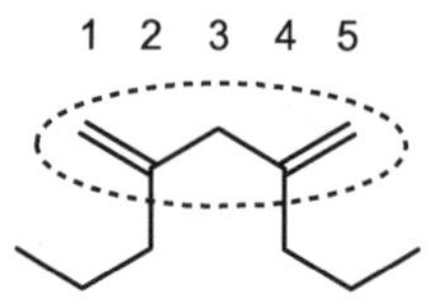

Lesson III.1.3. Alkene Priority in Heteroatom-Containing Compounds

If an alkene also contains an OH group, then that OH group takes priority over the C=C bond. Thus, we must select as parent the chain that contains an OH group, even if that results in fewer C=C bonds being included. Similarly, we must number the parent chain to give the OH group the lowest possible substituent number, regardless of what substituent numbers that assigns to the C=C bonds. To assign a name to the molecule, the "ene" suffix becomes "en" and is listed before the "ol" suffix which ends the name. A 6-carbon chain with one C=C bond and one OH group would thus be a "hexenol". For a complex molecule such as this, the substituent numbers can be placed immediately before the suffixes to enhance clarity.

Example III.1.2

Provide the unambiguous systematic name for the molecule shown below:

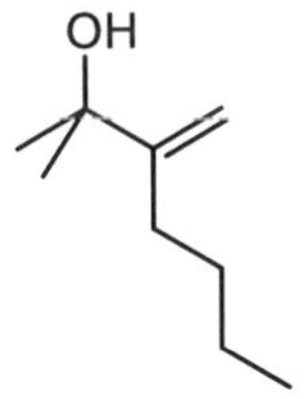

Solution III.1.2

The longest chain in this molecule has 7 carbons, but we must pick the longest one that has both the OH group and the C=C bond. The longest chain that meets these criteria has only 4 carbons, so the root is "butenol". We have to number from left to right because that gives the OH group a substituent number of 2 (going from right to left it would have a 3). As a result, the C=C bond begins at carbon 3, so the parent molecule is "3-buten-2-ol". We next need to identify the substituents and their positions in this molecule, which are methyl at carbon 2 and an *n*-butyl at carbon 3. The complete systematic name is thus "3-*n*-butyl-2-methyl-3-buten-2-ol".

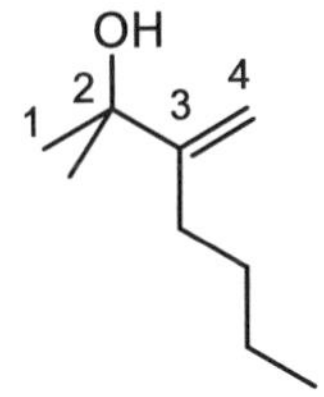

Lesson III.1.4. Hindered Rotation about C=C Bonds

As we learned in Lessons I.14 and I.15, rotation about carbon–carbon single bonds in alkanes and cycloalkanes can give rise to multiple conformational isomers, which readily interconvert at room temperature. In contrast, carbon–carbon double bonds do not undergo bond rotation. Rotating around a π-bond would require that bond to break. This means that alkenes can have configurational isomers (defined in Lesson I.17).

Example III.1.3

Indicate whether the following pair of molecules is identical or non-identical:

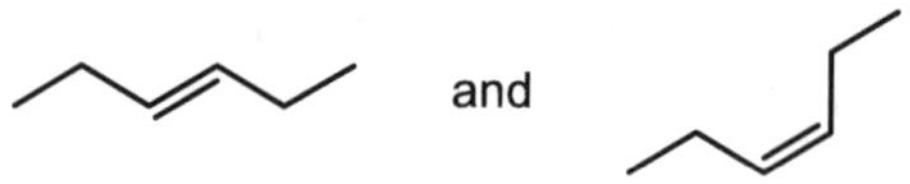

Solution III.1.3

The easiest way to approach this problem is to draw the Newman projection along the C(2)–C(3) bonds. For the molecule on the left, the ethyl substituents on carbons 2 and 3 are separated by 180° in the Newman projection, whereas the ethyl substituents in the molecule on the left are separated by 0°. When we were working with alkanes, we could freely rotate the substituents on the front and rear carbons in the Newman projection, because C–C σ-bonds can freely rotate at room temperature (and much lower!). With alkenes, there is a σ- and a π-bond present, which prevents free rotation about the C(2)–C(3) bond (we would need to break the π-bond before the σ-bond would be able to rotate freely). Because the C=C bond cannot freely rotate, these two molecules are non-identical (configurational isomers, to be precise).

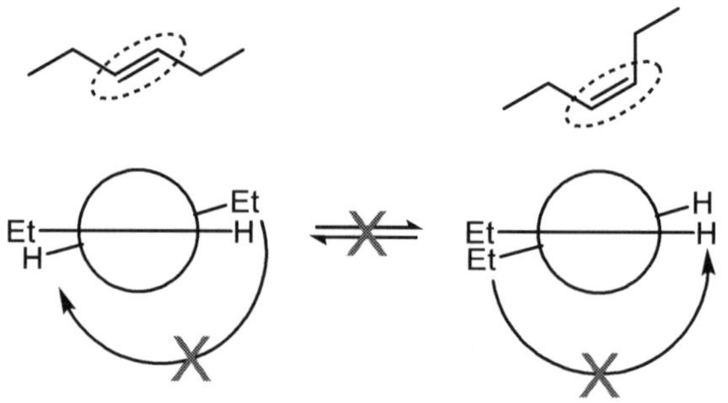

Lesson III.1.5. The cis-/trans- Prefix Convention

The two isomers of 3-hexene shown in Example III.1.3 differ only by whether the two ethyl substituents are on the same side or opposite sides of the C=C bond. Rephrasing this in more general terms, the two isomers differ only by the spatial orientation of their substituents, and the two isomers cannot be interconverted without breaking any bonds. Thus, based on the definition in Lesson I.20, we would classify the two isomers of 3-hexene as stereoisomers. Because the presence of a C=C bond may give rise to multiple stereoisomers, we would classify that C=C bond as stereogenic. Note, however, that it is only when each carbon in a C=C bond has two non-identical substituents that the C=C bond becomes stereogenic; if two of the substituents on one C atom are identical, then switching them will not yield a different stereoisomer.

The simplest type of stereogenic C=C bond is one in which each carbon in the C=C bond has one hydrogen and one non-hydrogen substituent, and the term for this type of molecule is a disubstituted, internal alkene. In contrast, a disubstituted, external alkene is a molecule in which one carbon in the C=C bond has two hydrogen substituents and the other carbon and has two non-hydrogen substituents.

If we consider 2-methyl-1-pentene (see below), we can see that switching the methyl and *n*-propyl substituents does not produce a new molecule, so 2-methyl-1-pentene does not have multiple stereoisomers. For similar reasons, a monosubstituted alkene like 1-hexene (another constitutional isomer of 3-hexene) does not have multiple stereoisomers.

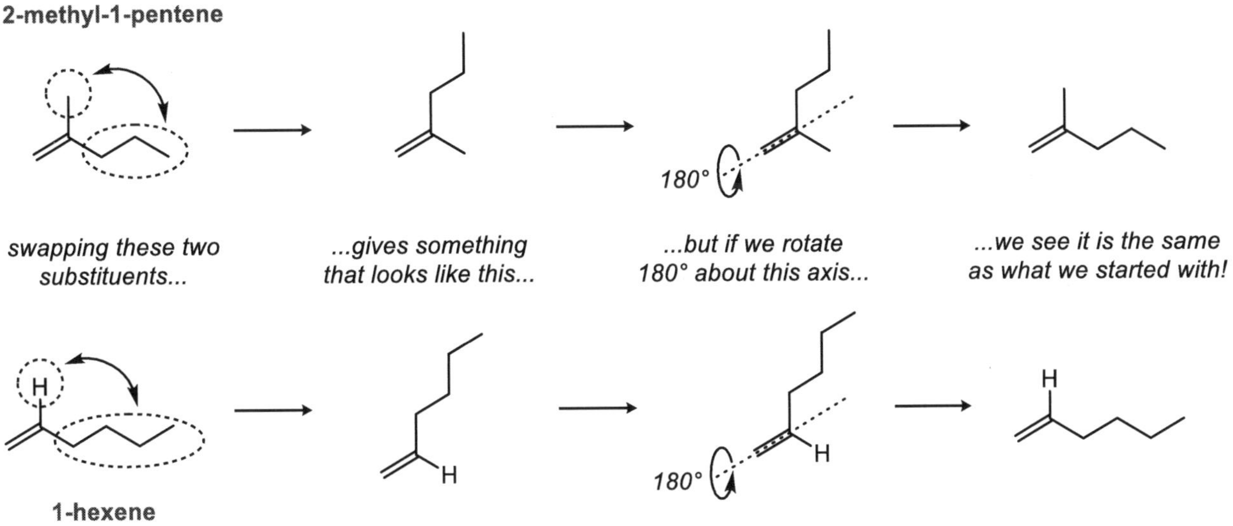

For an internal disubstituted alkene, if both substituents are on the same side of the line passing through the C=C bond, we apply the prefix "*cis*-". If the substituents are on opposite sides of the line passing through the C=C bond, we apply the prefix "*trans*-". Consider again the two configurational isomers of 3-hexene as an example:

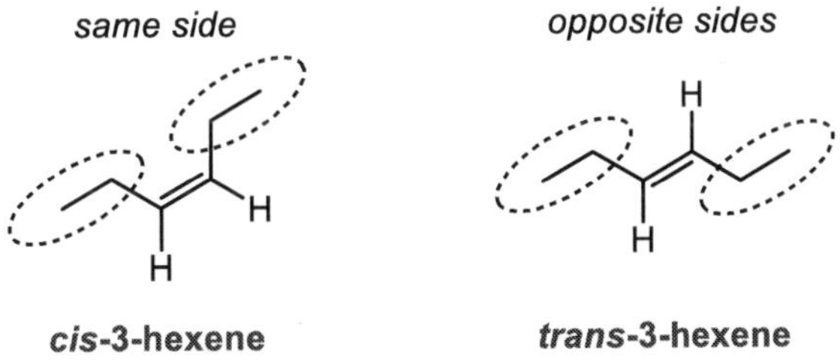

Note that the "*cis*-" and "*trans*-" prefixes belong to a convention developed specifically for disubstituted, internal alkenes and they can only be used for disubstituted, internal alkenes and cycloalkanes (as described in Lesson II.17)!

Example III.1.4

Provide the appropriate name for the molecule shown below:

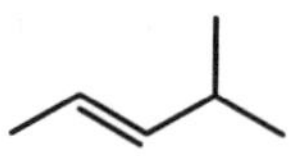

Solution III.1.4

The alkene carbon on the left has one non-hydrogen substituent, a methyl group. The carbon on the right also has one non-hydrogen substituent, an isopropyl group. Because these two non-hydrogen substituents are on opposite sides of the C=C bond with respect to each other, the proper stereochemical prefix for this molecule would be "*trans-*".

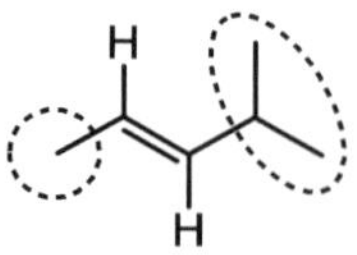

The complete name of the molecule is thus *trans*-4-methyl-2-pentene.

If a C=C bond is either (a) disubstituted and external or (b) monosubstituted, then it is not stereogenic and therefore no stereochemical prefix is necessary for an unambiguous name. Because the "*cis-*" and "*trans-*" convention was developed specifically for disubstituted, internal alkenes, stereogenic tri- and stereogenic tetrasubstituted C=C bonds require a different set of stereochemical prefixes, as described in Lesson III.1.6.

Lesson III.1.6. Tri- and Tetrasubstituted Alkenes and the E-/Z- Prefix Convention

When an alkene is tri- or tetrasubstituted, the *cis-/trans-* convention is inapplicable, but its C=C bond may still be stereogenic, so we need to have a way to describe the necessary stereochemical information in that molecule's name. For example, if we add a methyl substituent to the 3-position of 3-hexene, the C=C bond is still stereogenic: in one stereoisomer, the two ethyl substituents are on the same side of the C=C bond, and in the other stereoisomer, the two ethyl substituents are on opposite sides.

The terms "*zusammen*" (German: "together") and "*entgegen*" (German: "opposed") are used to describe stereoisomers of tri- and tetrasubstituted alkenes. The way that "*zusammen*" (abbreviated as the prefix "*Z-*") and "*entgegen*" (abbreviated as the prefix "*E-*") are used is to assign the Cahn-Ingold-Prelog (CIP) priority (see Lesson I.18) to the two substituents on each carbon in a C=C bond (i.e., each carbon has a "higher" and a "lower" priority substituent), and then we compare the relative spatial orientation of the two "higher" priority substituents. If the two "higher" priority substituents are on the same side of the C=C bond, then we say they are "together", or "*zusammen*", and we use the prefix "Z-

". If the two "higher" priority substituents are on opposite sides of the C=C bond, then we say they are "opposed", or "*entgegen*", and we use the prefix "*E*-".

Example III.1.5

Provide the appropriate stereochemical prefix for the molecule shown below:

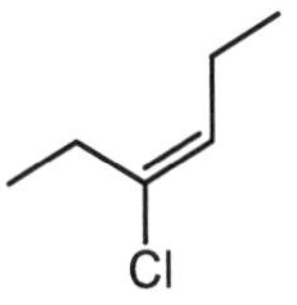

Solution III.1.5

The left carbon in the C=C bond has an ethyl substituent and a chlorine atom, and the Cl has higher priority. The right carbon in the C=C bond has an ethyl substituent and a hydrogen atom, and the Et has higher priority. Because the two "hi" substituents are on opposite sides of the C=C bond, this molecule would require a stereochemical prefix of "(*E*)-".

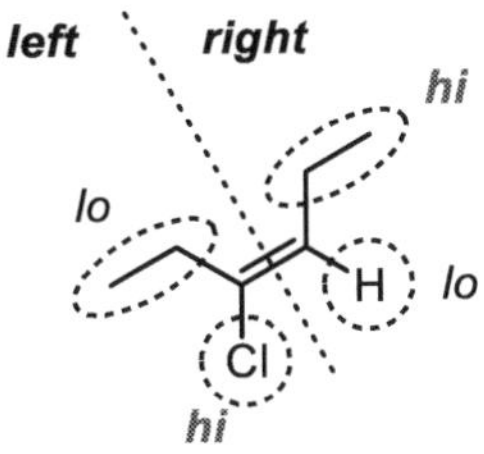

Lesson III.1.7. Alkene Priority in Molecular Nomenclature

If we look at the structure of 2,3-dimethyl-3-penten-2-ol (below left), we can see that the C=C bond is stereogenic, so the name must include a stereochemical prefix to be unambiguous, which in this case would be "(*E*)-". We would thus use the rules from Lesson I.14 to assemble the core name and place the label of configuration in front of the name in parentheses, in the same manner as we did with the *R*- and *S*- labels in Lesson I.18.

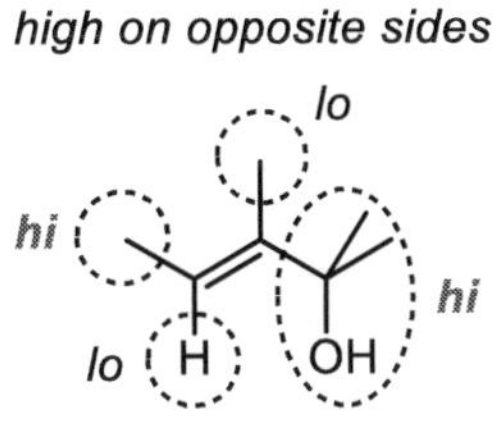

(*E*)-2,3-dimethyl-3-penten-2-ol

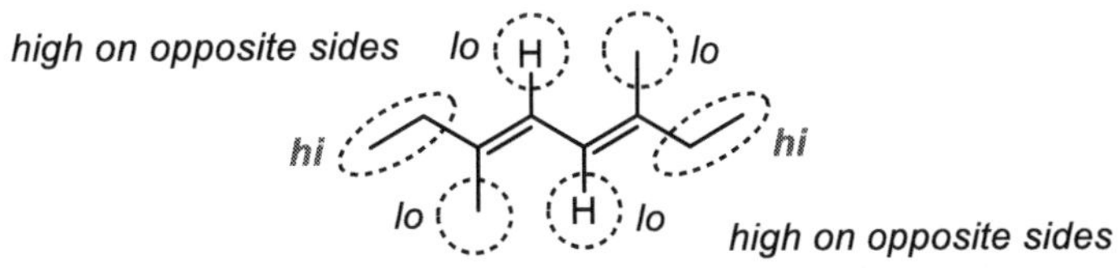

(3*E*,5*E*)-3,5-dimethyl-3,5-octadiene

What happens if a molecule has more than one stereogenic C=C bond? In this case, we would include a number with the stereochemical prefix to indicate which C=C bond has which stereochemistry. So, for the stereoisomer of 3,5-dimethyl-3,5-octadiene shown above, the complete prefix would be "(3*E*,5*E*)-".

Example III.1.6

Provide the unambiguous name for the molecule shown below:

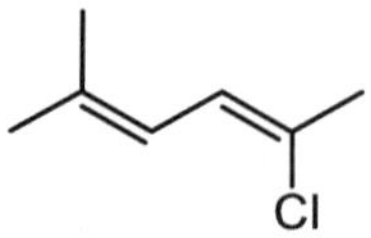

Solution III.1.6

The parent of the molecule is the 6-carbon chain containing the two C=C bonds, and we number from right to left to give a lower position number to the substituent that comes earlier in the alphabet ("chloro" vs. "methyl"). Although carbons 4 and 5 are connected by a C=C bond, this C=C bond is not stereogenic because carbon 5 has two identical substituents. The double bond between carbons 2 and 3 is stereogenic: on the left carbon, the hydrogen is lower priority than the carbon chain, and on the right carbon, the methyl is lower priority than the chlorine. Because the two "hi" substituents are on the same side of the C=C bond, the prefix should be "(*Z*)-". The proper name for the molecule would thus be "(*Z*)-2-chloro-5-methyl-2,4-hexadiene".

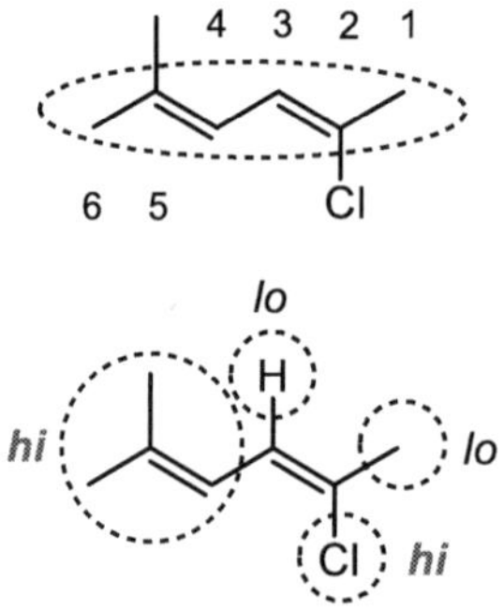

Lesson III.2. Properties of Alkenes and Hydrogenation of Alkenes

Lesson III.2.1. Saturated versus Unsaturated Molecules

Hydrocarbons that have π bonds are called **unsaturated**, because they are not “saturated” with as many H atoms as could be bound to the C atoms if all of the bonds were single bonds. You may be familiar with the terms “saturated fat” (generally solid fats like shortening and lard) and “unsaturated fat” (things like vegetable oil). Early chemists were able to determine the formula of a compound but did not always know the exact structure, so the extent to which a molecule was “saturated” with as many H atoms as the carbons could hold became an important property to determine. Consider these molecules:

C_5H_{12}
saturated

C_5H_{10}
one degree of unsaturation

C_5H_{10}
one degree of unsaturation

Notice that, compared to the saturated alkane having the same carbon count, each π bond or ring that we add requires us to remove two H atoms, or one H_2 molecule, compared to the saturated structure. For this reason, each ring or π bond in a structure is referred to as a **degree of unsaturation** (DU) or one unit on the **hydrogen deficiency index** (HDU). By looking at a structure it is relatively easy to determine the degree of unsaturation, as illustrated by this example:

Example III.2.1.

How many degrees of unsaturation are present in a molecule of Lexapro:

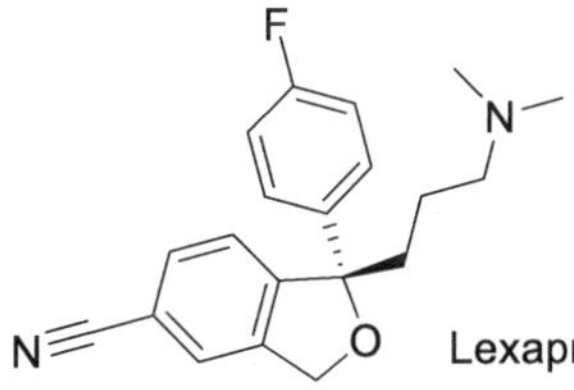

Lexapro

Solution III.2.1.

We count up a total of eight π bonds and three rings, for a total of eleven degrees of unsaturation.

It is also possible to determine the DU for a molecule given only its formula. A general formula for this is:

$$\text{Degree of unsaturation} = \frac{\mathbf{2(\#\ of\ C) + 2 - (\#\ of\ H) - (\#of\ X) + (\#\ of\ N)}}{\mathbf{2}}$$

Where "# of X" is the total number of all halogens. Note that if any of the atom types are not found in the formula, those terms will just be zero in the formula. Another notable point is that if there are O atoms in the formula, their presence does not influence the DU, so "# of O" does not show up in the equation.

Example III.2.2.

How many degrees of unsaturation are present in a molecule having each of the following formulae?

A. $C_{13}H_{16}BrCl$

B. C_6H_7N

C. $C_{13}H_{15}NO$

Solution III.2.2.

A. Using the formula, we get: DU = [(2*13) + 2 – 16 – 2 + 0]/2 = 5
B. Using the formula, we get: DU = [(2*6) + 2 – 7 – 0 + 1]/2 = 4
C. Using the formula, we get: DU = [(2*13) + 2 – 15 – 0 + 1]/2 = 7

Lesson III.2.2. Hydrogenation of Alkenes to Alkanes

In Lesson I.11.1, we saw how the thermodynamics of alkene hydrogenation reactions can be used to infer stability, but we have not yet discussed the mechanism of the hydrogenation reaction. The typical conditions used to perform alkene hydrogenation reactions are H_2 (or D_2) in the presence of a group 10 transition metal (Ni, Pd, or Pt) as a catalyst. Under these conditions, a molecule of H_2 first reacts with the metal surface, then the alkene reacts via a series of steps shown below to form an alkane:

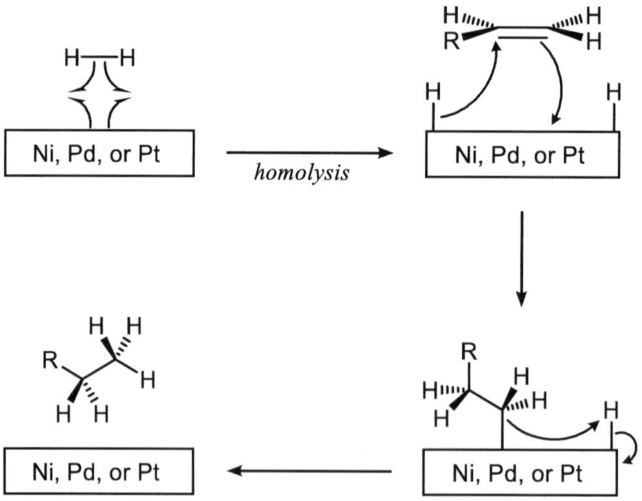

Because both H atoms added to the alkene come from the metal surface, both H atoms will add to the same face of the C=C bond. **A process that adds two units to the same side of an alkene is called**

***syn*-addition**. This reaction does not proceed through any achiral intermediates or intermediates that can rearrange, so stereochemistry will be preserved. Note that each alkene has two faces (for example, a "front" and "back" face if it is drawn flat in the plane of the page/screen). If there is no difference in energy between adding to one face versus the other, each face will be substituted in equal abundance and a 1.1 mixture of stereoisomers will be formed.

Example III.2.3

Draw the major product of the reaction shown below:

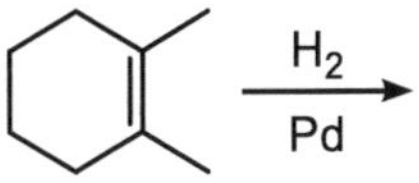

Solution III.2.3

In the presence of Pd metal, each molecule of H_2 undergoes homolysis to form 2 Pd–H bonds. When the reactant, 1,2-dimethyl-1-cyclohexene, then binds to the metal surface, both H atoms are transferred to the same face of the π-bond. The hydrogens add *syn* with respect to each other, which restricts the 2 methyl groups to be *syn* with respect to each other. An alkene can bind to the metal surface with either face of the π-bond, but *cis*-1,2-dimethylcyclohexane, the product of this reaction, is a meso compound, so only one stereoisomer is possible.

Lesson III.3. Electrophilic Addition to Alkenes and Carbocation Rearrangement Revisited

Lesson III.3.1. The C=C Bond is a Nucleophile

We are familiar with lone pairs acting as nucleophiles. We learn in general chemistry, for example, that a proton coordinates to a water molecule in aqueous solution, forming the hydronium ion (H_3O^+). In this case, the proton is the electrophile, and water is the nucleophile:

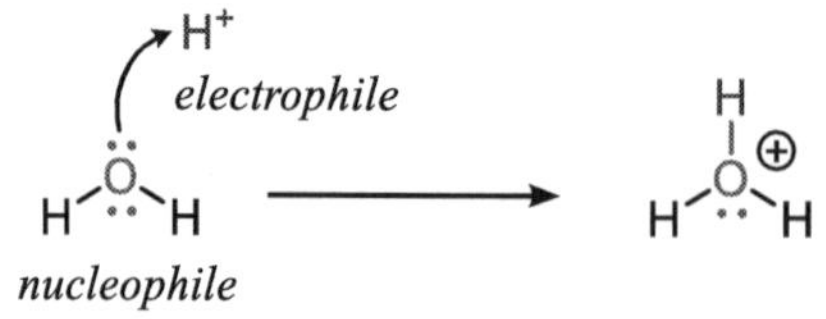

Similarly, the first step of reaction between an alcohol and a strong acid is protonation of the OH group of the alcohol. We have also seen many S_N1 and S_N2 reactions in which nucleophilic species act by donating lone pair electrons:

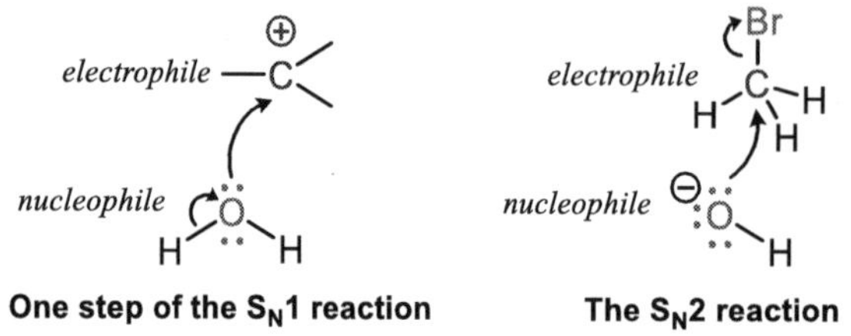

Lone pair electrons are generally held less tightly than are bonding electrons, so lone pairs are more easily removed by an electrophile. A C=C bond can also be a nucleophile, because π-bonding electrons are less tightly bound than are σ-bonding electrons. Consequently, electrophiles can pull the π-bonding electrons away from a double or triple bond:

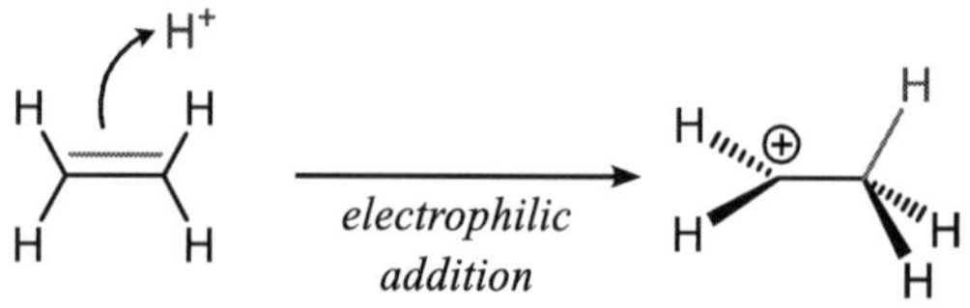

When the proton adds to the alkene, the carbon on the right is a normal, neutral, sp^3-hybridized carbon. The carbon on the left only has three bonds, so it is still sp^2-hybridized and it has a formal charge of +1.

Sometimes, the doubly-bound carbon atoms are not identical. We could possibly form two different cations in such a case. Consider the reaction of 1-propene with H^+:

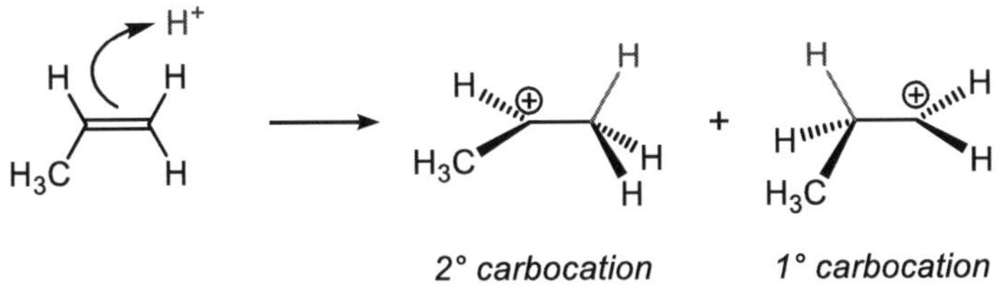

In this case, if the proton adds to the carbon on the right, it leaves only three bonds on the other carbon, making a secondary carbocation. If the proton adds to the carbon on the left, then a primary carbocation will form. We know from Lesson I.11 that a 2° carbocation is more stable than a 1° carbocation. We also know from Lessons I.11-12 that a more stable product most often forms faster than a less stable product. For these reasons, **the major product of nucleophilic addition is the more stable carbocation**. We also know that whenever a carbocation is formed it will spontaneously rearrange if it can form a more stable carbocation by doing so, as the following example illustrates.

Example III.3.1

In each reaction shown below, what would be the ultimate carbocation intermediate formed prior to any steps that may follow?

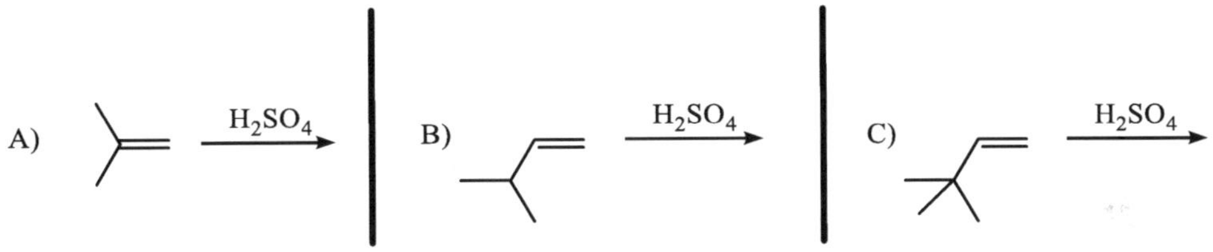

Solution III.3.1

Sulfuric acid, H_2SO_4, is a strong acid, so it provides a proton to add to the alkene in each case. Our process to solve this problem is: 1) add the proton to the less substituted side to give the more substituted carbocation. 2) check to see whether the carbocation spontaneously rearranges; if so, show the rearranged product. In order to keep good track of where the positive charge will be (the C with only three bonds), it is helpful to fill in any H atoms on C=C unit.

Solution for A). Make the tertiary carbocation.

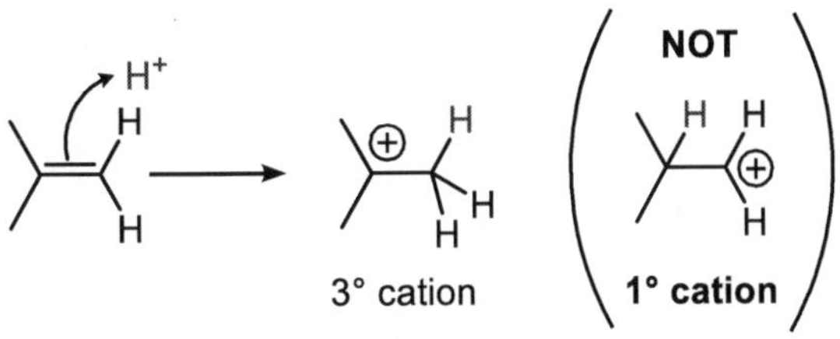

The tertiary carbocation will not rearrange, so this solution is complete!

Solution for B). Initially form the secondary carbocation:

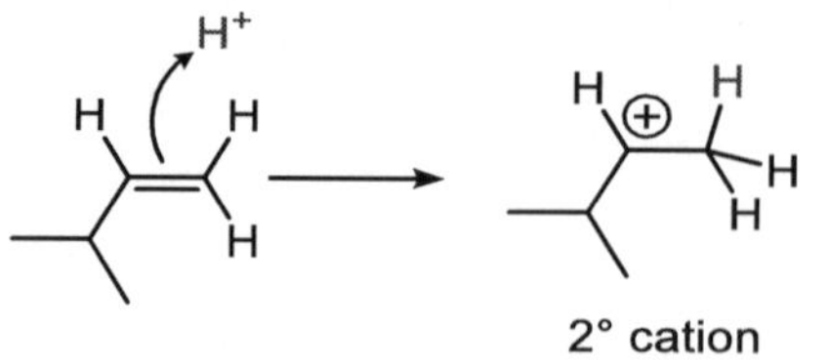

Check the sites adjacent to the cationic carbon to see if the carbocation can rearrange to a more stable site. Here, there is a tertiary site right beside the cationic C, so a rearrangement occurs. The tertiary C has an H on it. **Remember it is easier to move a lightweight H than a heavier alkyl group**. So, the easiest spontaneous rearrangement to give the tertiary carbocation is:

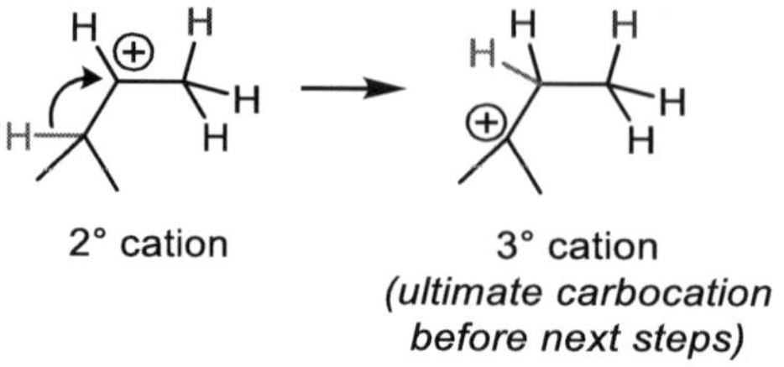

Solution for C). As for part B), initially form the secondary carbocation:

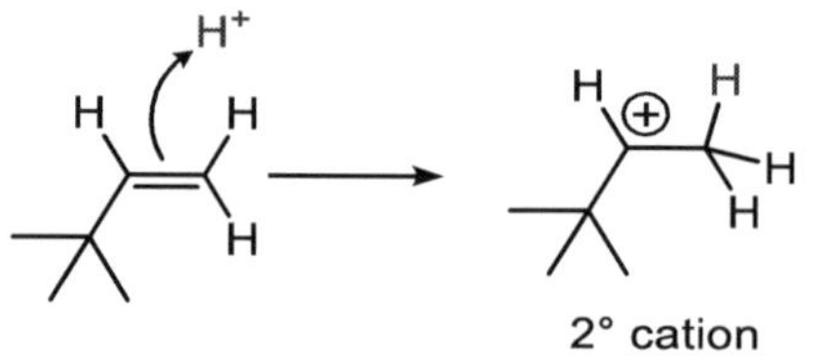

Check the sites adjacent to the cationic carbon to see if the carbocation can rearrange to a more stable site. Here, there is a quaternary site right beside the cationic C, so a rearrangement occurs. The quaternary C has only methyl groups on it, so we have to move a methyl group in the spontaneous rearrangement to give the tertiary carbocation is:

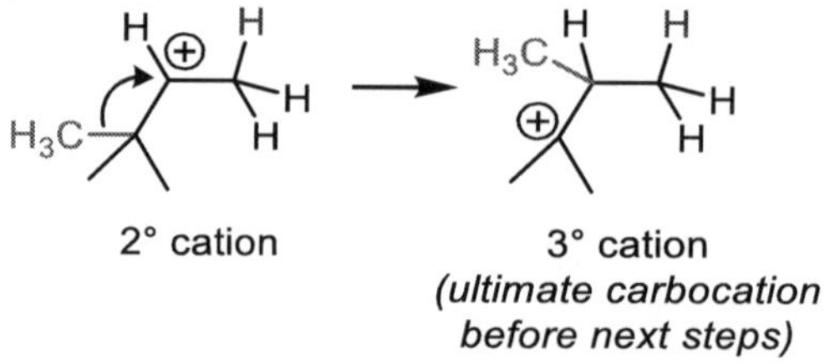

Lesson III.4. Introducing Addition to Alkenes

Lesson III.4.1. Addition to 1-Propene

In Lesson III.3, we saw that the π-bonding electrons in a C=C bond can be pulled away from an alkene to form a carbocation. We know from our study of the S_N1 (Lesson II.3) that, once a carbocation is formed, a nucleophile like water can add to that carbocation. Likewise, a nucleophile can add to a carbocation that forms from addition of a proton to an alkene:

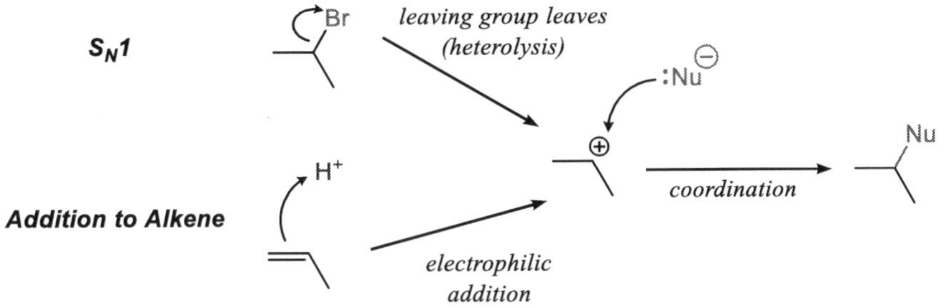

Of course, the nucleophile does not react differently with a carbocation depending on how the carbocation was formed: nucleophile coordination to a carbocation is a thermodynamically favorable reaction step. The net addition reaction to 1-propene was H to the less substituted side of the double bond and the nucleophile to the more substituted side of the alkene. Remember also that the more electronegative elements (N, O, Cl, Br, I) tend to have the negative charge in most nucleophiles. So, what is usually observed in addition reactions to alkenes is that **the more electronegative atom adds to the more substituted carbon of the C=C bond**. This general observation became known as **Markovnikov's Rule**. There are only two exceptions to this rule that we will cover in this text.

Now that we know Markovnikov's Rule, we can start to learn some of the specific nucleophiles that we can add to alkenes, watching for carbocations that could rearrange ("C+? column"). We will start with the simplest reactions. Each reaction will be studied in more detail and a few other reactions will be added in future lessons. The first nine alkene addition reactions are summarized in the table below:

TABLE III.4.1

	Reagents	Add to **more** subst. C	Add to **less** subst. C	C+?	*syn-*, *anti-*, or mix
1	HX (X = Cl, Br or I)	–X	–H	Yes	Mix
2	H_2SO_4, H_2O	–OH	–H	Yes	Mix
3	1. $Hg(OAc)_2$, H_2O; 2. $NaBH_4$	–OH	–H	No	Mix
4	X_2 (X = Cl or Br)	–X	–X	No	*Anti-*
5	X_2 , H_2O (X = Cl or Br)	–OH	–X	No	*Anti-*
6*	**HBr, ROOR**	**–H**	**–Br**	**No**	**Mix**
7*	**1. BH_3; 2. H_2O_2, NaOH**	**–H**	**–OH**	**No**	***Syn-***
8	H_2, Pd	–H	–H	No	*Syn-*
9	1. OsO_4; 2. $S(CH_3)_2$ or $NaHSO_3$	–OH	–OH	No	*Syn-*

**These reactions lead to non-Markovnikov addition*

Although many different mechanisms are involved in these reactions (you may notice that only two of them involve making a carbocation), notice that most of these reactions do follow Markovnikov's rule: the more electronegative element bonds to the more substituted C of the C=C. For some of the reactions, the same atom adds to each side of the C=C. The two reactions shown in bold in **TABLE III.4.1** show the opposite of what is expected. If the **more electronegative atom adds to the less substituted side**, this is called a **non-Markovnikov** (or anti-Markovnikov) reaction, giving the non-Markovnikov product. We will address the two other columns shortly, but if we know just what adds to the more substituted and what adds to the less substituted C of a C=C bond, we can do a lot of problems, as illustrated in this example:

Example III.4.1

Using the **TABLE III.4.1**, provide the major product for each of these reactions, all of which use 1-propene as the starting alkene.

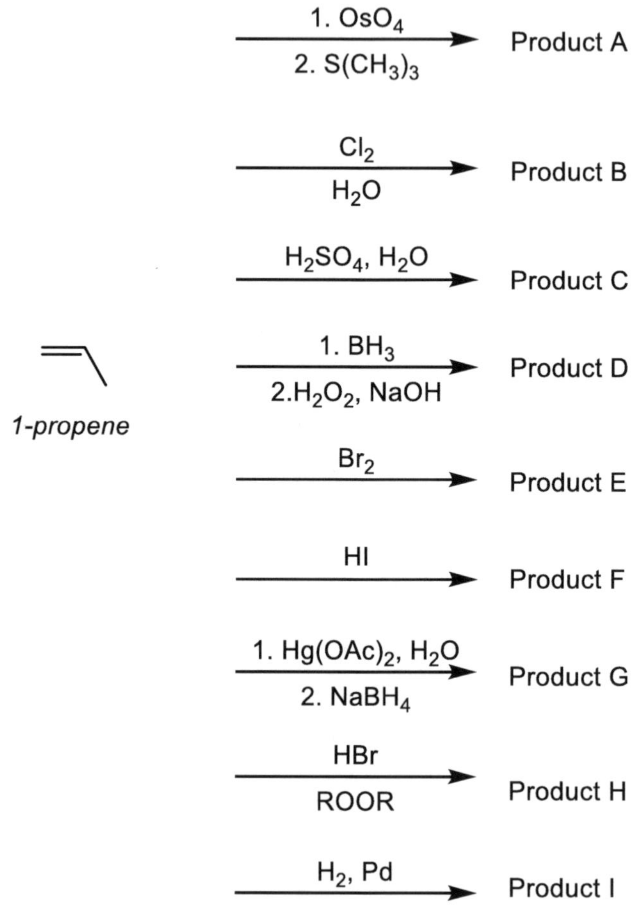

Solution III.4.1

For each product, the π-bond is gone and the groups listed in **TABLE III.4.1** add to more/less substituted C that had been in the double bond. The added groups are shown. The number beside each product refers to the reaction number in the leftmost column of the table. As we know from Lesson II.1, if we generate any chiral materials from the achiral alkene it will be formed as a racemic mixture.

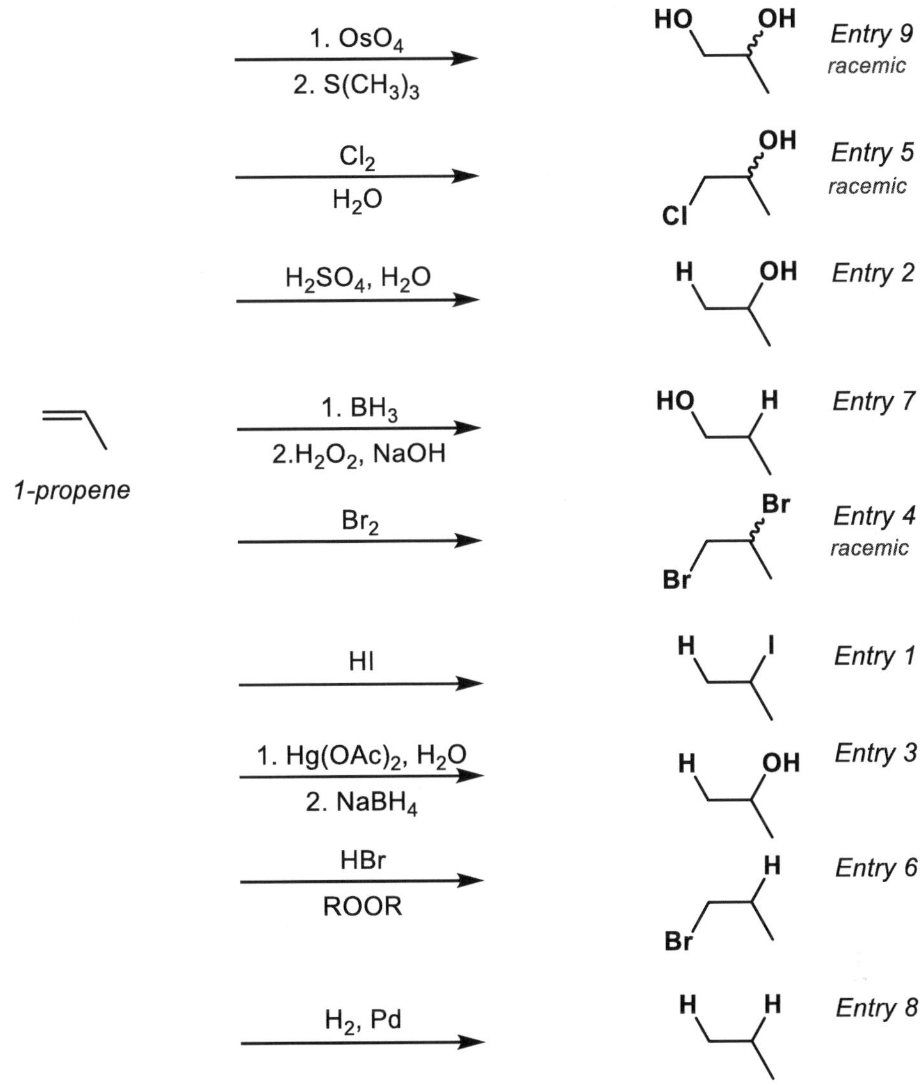

Lesson III.4.2. Carbocation Intermediates; Syn- and Anti-Addition

We know that a proton will undergo electrophilic addition to an alkene to form a carbocation. Strong acids (HX or H_2SO_4) produce protons, so a carbocation intermediate will be involved in these cases (see the column labelled "C+?" in **TABLE III.4.1**). Whenever a carbocation intermediate forms, we need to look out for carbocation rearrangement prior to any subsequent steps. As an example:

Reaction:

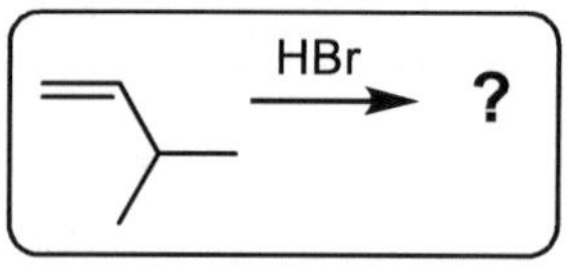

Process:

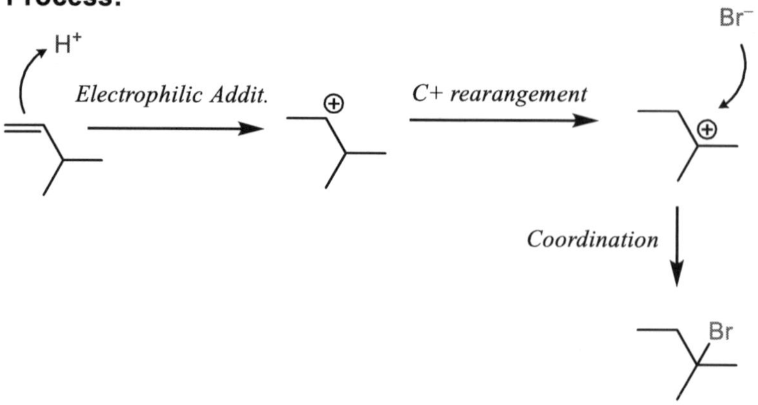

Other than hydrogenation, which we saw in Lesson III.2, we do not yet know the mechanistic details for the reactions is our alkene reaction table. We will learn the mechanisms in following lessons. For now, it is worth noting that there are two possible geometries with which the two added groups may add to the alkene. Two groups may add so that they end up pointing the same way as one another, a process called ***syn*-addition** (like we saw for hydrogenation in Lesson III.2). Alternatively, the two groups can add so that they point in opposite directions, a process called *anti*-addition. Some reactions are not specific for either *syn*- or *anti*-addition, in which cases a mix of the two addition modes is observed:

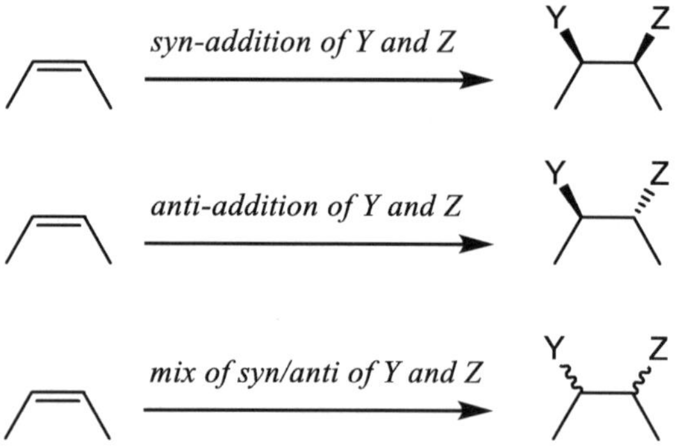

For any of these three reactions
If a chiral compound forms from achiral reactants, the product will form as a racemic mixture!

The rightmost column in **TABLE III.4.1** lists the stereochemistry of addition for each reaction. The Example on the following page illustrates products in which it is important to specify whether a *syn-*, *anti-* or a mix of addition is observed. **The information in TABLE III.4.1 should be committed to memory as soon as possible!**

Example III.4.2

Use the information in TABLE III.4.1 to help you provide the major product of each reaction. Label each as achiral or racemic. If an achiral compound is also a meso compound, label it as meso.

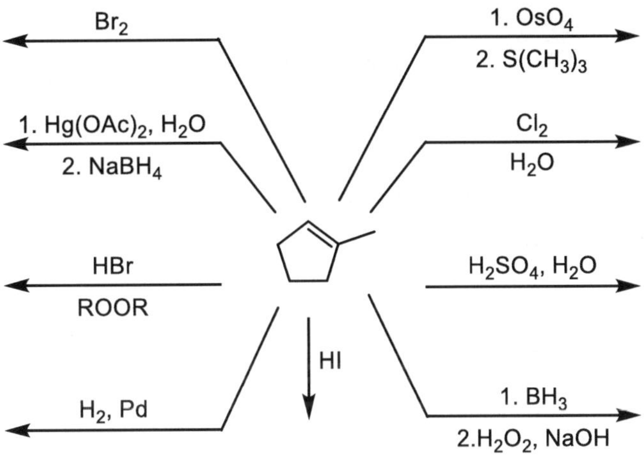

Solution III.4.2

The added groups are shown in bold. A squiggly line is used to indicate that the added group could point either way, towards or away from the viewer.

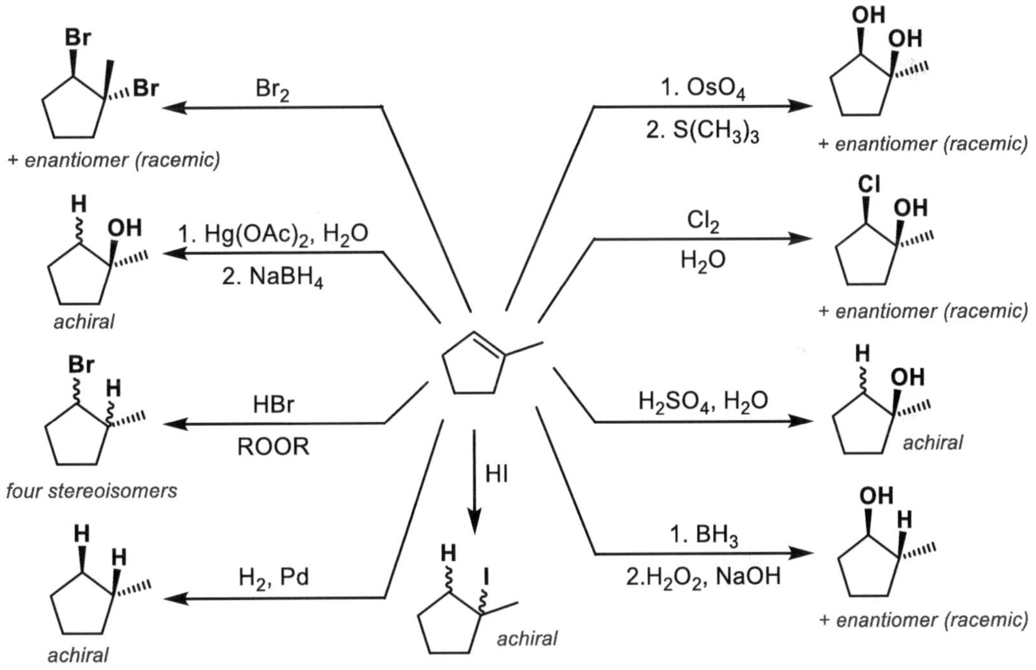

Lesson III.5. Hydrohalogenation and Hydration of Alkenes

Lesson III.5.1. Hydrohalogenation – Electrophilic Addition of H$^+$, then Coordination of X$^-$

Hydrohalogenation refers to the addition of H–X (X = Cl, Br, I) across a C=C bond. H–X is a strong acid, so it will dissociate to H$^+$ and X$^-$ in solution, so we expect electrophilic addition of the proton to the C=C bond as the first step. Remember that we expect to make the more substituted, more stable carbocation in the electrophilic addition step. If rearrangement leads to a more stable carbocation, then rearrangement happens rapidly before the next step. We saw examples of this pathway in the previous lesson.

Note that a positively-charged carbon cannot be a chiral atom because it is only bound to three groups and has no lone pairs. Once coordination occurs, adding a fourth bond, that C might become a chiral center. Remember the general rule (Lesson II.1) that a chiral center generated from an achiral starting material leads to a racemic mixture. As we saw in Lesson II.5, coordination to a carbocation is equally likely from either face of the trigonal planar, positively-charged C, leading to an equal amount of each configuration, as further illustrated in the example below.

Example III.5.1

Draw the major product of the reaction shown in the box:

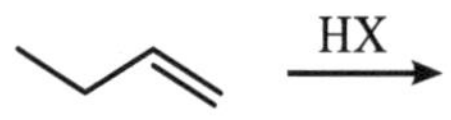

Solution III.5.1

In the first step of the reaction, the electrophile H$^+$ adds to the π-bond and forms a C–H bond at the less substituted alkene carbon, which allows the carbocation to form at the more substituted carbon. There is no way to form a 3° carbocation, so no rearrangement occurs. After the carbocation has formed, X$^-$ will then coordinate to the carbocation. Because there is no difference in energy between X$^-$ coming from the top or bottom, a 1.1 mixture of the two enantiomers will be formed.

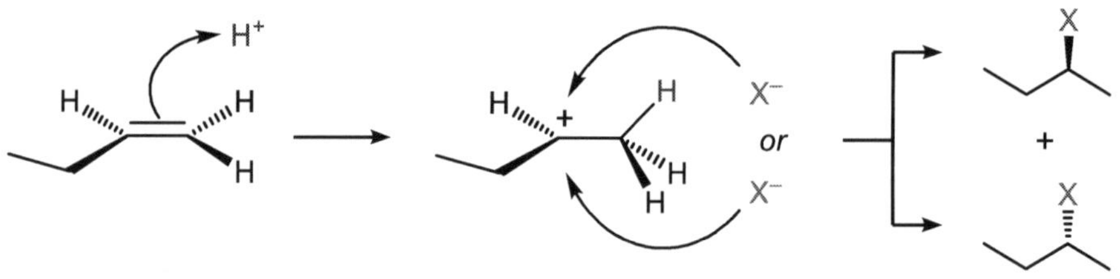

Hydration follows exactly the same mechanism as hydrohalogenation: electrophilic addition of a proton then coordination of a nucleophile. The only differences here are that the proton is supplied by sulfuric acid (H_2SO_4) and the nucleophile is water:

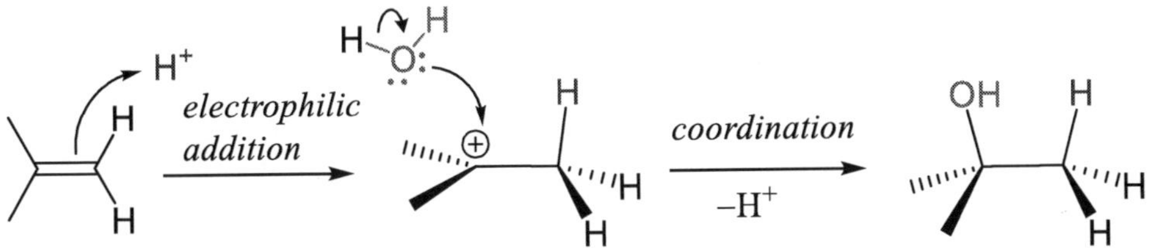

Example III.5.2

Draw the major product of the reactions shown below:

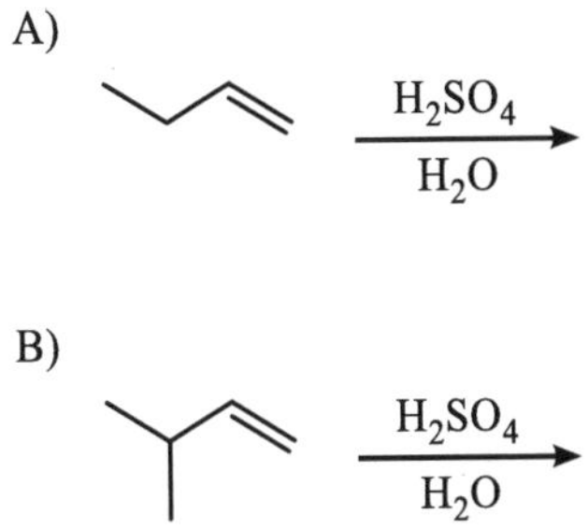

Solution III.5.2

Solution for A): The first step is to make the more stable carbocation and to check whether it rearranges. If it does not, we can just coordinate the nucleophile:

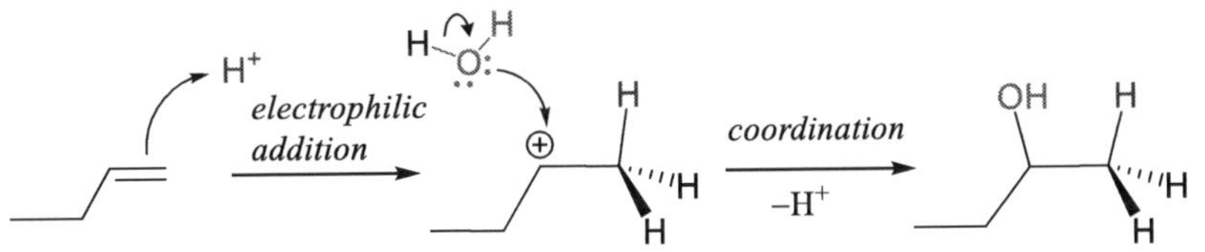

Finally, check the stereochemistry. In this case, we generated one stereocenter from an achiral compound, so the major product will be a racemic mixture of both enantiomers:

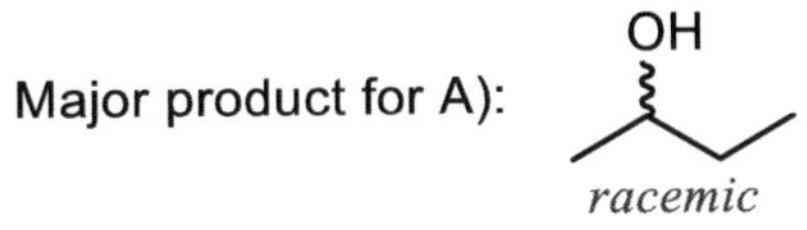

Solution for B): The first step is to make the more stable carbocation and to check whether it rearranges. Here it does rearrange, so we rearrange before coordination:

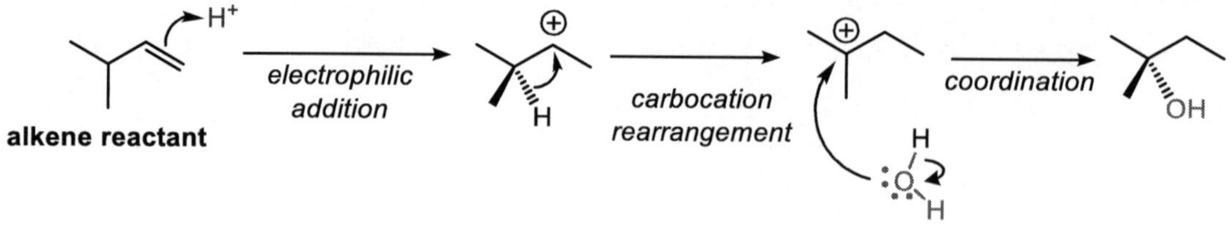

Finally, check the stereochemistry. In this case, there are no chiral centers, so we are done.

Lesson III.6. Alkene Halogenation and Halohydrin Formation

Lesson III.6.1. The Halonium Intermediate

In the previous Lesson, we considered the reaction of an alkene with H^+ as the electrophile, which forms a carbocation intermediate. Other electrophilic reagents that do not lead to carbocation intermediates can also add to alkene π-bonds, as we saw in (see Table III.4.1 in Lesson III.4). A halogen, X_2 (X = Cl or Br) is one such reagent. A halogen might not initially look like an electrophile because an electrophile has a deficiency of electrons or a positive charge, like a partial positive end of a polar bond. A symmetric halogen is not a polar compound, but it is **polarizable**. When X_2, surrounded by the large clouds of electrons from the six lone pairs, comes close to the π-bond of the alkene, the alkene electrons repel the halogen electrons, creating a polarized X–X:

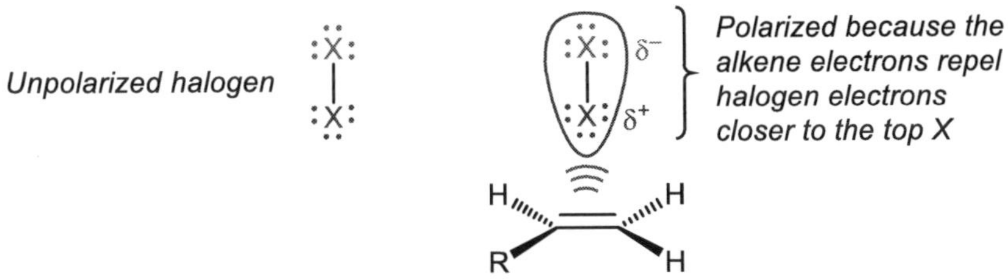

A solvent can also interact with the halogen to polarize it in the same way, so polarized halogens, $^{\delta+}X–X^{\delta-}$ are present in our reaction mixture. The positive end of a polarized halogen attracts alkene π-bonding electrons just as a proton would. However, rather than a carbocation forming, the lone pair on the halogen is attracted to the positive carbon to form a three-membered ring structure:

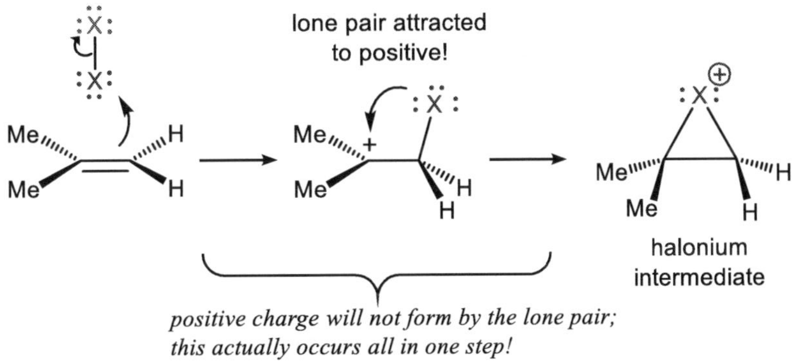

Because the lone pair is attracted to the positive charge as it begins to form, it actually donates to the more substituted carbon before a carbocation forms So, **the proper arrow-pushing mechanism shows all three arrows in one, concerted step to make a halonium intermediate:**

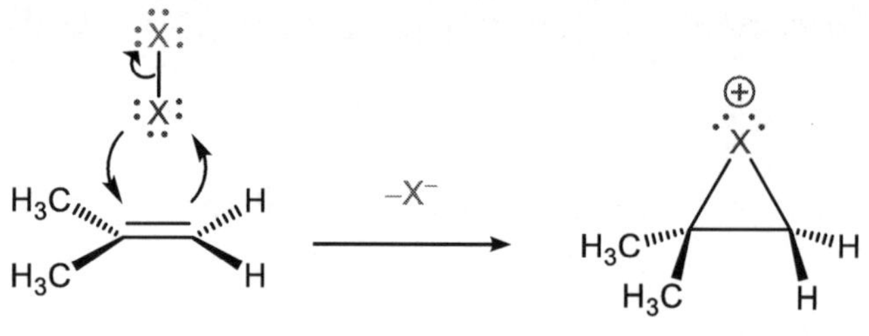

Halonium intermediate
X = Cl: chloronium
X = Br: bromonium

It is important to note that, among reactions in this book, only carbocation intermediates rearrange. Halonium intermediates always include the three-membered ring of the X and the two carbon atoms that were in the double bond, and **halonium intermediates do not rearrange**.

Lesson III.6.2. Halogenation of Alkenes

The halonium ion is just an intermediate; it is not the final product of the reaction of X_2 with an alkene. We know from Table III.4.1 in Lesson III.4 that the overall reaction will give *anti*-addition of two X atoms to the C=C. How can we explain this mechanistically? Well, once formed, the halonium intermediate undergoes ring-opening by reaction with a nucleophile. This step is similar to the ring opening of the epoxide that we saw in Lesson II.15. **For the halonium, the more substituted C is selectively attached by the nucleophile** because it has a greater δ+. The nucleophilic substitution occurs via an S_N2-like mechanism with inversion of configuration. When this transformation is complete, the net resultant is *anti*-addition of X and Y to the C=C bond:

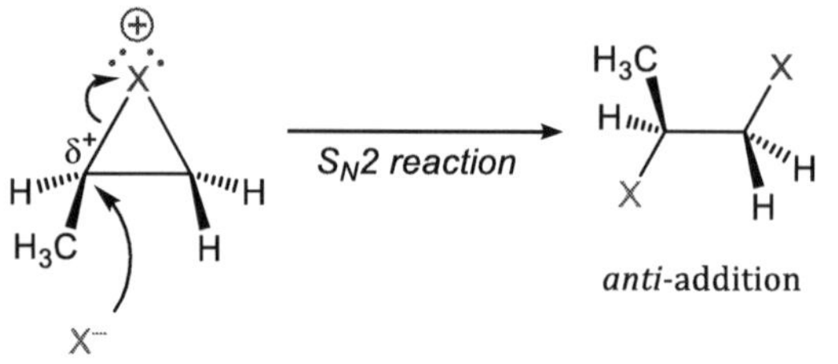

We see that we generated a chiral product, but we started with achiral reagents, so we should get a racemic mixture as the major product. So, the complete net reaction of 1-propene with Br_2 is:

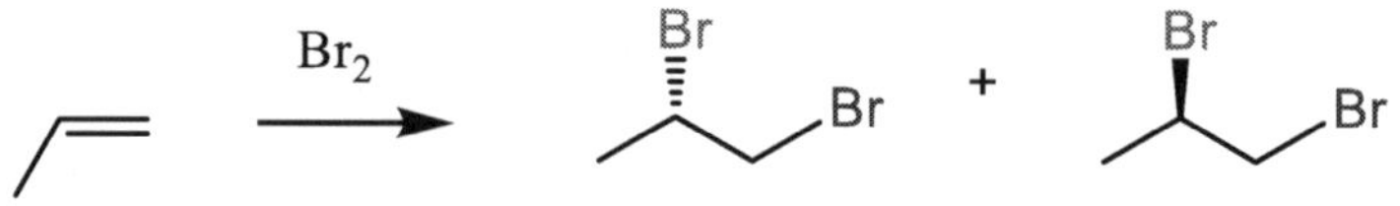

50:50 (racemic)

Lesson III.6.3. Halohydrin Formation – Nucleophilic Addition of H_2O to the Halonium Intermediate

In Lesson III.5, we saw that a carbocation intermediate can react with either X^- or H_2O nucleophiles. The halonium intermediate can also react with either X^- or H_2O as the attacking nucleophile. The reaction with X_2/H_2O (entry 5 in Table III.4.1) leads to addition of an X– to the less substituted and HO– to the more substituted C:

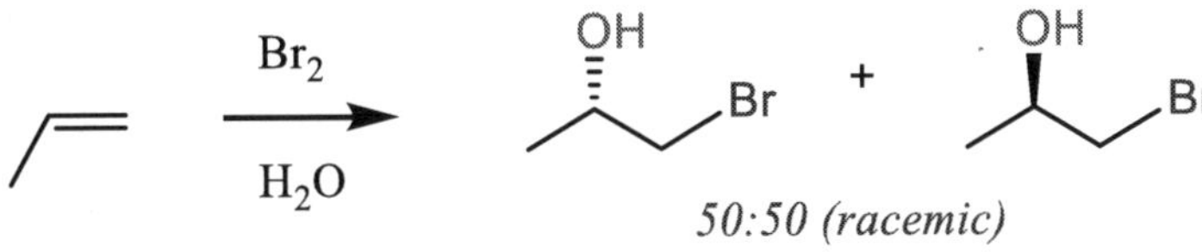

A molecule having a halogen and an OH on adjacent carbons is called a halohydrin, so this reaction is called halohydrin formation. The mechanism of this reaction is identical to that of the halogenation, but using water as the nucleophile that adds to the halonium intermediate:

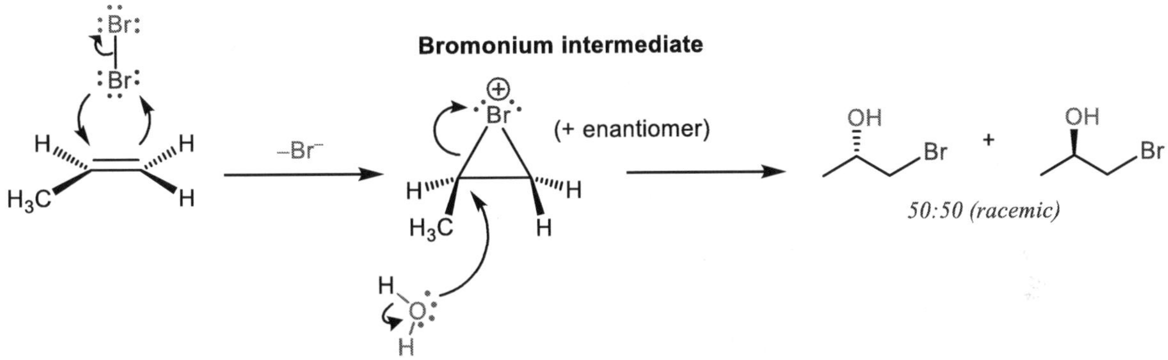

Example III.6.1

Draw the major product for each of the reactions shown below. Clearly indicate the stereochemistry for each product. Refer back to Table III.4.1 if you need to refresh your memory on some of the reactions.

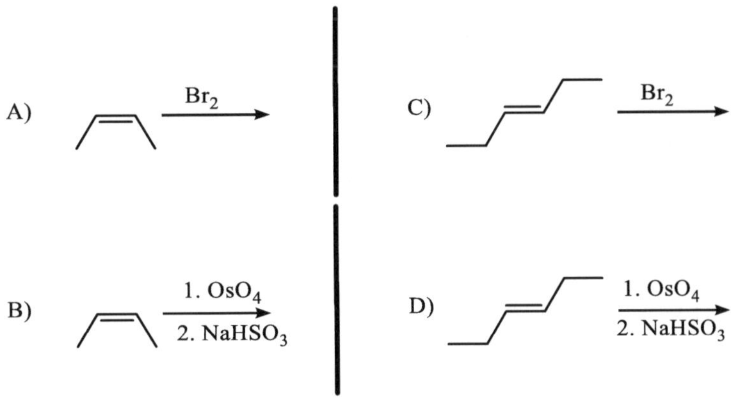

Solution III.6.1

Reactions A and B both start with the same alkene. Neither involves a carbocation intermediate, so no need to check for rearrangement. In such cases, we just need to know 1) what adds and 2) are the added units *syn-*, *anti-*, or a mix with respect to each other. For A), it is simply anti-addition of two Br. For B, it is *syn*-addition of two OH:

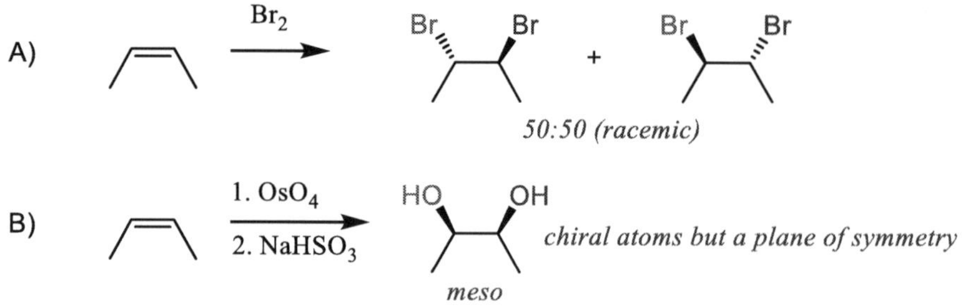

Note that the carbon chain keeps its initial shape in the course of addition. Of course, once formed there can be rotation around the single bonds, so one could draw it in different ways. To assess the stereochemistry, we first see how many chiral centers are formed. In both A) and B), we form two new chiral centers. The product of reaction A) is a chiral molecule made from an achiral starting material, so it is formed as a racemic mixture. The product of reaction B) has a plane of symmetry so it is an achiral, meso compound.

Next, we do the same analysis for reactions C) and D). We gain generate two new chiral centers in each case. When checking for symmetry (to see if there is a meso compound) for product D), the OH groups line up with each other but the ethyl groups do not. It is easy to see that there is no symmetry plane in D), so it is formed as a racemic mixture. When we evaluate product C), however, none of the groups initially line up. So, we have to rotate around the single bond joining the two chiral carbons. When we do this, we see that there is a plane of symmetry and this is an achiral, meso compound:

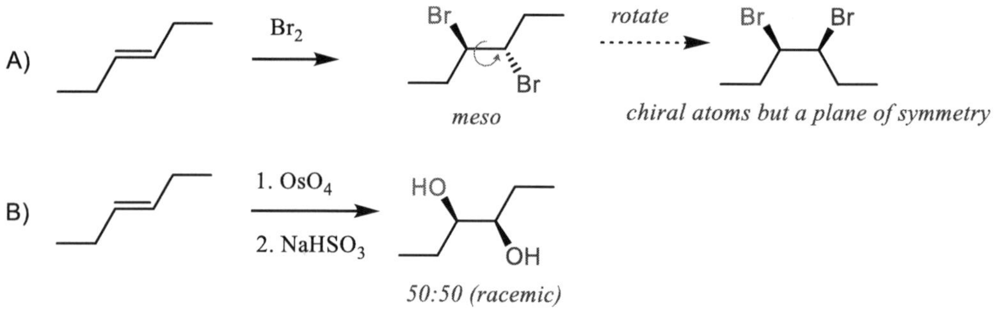

.7. Oxymercuration/Reduction of Alkenes

II.7.1. The Mercurinium Intermediate

Lesson III.6, we saw that X_2 can react with an alkene to make a cationic, three-membered ring rmediate called a halonium intermediate. Several other electrophiles can likewise react with an kene to form a cationic three-membered ring intermediate. We will only look at one more in this book. When mercuric acetate ($Hg(OAc)_2$) and water react with an alkene, a **mercurinium intermediate** initially forms and then water acts as the nucleophile to do an S_N2 reaction on the more substituted site. In terms of the required arrow-pushing, this mechanism is identical to that for halohydrin formation:

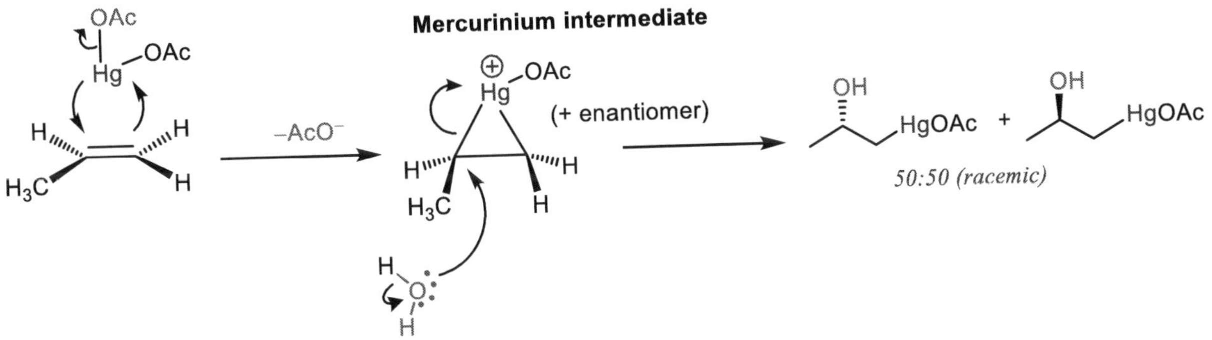

The two-step mechanism shown above is called **oxymercuration** because an oxygen and a mercury form bonds to the carbons. This is not the product that is generally isolated though. As we saw in Table III.4.1, the oxymercuration step is followed by reaction with $NaBH_4$. **Compounds having boron-hydrogen bonds are often good at transferring a hydrogen to a carbon.** In this case, an H replaces the –HgOAc, to give an alcohol as the final product:

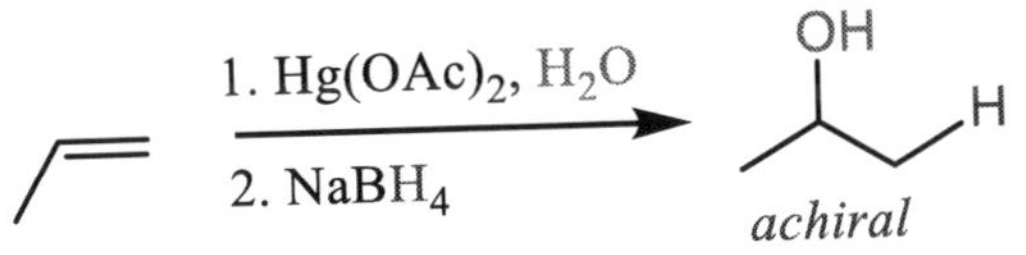

Because adding a hydrogen to a carbon atom is a reduction reaction (Lesson II.12), the overall reaction shown above is called **oxymercuration/reduction**. The mechanism for the reduction step is rather complex and is not shown here, but the reduction is not specific regarding to which side the H adds, so the product shows a **mix of *syn*- and *anti*-addition.** Although a Markovnikov alcohol is formed by both oxymercuration/reduction and hydration (Lesson III.5), different major products can result. This is because hydration has a carbocation that can rearrange but oxymercuration does not, as illustrated by the following example:

Example III.7.1

Draw the major product for each of the following reactions:

A) 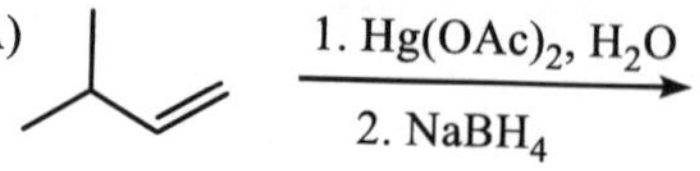

B) 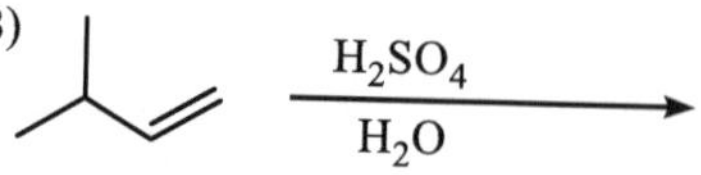

Solution III.7.1

For reaction A, we add –OH to the more substituted and –H to the less substituted C in the C=C bond and there is no carbocation, so no need to worry about rearrangement. Then we check for stereochemistry and see that we have made one new stereocenter from an achiral starting material, so a racemic mixture is the major product:

A) 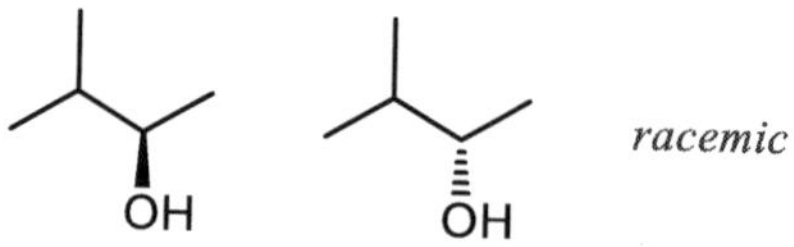

For reaction B), we have to be more cautious, drawing the carbocation to see whether it will rearrange. Here, the carbocation does rearrange, and we end up with a different alcohol than we got from reaction A!

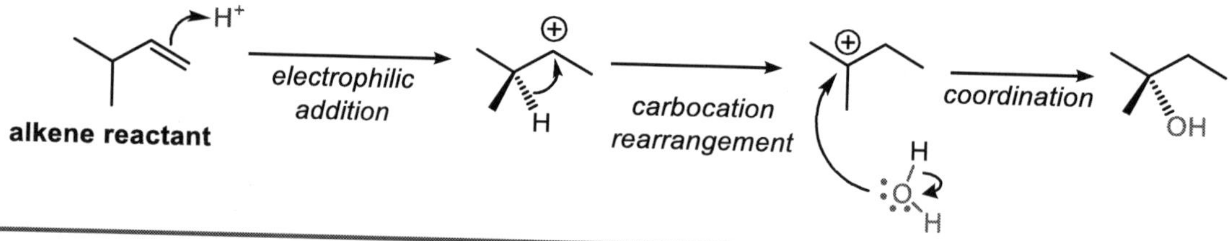

Lesson III.8. Epoxidation of Alkenes

We learned in Lesson II.15 about ring-opening of epoxides, but we did not discuss how epoxides are made. We might envision adding an "O" to an alkene to make a three-membered ring similar to the rings formed to make the halonium or mercurinium intermediates we saw in the previous two lessons. Peroxyacids (molecules of the form RCO_3H) are a good source of an oxygen atom to add to an alkene. The reaction, called **epoxidation, is concerted:**

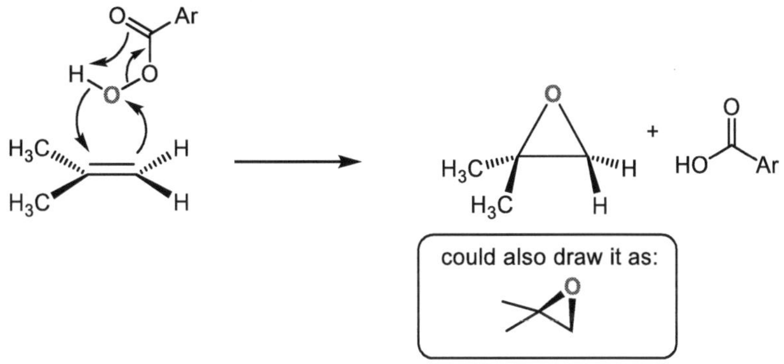

One particular, widely-used peroxyacid that you should know is *m*-chloroperoxybenzoic acid, abbreviated as mCPBA. Its structure is:

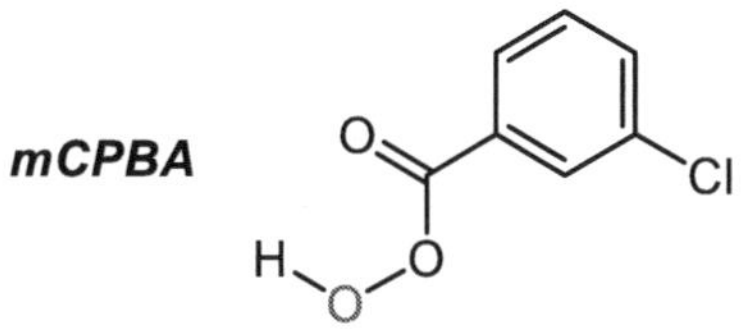

Note that, since both C atoms from the C=C bond attach to the same atom in the product, both bonds have to point in the same direction, so **epoxidation shows *syn*-addition.** This is the first completely new alkene addition reaction we have learned that is not shown in Table III.4.1.

Because epoxidation is concerted step, if we begin the reaction with a *trans*-alkene, the substituents on the epoxide ring will also be *trans* to each other. As always, we will have to check the stereochemistry of our product and if a chiral product is made from an achiral starting material, it will be made as a racemic mixture.

Example III.8.1

What is the major product of each of the reactions shown?

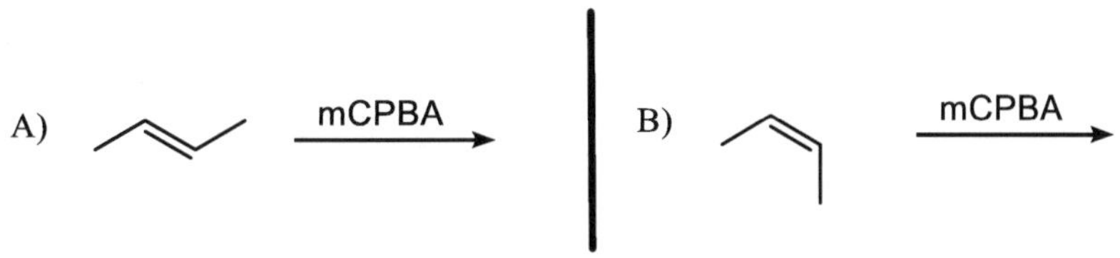

Solution III.8.1

Epoxidation is concerted, so we start by drawing the *trans*-epoxide product from the *trans*-alkene and the *cis*-epoxide product from the *cis*-alkene:

A)

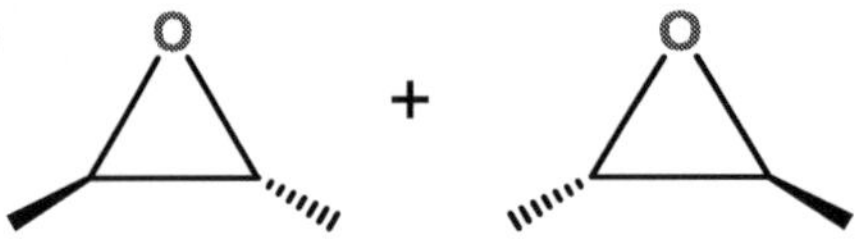

Chiral molecule, so its enantiomer will also form in equal amount. This is a racemic mixture!

B)

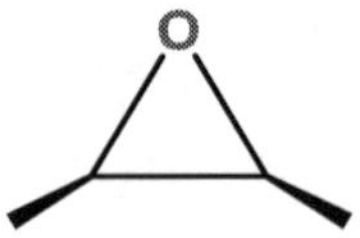

There are chiral atoms but the molecule as a whole is not chiral because it is symmetric. This is a meso compound!

Finally, we check for chiral centers to deduce that product A) is formed as a racemic mixture, while product B) is an achiral, meso compound.

Lesson III.9. Hydroboration/Oxidation of Alkenes

Lesson III.9.1. Hydroboration – Concerted Addition of B– and H– to a C=C Bond

The alkene reactions we have studied so far have given us Markovnikov products, but we know from Table III.4.1 that there are two non-Markovnikov (sometimes called anti-Markovnikov) reactions we will learn. The first example we will study involves reaction of an alkene with a borane (something with a B–H bond) like BH_3, followed by addition of H_2O_2/NaOH. **Compounds with a B–H bond often transfer a H to a carbon**. First, consider what happens when an alkene comes in contact with a borane, R_2BH. There are two notable things R_2BH: 1) boron is less electronegative than H, so a δ– is on H and 2) the B has only six valence electrons, so it has a strong pull on electrons to fill its octet. Because of these two features of a borane, a **concerted**, **hydroboration** step occurs. The net result of this step is ***syn*-addition of the H– to the more substituted side and of R_2B– to the less substituted side:**

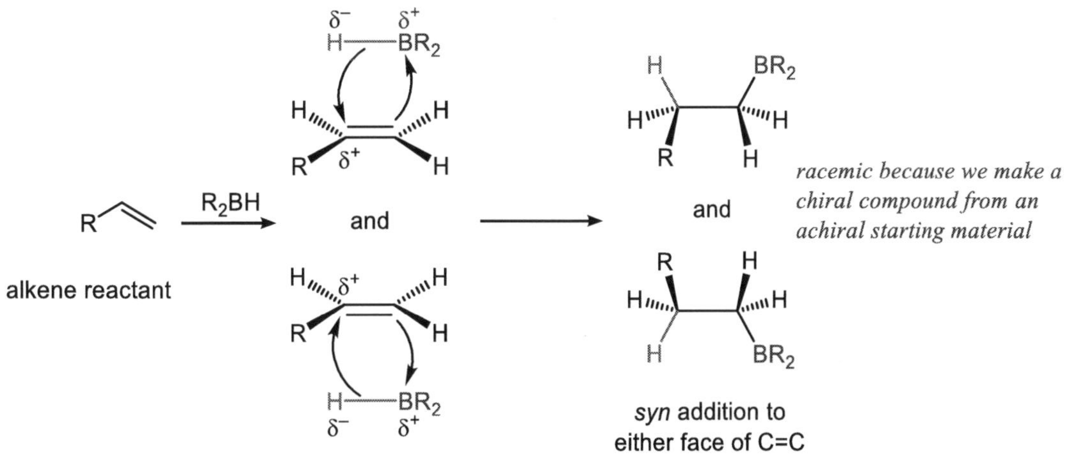

Lesson III.9.2. Oxidation – Replacing BH_2 with OH

Following hydroboration, we add H_2O_2/NaOH(*aq*) to the reaction. This second step replaces the BH_2 group with an OH group. This oxidation occurs via a series of steps that retain stereochemistry at the C, meaning that the –OH will still be *syn*- to the added H. The mechanism is not shown here, but is presented in the companion book (mentioned in the Preface) if you are interested. The combination of **hydroboration/oxidation results in Non-Markovnikov addition of H and OH across a C=C bond with *syn*- stereochemistry.**

Example III.9.1

What is the major product of the reaction shown below?

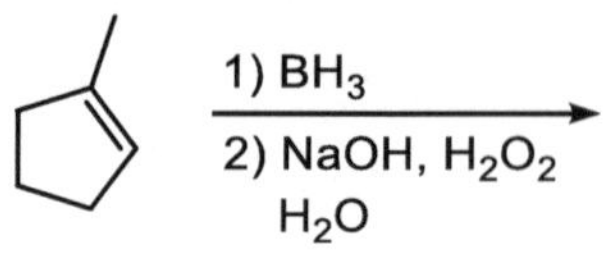

Solution III.9.1

The reagent BH_3 facilitates concerted *syn*-addition of –H and $–BH_2$. The oxidation replaces the BH_2 group with an OH group with retention of stereochemistry. The net hydroboration/oxidation forms two chiral centers, giving a racemic mixture of the two possible isomers:

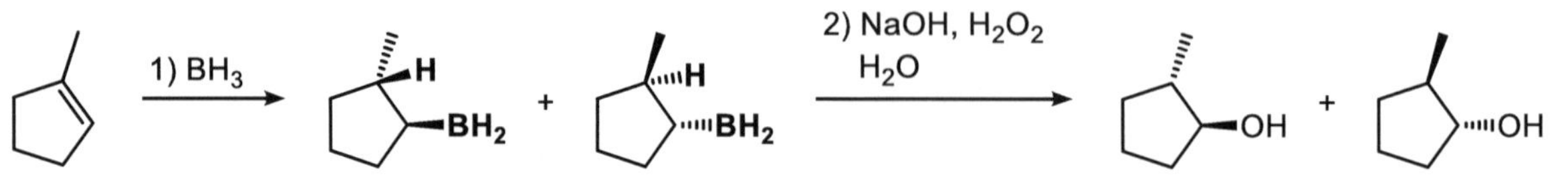

trans-products formed as a racemic mixture!

Note that in the final products we did not draw in the added H atoms, only the methyl and the OH substituents.

Lesson III.10. Ozonolysis and Preparation of Vicinal Diols from Alkenes

Lesson III.10.1. The Ozonide Intermediate

In the Lewis dot structure of ozone (O_3), the very polar O–O bonds (there is a formal +1 charge on the central O) allows the flow of electrons between ozone and an alkene to occur as follows to form an unusual 5-membered ring termed a "molozonide", which rearranges (over complex steps) to give a more stable ozonide:

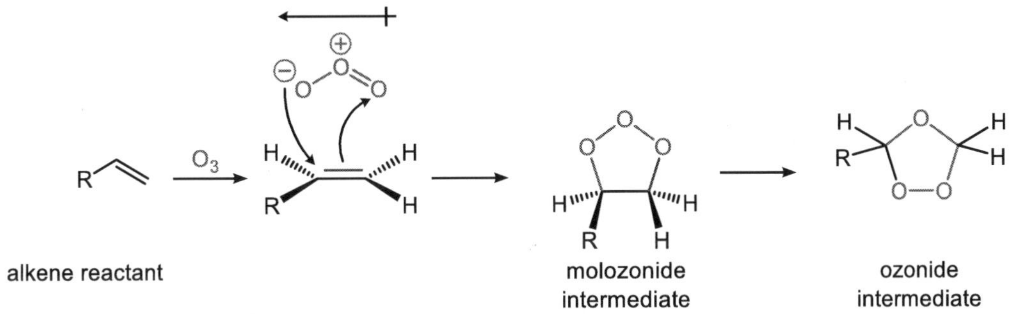

Lesson III.10.2. Ozonolysis – Replacing C=C with C=O Bonds

Ozonolysis is an alkene oxidation reaction that breaks one C=C bond and replaces it with two C=O bonds. In the first step of ozonolysis, an alkene reacts with O_3 to form an ozonide as described in the previous section. In the second step, the ozonide is subjected to reducing conditions (common reductants are R_2S or Zn/H_2O).

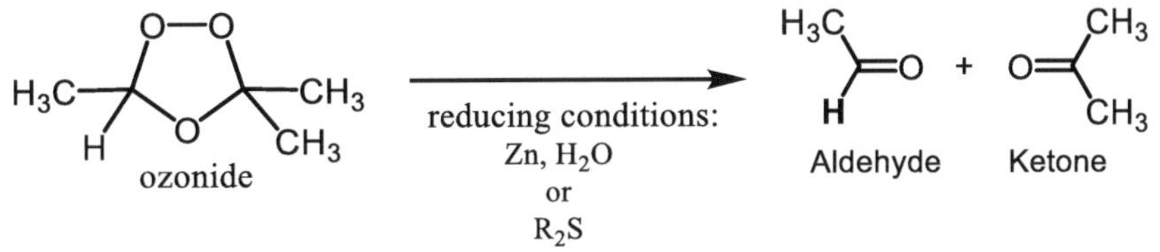

Example III.10.1

What is the major product of the reaction shown below?

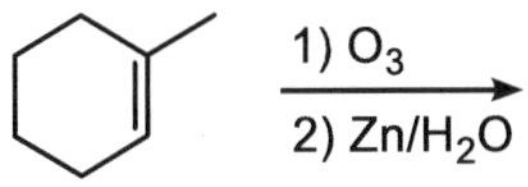

Solution III.10.1

The two reactions in this Example have the net effect of cleaving a C=C bond and replacing it with two C=O bonds. In this example, useful problem-solving strategy is illustrated. First, if you number the C atoms in the ring in the starting material, it will be easier to keep track of each carbon's fate in the product. Second, you can just replace the C=C with C=O bonds and keep the structure in about the same shape as the starting material to readily get the right product. Finally, you will have to redraw your structure as a chain. This is made easier because you have numbered your carbon atoms. You have the six carbon atoms from the original ring, and you know to put a methyl group coming off of C1, and to put the C=O bonds on carbons C1 and C6. Now you have the right product.

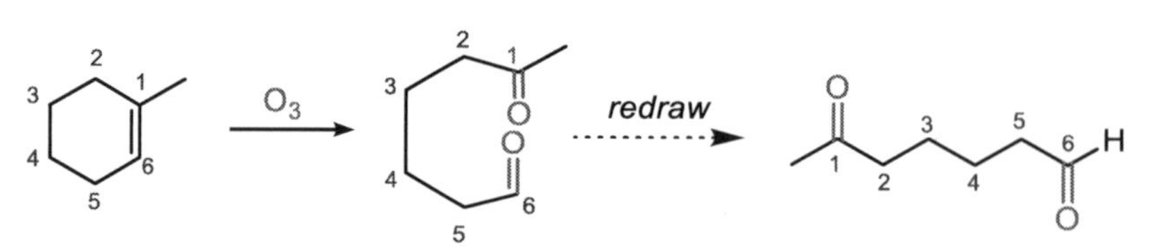

Lesson III.10.3. Preparation of Vicinal Diols – Adding "OH" to Each Carbon in a C=C Bond

We have seen that ozonolysis is a way to cleave the C=C bond and add two doubly-bound O atoms. Other reactions have been developed to add two singly-bound O atoms across the π bond of an alkene. Reaction of an alkene with OsO_4 (osmium tetroxide) is one example. The osmium in OsO_4 is electron deficient due to the presence of four oxygen atoms around it, so it will want to react in such a way to gain electron density by sharing fewer pairs of electrons with O. This facilitates the flow of electrons between an alkene and OsO_4 in a way that pushes an electron pair onto the osmium atom:

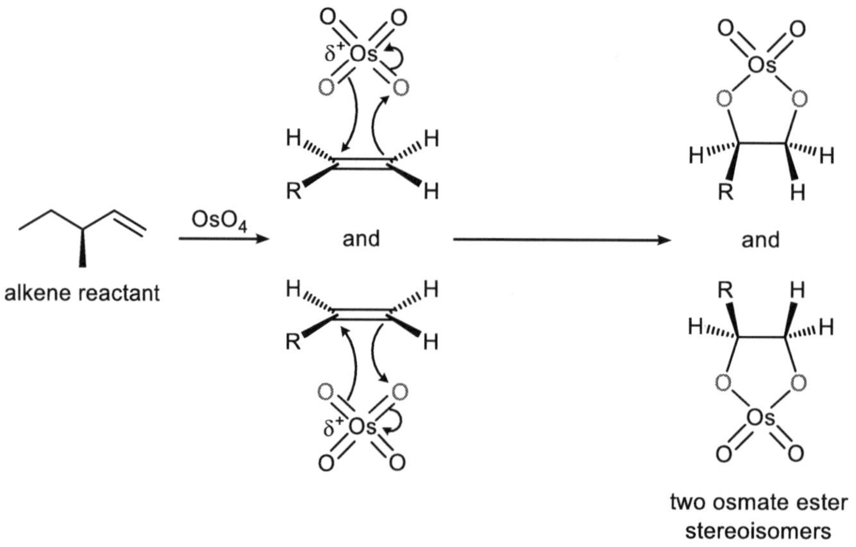

Once the osmate ester forms, we add the second set of reagents listed in Table III.4.1 (Lesson III.4), and this causes the Osmium to break away and protonating the oxygen atoms that we added to the alkene carbons:

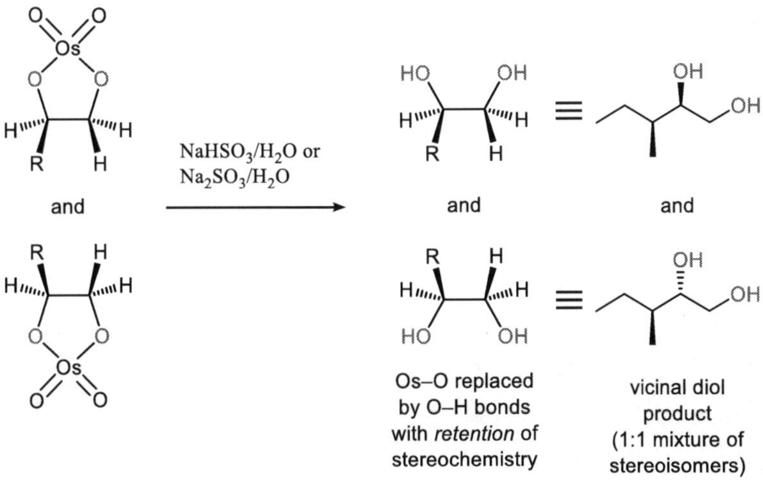

The formation of the osmate ester is concerted. Note that both oxygen atoms have added to the same face of the alkene (i.e., *syn* addition). Remember that OsO_4 can add from either above or below the C=C plane, and because OsO_4 is achiral, if the alkene lacks chiral centers, a 1.1 mixture of stereoisomers will be produced. The net reaction is the conversion of a C=C group into a vicinal diol group (i.e., two C–OH groups adjacent to each other). Review the example problem in Lesson III.6 to review stereochemical outcomes from formation of vicinal diols from *cis*- versus *trans*- alkenes.

Lesson III.11. Naming and Making Alkynes

Lesson III.11.1. The "-yne" Suffix

When naming alkynes, we make use of the core nomenclature rules found in Lesson I.14. For an alkyne, we replace the "-ane" suffix with "-yne". So, if we have a 5-carbon chain with one C≡C bond, it is a "pentyne"; a 9-carbon chain with one C≡C bond is a "nonyne". If two or three C≡C bonds are present, we use "-diyne" or "-triyne", and we add an "a" to the end of the root. A 5-carbon chain with two C≡C bonds is a "pentadiyne", and a 9-carbon chain with three C≡C bonds is a "nonatriyne". Note that since the triple bond is linear and has only one group on each side of the C≡C, we do not need any *cis*-, *trans*-, *E*-, or *Z*- labels as we did for alkenes.

Lesson III.11.2. Alkyne Priority in Hydrocarbons

To determine the name for an alkyne, we first pick as parent the longest chain that contains the C≡C bond. If multiple C≡C bonds are present, we select as parent the chain that contains all the C≡C bonds (we will not deal with examples where multiple branches all have alkyne units). We then number the parent chain to give the C≡C bonds the lowest possible substituent numbers. If both C=C and C≡C bonds are present, we number the parent chain to give the multiple bond substituents the lowest possible numbers, regardless of whether they are C=C or C≡C bonds. If, and only if, a C=C bond and a C≡C bond would have the exact same number, do we employ the alphabetization rule, and assign priority to the C=C bond over the C≡C bond ("ene" comes before "yne" alphabetically). When both a C=C bond and a C≡C bond are present in a molecule's structure, the suffix for the C=C bond becomes "en" and it is listed before the C≡C suffix "yne". So, a 5-carbon chain with one C=C bond and one C≡C bond would be a "pentenyne". For complex molecule names such as these, it is most convenient to list the substituent numbers immediately before the suffixes. The following example illustrates how to apply these rules:

Example III.11.2

Provide the proper systematic name for the molecule shown below:

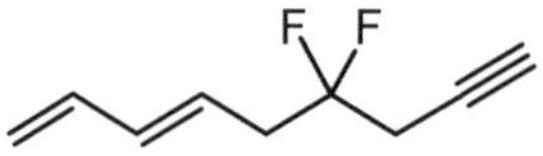

Solution III.11.2

In this molecule, we can see that the parent chain contains 9 carbons, two C=C bonds, and one C≡C bond, so the base name will be "nonadienyne". We have to number from left to right to give the multiple bonds the lowest possible substituent numbers, which are 1, 3, and 8.

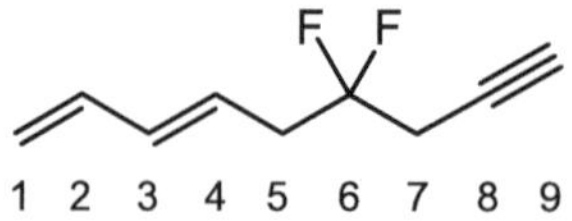

Numbering from right to left would have given 1, 6, and 8. There are 2 fluorine substituents at carbon 6, and the C=C bond beginning at carbon 3 is stereogenic (the two non-hydrogen substituents are *trans*). The systematic name for this molecule is "*trans*-6,6-difluoronona-1,3-dien-8-yne".

Lesson III.11.3. Alkyne Priority in Alcohols

If an alkyne also contains an OH group, then that OH group takes priority over the C≡C bond. Thus, we must select a parent chain that contains the OH group and give the OH group the lowest possible substituent number. To assign a name to the molecule, the "yne" suffix becomes "yn" and is listed before the "ol" suffix which ends the name. A 5-carbon chain with one C≡C bond and one OH group would thus be a "pentynol". Again, for a complex molecule such as this, we would list the substituent numbers immediately before the suffixes.

Example III.11.3

Provide an unambiguous systematic name for the molecule shown below:

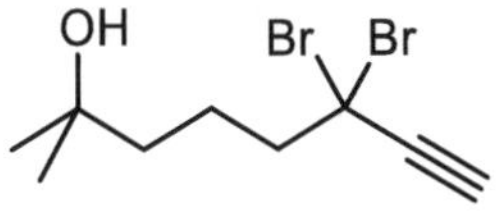

Solution III.11.3

The parent chain for this molecule contains 8 carbons, 1 C≡C bond, and 1 OH group, so the base name will be "octenol". We number this chain from left to right to place the OH group at carbon 2 (going right to left would place it at 7) and the C≡C begins at carbon 7. There is 1 Me group at carbon 2, 2 Br substituents at carbon 6, and no stereocenters or stereogenic bonds. The systematic name for this molecule is "6,6-dibromo-2-methyloct-7-yn-2-ol".

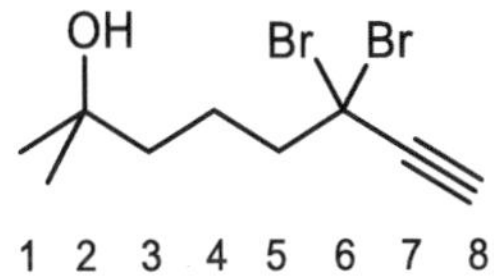

Lesson III.11.4. Making Alkynes by Double E2 Reaction

We know that the E1 and E2 reactions are good routes to make π-bonds (Lessons II.6–9) by elimination of a leaving group and an H on adjacent carbons. If we want to prepare an alkyne, we need to make *two* π-bonds, so we need to have *two* leaving groups and *two* H atoms to eliminate from adjacent carbons. We can envision this double elimination reaction if we start with two leaving groups on adjacent carbon atoms (a molecule having this feature is called a **vicinal** dihalide) or we can have two leaving groups on one C (a molecule having this feature is called a **geminal** dihalide). The preparation of alkynes is usually done by a double E2 reaction using either $NaNH_2$ or NaH as the base:

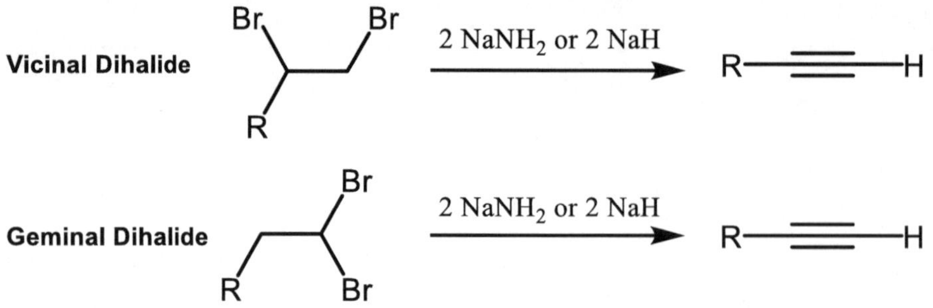

It is possible to deprotonate the terminal alkyne if too much base is added. This is because the alkyne C is *sp*-hybridized and is nearly as electronegative as a N atom, so it can stabilize an anion relatively well (Lesson I.6.2). So, if an excess of the base is used, you initially get the deprotonated alkyne. Water can be added to deprotonate the anion to obtain the neutral alkyne:

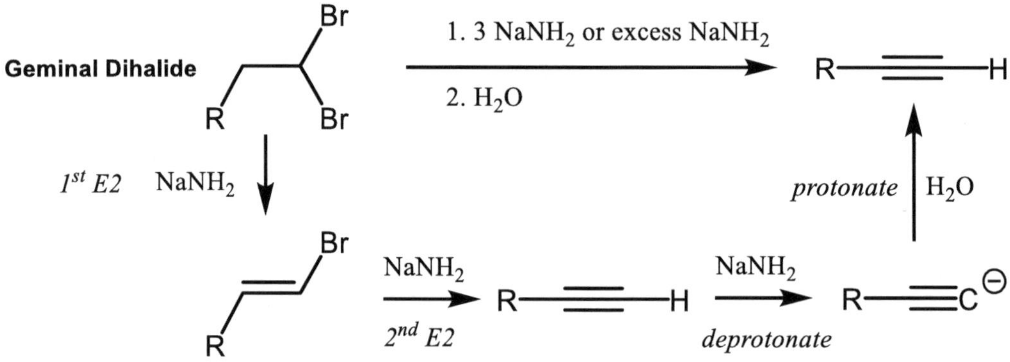

Lesson III.12. Hydrohalogenation and Halogenation of Alkynes

Lesson III.12.1. Hydrohalogenation of Alkynes

In Lesson III.3, we demonstrated that the π-bond in an alkene can function as a nucleophile. An alkyne has two π-bonds, thus the C≡C bond in an alkyne can also function as a nucleophile. However, the mechanisms of alkyne reactions exhibit important differences from those of alkenes. For example, in hydrohalogenation (reaction of with HX) a proton undergoes electrophilic addition to an alkyne π-bond to form a cation selectively on the more substituted carbon. This happens in hydrohalogenation of alkynes just as we saw for electrophilic addition of a proton to an alkene:

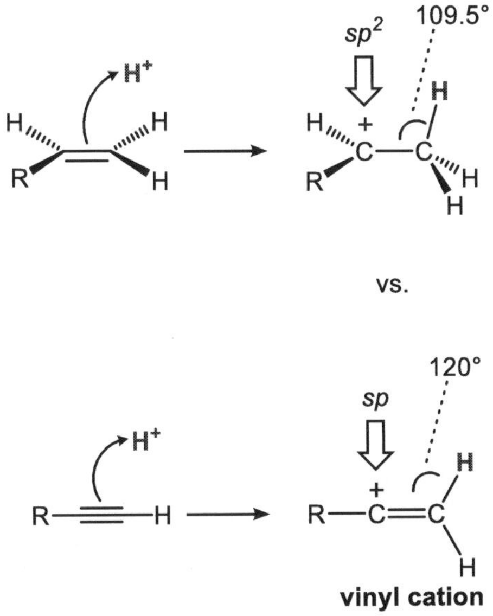

Neighboring H atoms are farther away in the vinyl cation than in the sp²-hybridized cation above because the angle is bigger.

The neighboring groups are too far away to move to the cation, so the carbocation will not rearrange.

Note, however, that the carbocation that forms is sp^2-hybridized. Because the carbon adjacent to the carbocation is sp^2-hybridized, its C–H and C–C bonds will be pointed further away from the empty *p*-orbital than those on an sp^3-hybridized carbon (i.e., 120° vs. 109.5°), so the neighboring groups are too far away to shift over to the positively-charged carbon. As a result, **the carbocation made by electrophilic addition to an alkyne cannot rearrange**. Once the carbocation forms, the nucleophile will coordinate just as we saw for alkenes:

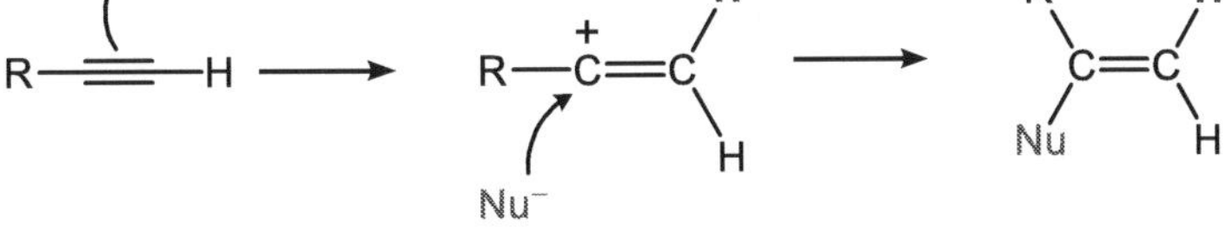

Another important differentiation between alkenes and alkynes is that the alkyne has *two π-bonds* in its triple bond, so you can actually do *two additions* to the alkyne C≡C bond if there are two or more equivalents of the addition reagent:

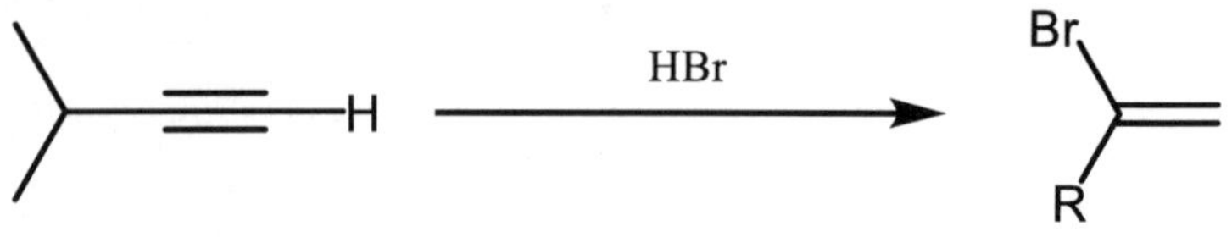

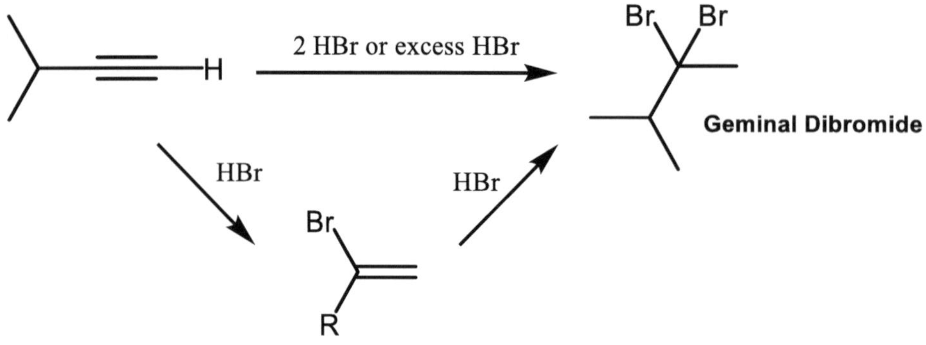

Example III.12.1

Draw the major product of the two-step reaction sequence shown below:

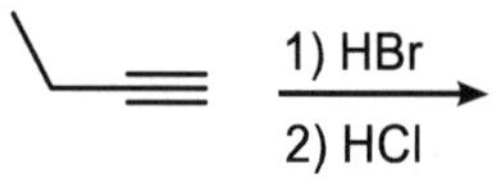

Solution III.12.1

We know that the first reaction, with HBr, will lead to Markovnikov addition of H and Br:

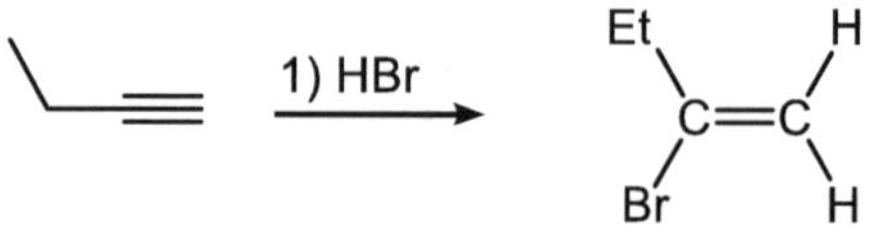

The second step, reaction with HCl, will result in Markovnikov addition of H and Cl. This generates a chiral center, so a racemate results:

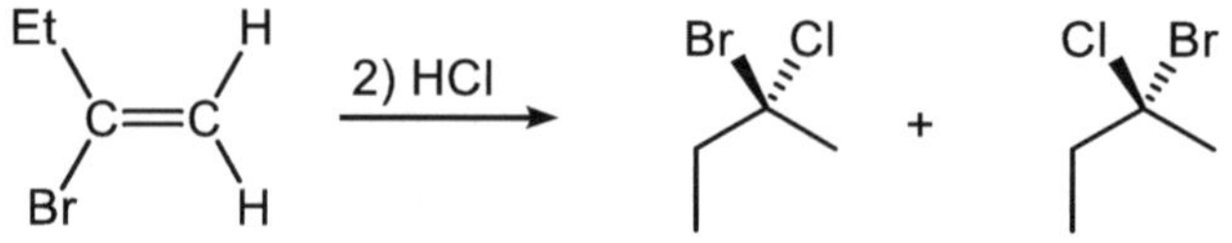

Lesson III.12.2. Halogenation - Electrophilic Addition by X^+, then Coordination by X^-

We saw in Lesson III.6 that X_2 (X = Cl or Br) will react with a π-bond in an alkene to give *anti*-addition of two X– units. The same reaction is possible with alkynes. Earlier in this lesson, we saw that a difference in bond angles in the alkyne-derived carbocation prevents nearby groups from moving over to the positively-charged carbon in alkynes. This effect also changed the intermediate for halogenation of alkynes compared to the mechanism we saw for akenes. Once the first X adds, it is too far from the positively-charges C to form the three-membered ring halonium intermediate (this happens for halogenation of alkenes; see Lesson III.6). However, **the net result of halogenation of alkynes is *anti*-addition of two X**, the same as the net result for halogenation of alkenes:

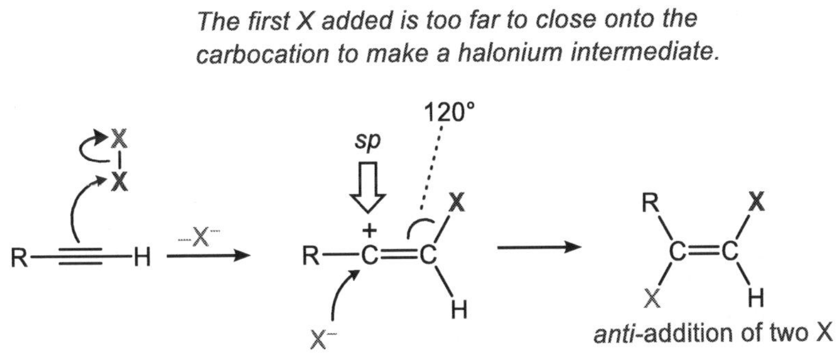

If two or more equivalents of X_2 are added, two X will be added to each side.

Example III.12.2

Draw the major product of the reaction shown below:

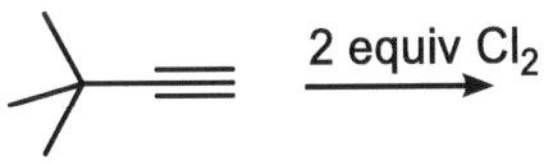

Solution III.12.2

We know that the first equivalent of Cl_2 will undergo *anti*-addition to one of the π-bonds of the alkyne:

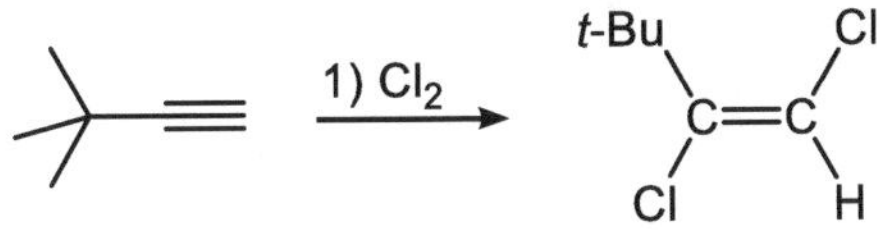

Next, the resulting alkene undergoes halogenation by the route we learned in Lesson III.6, first by forming the chloronium intermediate, then undergoing S_N2 ring opening to give the final product in which two Cl have added to each of the C in the original C≡C bond:

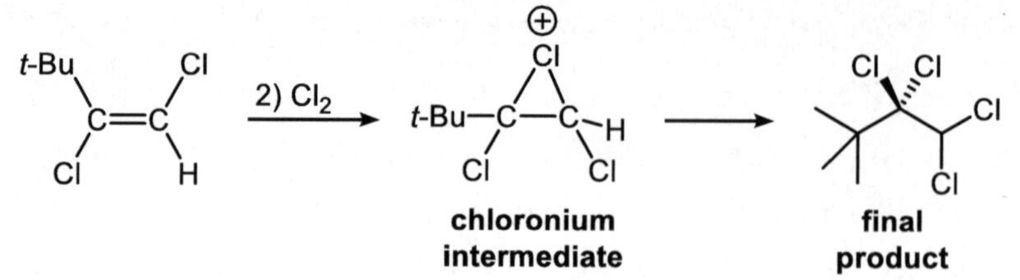
t-Bu
Cl
Cl
H
2) Cl2
t-Bu
Cl
H
Cl
chloronium intermediate
Cl
Cl
Cl
Cl
final product

Lesson III.13. Three Ways to Reduce Alkynes

Lesson III.13.1. Alkyne to Alkane

In Lesson III.2, we discussed hydrogenation of an alkene to an alkane. If we perform hydrogenation on an alkyne, we can reduce the π-bonds in the same way. One problem with hydrogen as a chemical reagent is that hydrogen is a gas, so it is not possible to just pour some into a container and weigh it at room temperature. So, **hydrogen (H_2) is always present in excess in these reactions**. The first equivalent of H_2 would convert the C≡C bond to a C=C bond. We know, however, that an alkene will be rapidly hydrogenated to the alkane since there is excess H_2 around. So, **the reaction of an alkyne with H_2/Pd will go all the way to an alkane**:

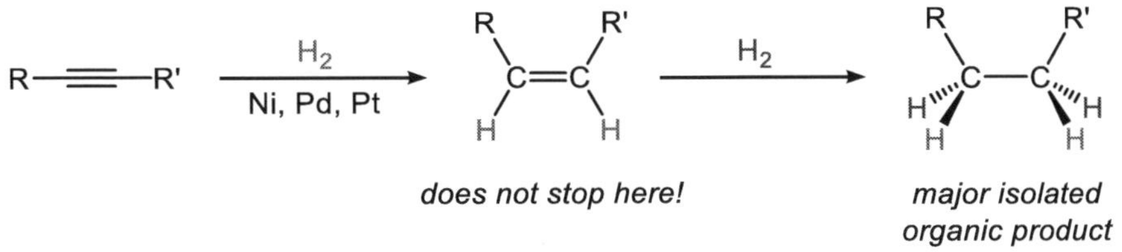

Lesson III.13.2. Syn-Addition of H_2 to Alkyne using the Lindlar Catalyst

If we want to take an alkyne and add only two hydrogen atoms, even though there is an excess of H_2, we can accomplish this by using a weak catalyst. Herbert Lindlar discovered that the addition of lead or sulfur compounds "poisoned" the Pd catalyst used for hydrogenation, hence the system named **Lindlar's catalyst will facilitate *syn*-addition of H_2 to an alkyne to form an alkene**.

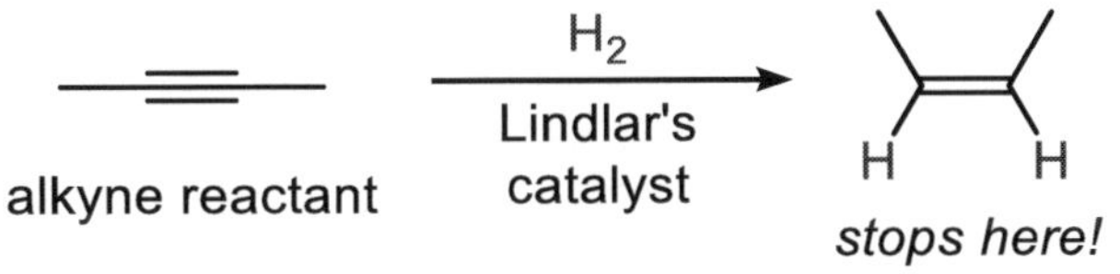

Example III.13.1

Draw the major product for each reaction shown:

A) 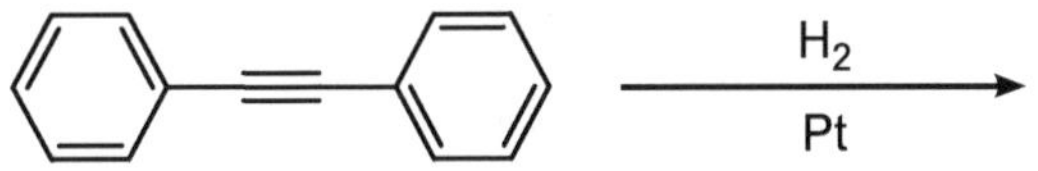

B) 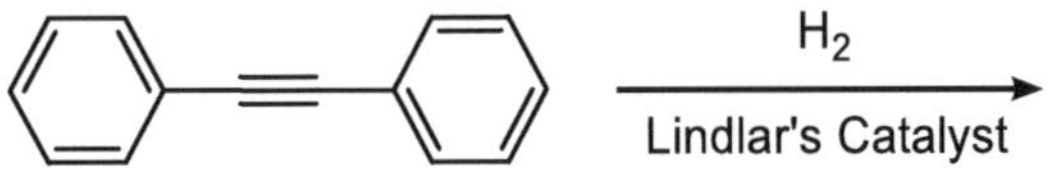

Solution III.13.1

For reaction A), the catalyst is very active and there is an excess of H_2, so both π-bonds of the alkyne are reduced, and two H add to each of the C in the C≡C:

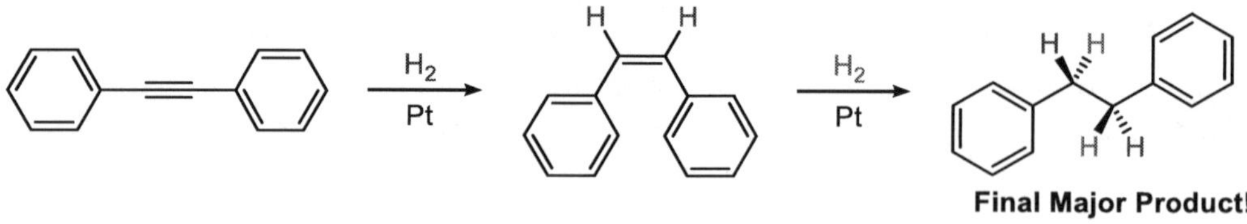

For reaction B), the catalyst is weaker, and will top after just one of the alkyne π-bonds has been reduced. The two added H atoms must be *syn*- to one another:

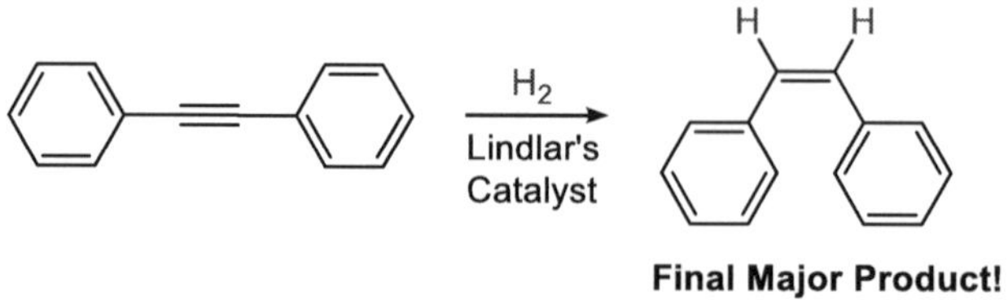

Lesson III.13.3. Anti Addition of H_2 to Alkyne

If *anti* addition of hydrogen to an alkyne is desired, we cannot use a Pd or Pt catalyst with hydrogen gas. Instead, we will need to use a different H source (NH_3) and a different metal (Na or K) to facilitate addition of the H to the alkyne. **Reaction of an alkyne to Na/NH_3(*l*) will result in *anti*-addition of two H atoms to an alkyne and stop at the alkene**:

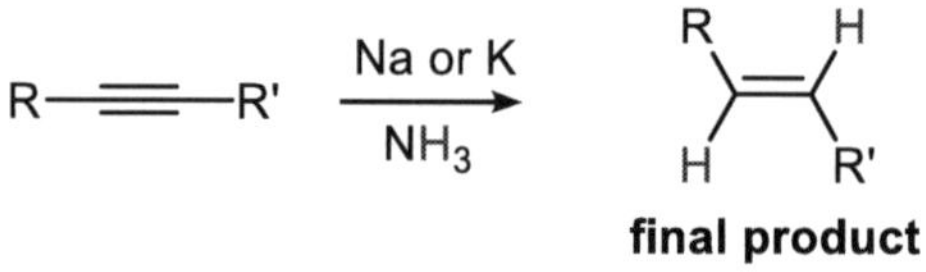

The mechanism of this reaction is complex, and is show in the companion book (mentioned in the Preface) for the interested reader.

Example III.13.3

Draw the major product for the reaction shown below:

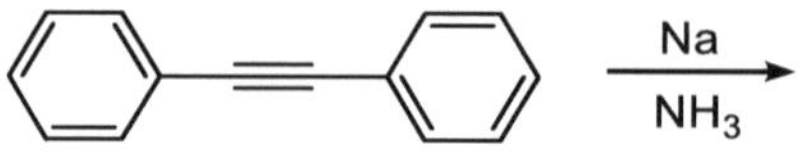

Solution III.13.3

The combination of Na and NH_3 leads to *anti*-addition of two H, in this case yielding *trans*-1,2-diphenylethylene. Phenyl is the name given to a $–C_6H_6$ (benzene) substituent:

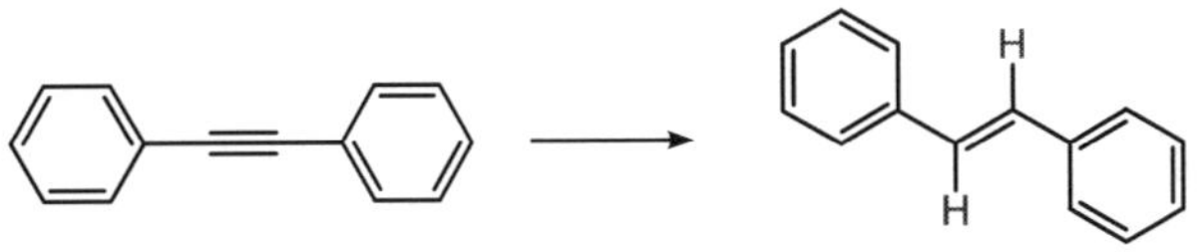

final product

Lesson III.14. Facile Constitutional Isomerism: Tautomerization

Lesson III.14.1. Enol and Keto Tautomers – A Special Class of Constitutional Isomers

We learned about constitutional isomers in Lesson I.14. Remember that constitutional isomers are molecules with identical molecular formulae, but different bond connectivity, and they cannot be interconverted without breaking one or more σ-bonds and reattaching them at a different atom. In general, σ-bond breakage does not occur readily and requires a significant amount of energy. However, there are special classes of constitutional isomers that can readily interconvert because the energy required is low. **Tautomers** are one such class. In the "enol" form, there is an OH group attached to a C=C bond. In the "keto" form, the bonds have rearranged so that there is an O=C–CH– unit where the HO–C=C– was in the enol form. These two species exist in chemical equilibrium with each other, but the keto form is usually the more stable one, so **the keto form predominates**.

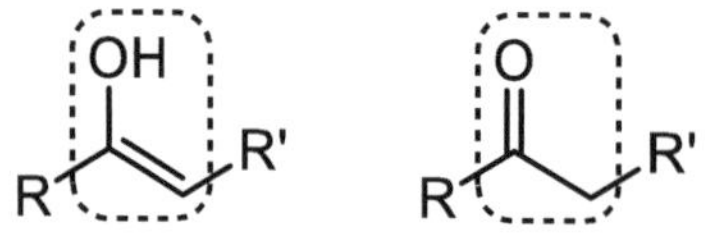

tautomers:
constitutional isomers that can easily interconvert

enol form **less stable**   keto form **more stable**

Lesson III.14.2. Keto–Enol Tautomerization and the Keto–Enol Equilibrium

Tautomerization refers to a chemical reaction that converts one tautomer to another. Keep in mind that this is NOT resonance: resonance only involves π-bonds and lone pairs. Any process that breaks a σ-bond (i.e., O–H) is a chemical reaction. The mechanism of tautomerization to convert an enol to a keto form is can be facilitated by an acid or a base:

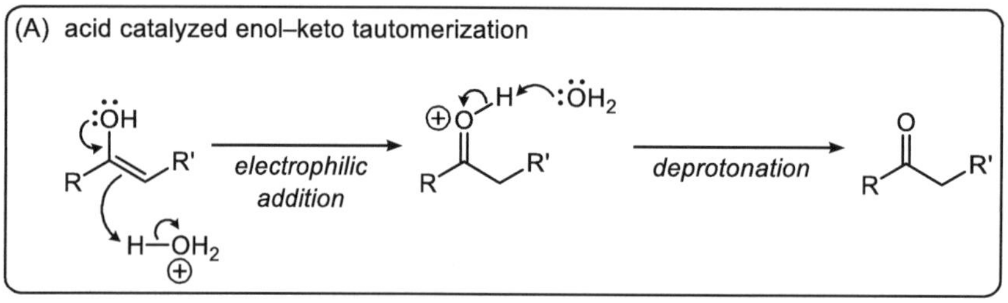

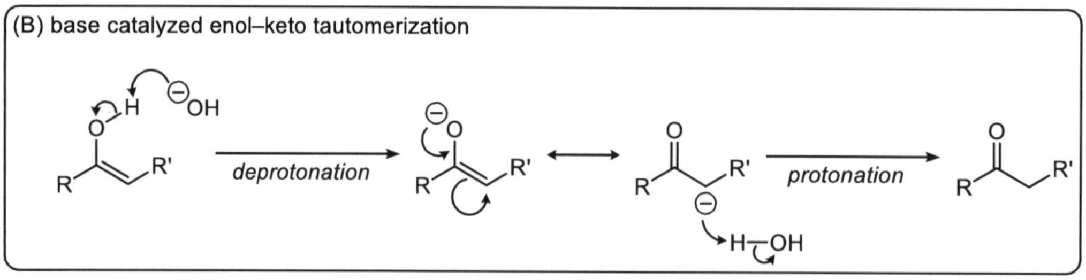

The way to represent the interconversion of two molecules via forward and reverse chemical reactions is with an equilibrium arrow. The specific chemical equilibrium that interrelates tautomers is termed the "keto–enol equilibrium". In general, the keto form because the bonds in the former (O–H and C=C) are weaker than the bonds in the latter (C–H and C=O). As a result, the keto form usually predominates over the enol form in solution.

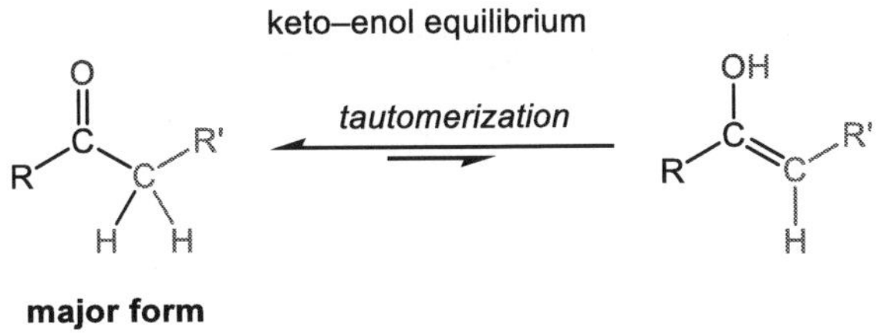

The take-home lesson for our purposes is that **any time an enol forms, it will spontaneously convert to the keto form as the major product**. In the next lesson, we will see why this concept is so important for determining the major isolated organic products for some alkyne reactions.

Lesson III.15. Preparing Carbonyls from Alkynes

Lesson III.15.1. Oxymercuration/Demercuration: Markovnikov Enol Formation

Let us examine what will happen if we subject an alkyne to oxymercuration/reduction conditions as we saw for alkenes in Lesson III.7. One difference in typical alkyne reactions is that $HgSO_4$ is generally used instead of $Hg(OAc)_2$, and H_2SO_4 is added to supply the H (as we will see in the mechanism). The reaction with the alkyne proceeds via a different mechanism than an alkene, because the 3-membered ring containing Hg and a C=C bond that would result in the alkyne is far too unstable:

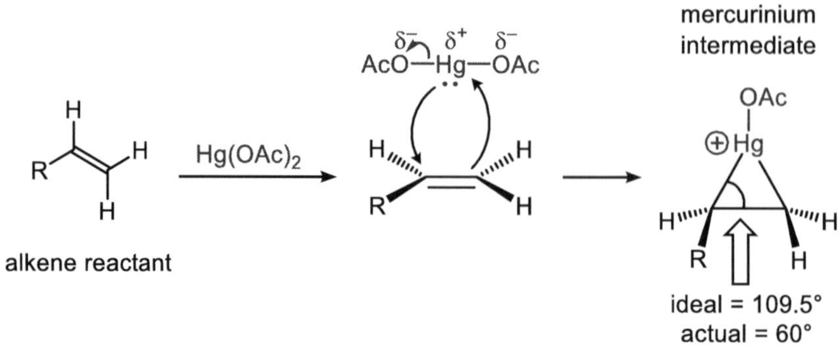

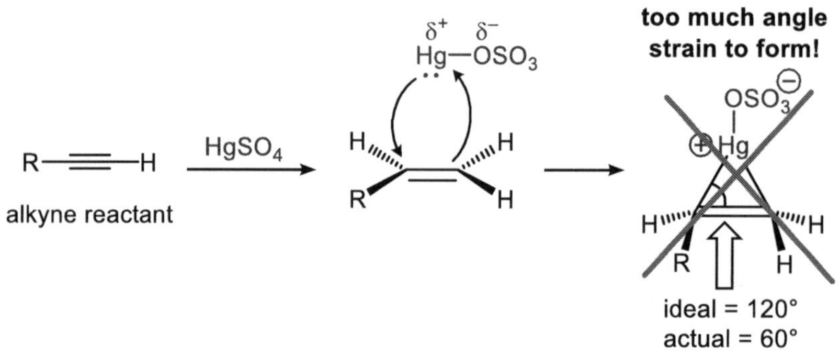

Instead, $HgSO_4$ makes a complex, similar to what we saw in Lesson III.12, with a greater δ+ on the more substituted carbon. Water then does nucleophilic attack at the more substituted alkyne carbon:

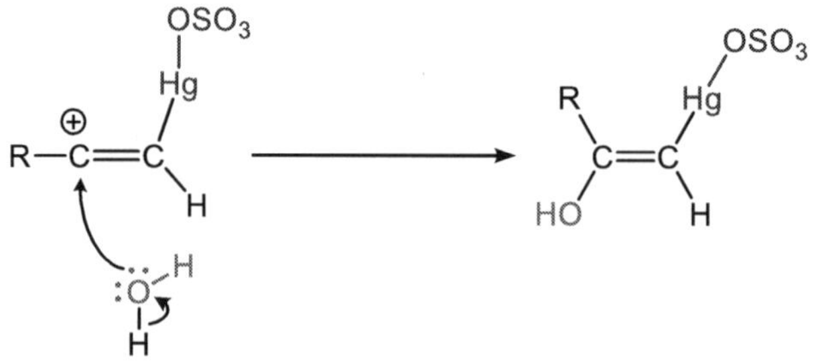

Whereas alkene hydration required oxymercuration followed by reduction with $NaBH_4$, alkyne hydration will continue from the oxymercuration product using the added sulfuric acid.

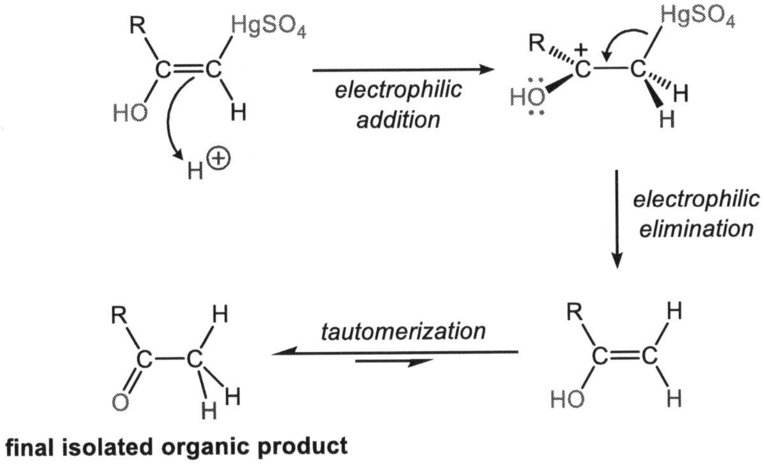

The net result of oxymercuration/demercuration is the conversion of an alkyne to a ketone with Markovnikov regioselectivity.

Example III.15.1

Draw the major product of each the reaction shown below:

A)

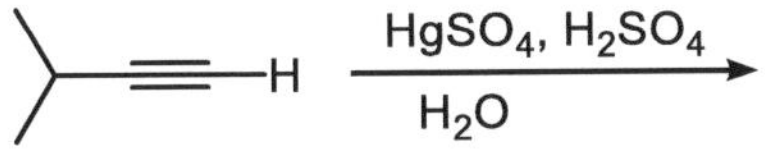

B)

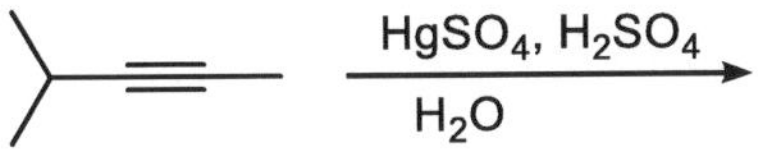

Solution III.15.1

For reaction A), we envision Markovnikov addition of OH and H to the alkyne. This gives an enol. We know that the enol tautomerizes to the keto form, so we do not actually isolate the enol at all. The major product will be the ketone:

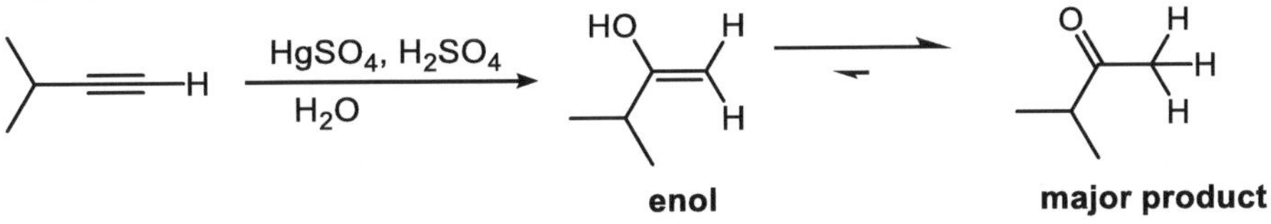

For reaction B, we again envision Markovnikov addition of an H and an OH. In this reaction, both sides of the triple bond are equally substituted, so a mixture of both initial products will result:

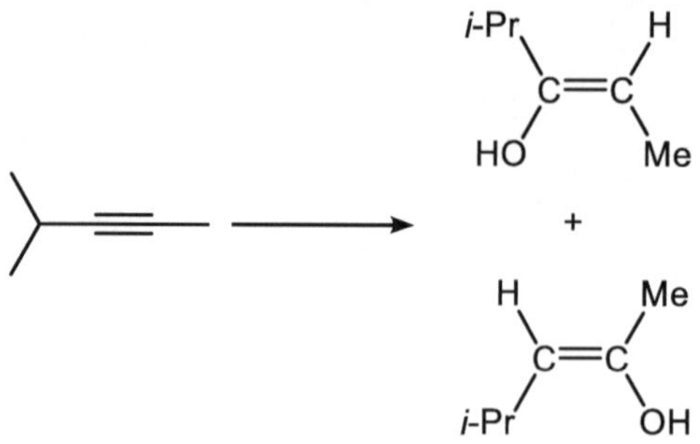

We recognize that these structures are **enols**, so they will favor the keto tautomers. In this case, we get two ketones in about equal yield.

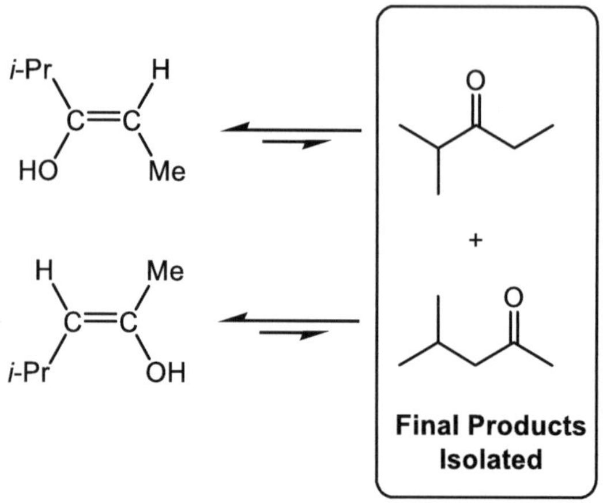

Lesson III.15.2. Hydroboration/Oxidation: Anti-Markovnikov Enol Formation

Hydroboration/oxidation of alkynes proceeds via a mechanism very similar to that of alkenes, but instead of BH_3, the boron reagent used is usually a bulky R_2B–H (R = secondary or tertiary alkyl group) such as a reagent abbreviated 9-BBN:

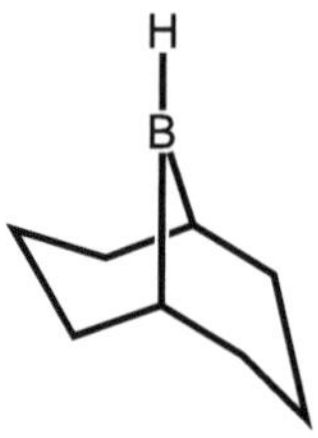

9-BBN
a reagent for hydroboration of alkynes

Whichever R_2B–H is used, the first step of the reaction is hydroboration, which is *syn*-addition of –H on the more substituted C and the $–BR_2$ on the more substituted C. The oxidation step replaces the $–BR_2$ with an –OH. This is exactly what we observe for hydroboration/oxidation of an alkene.

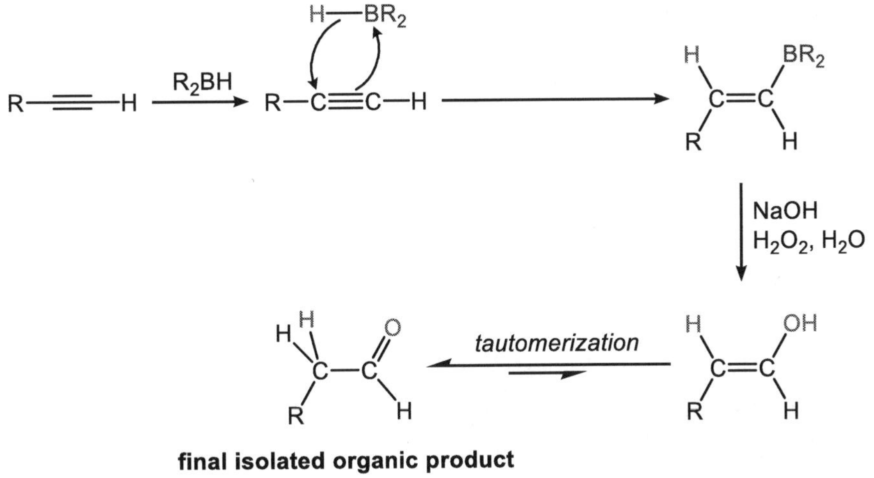

We know that the initially-formed enol is not stable, and will rapidly tautomerize to the keto form. So, in the reaction above **the isolated product is the aldehyde.**

Example III.15.2

Draw the major product of the reaction shown below:

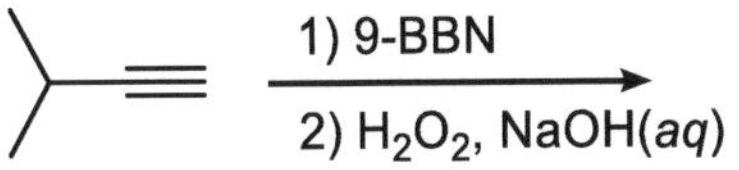

Solution III.15.2

We expect non-Markovnikov addition of an H and an OH, followed by tautomerization to give the final product:

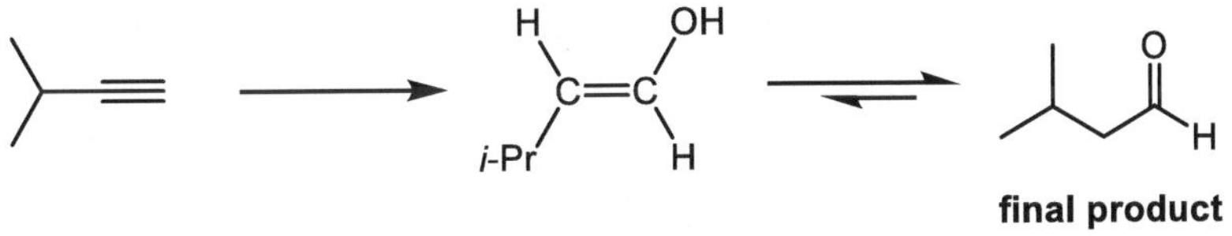

Lesson III.16. Acetylide Anion Preparation and its use as a Nucleophile

Lesson III.16.1. Atomic Orbital Hybridization and Effective Electronegativity

We know from Lesson I.6 that electronegativity of an atom increases with the *s*-character of its hybrid orbitals:

(more electronegative) $sp > sp^2 > sp^3$ (less electronegative)

In fact, this effect is so dramatic that a negative charge is more stable on an *sp*-hybridized carbon than on an sp^3-hybridized nitrogen. Because negative charge is stabilized on a an *sp*-hybridized C**, it is relatively easy to deprotonate a terminal alkyne if we use a strong base like NaH or $NaNH_2$** to form what is often referred to as an **acetylide anion**:

Once generated, the **acetylide anion is a good nucleophile and so can be used to do an S_N2 reaction**:

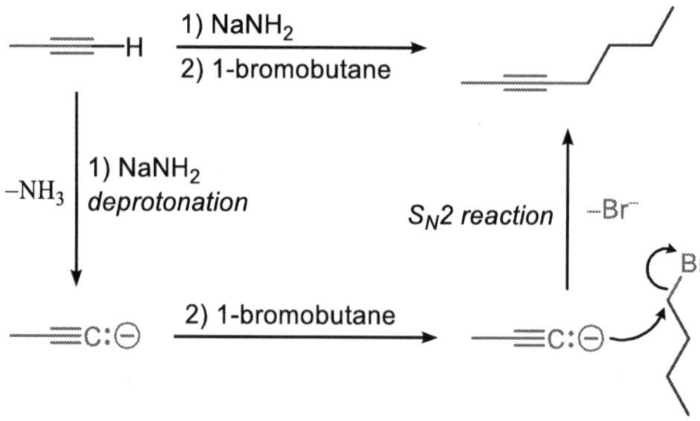

Like any other S_N2 reaction, if the center attacked is chiral, it will be inverted, so:

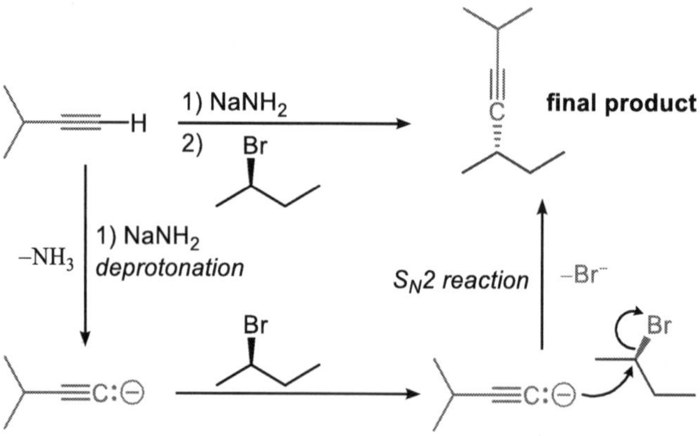

Example III.16.1

Propose a synthesis of the molecule shown below starting with acetylene and assume you have access to any acid, base, and alkyl halide you might need:

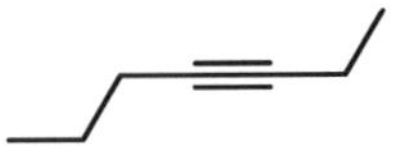

Solution III.16.1

We have seen in this Lesson that a primary way to make internal alkynes is by adding the required chains to the acetylide anion. So, we start with acetylene and deprotonate it to make the necessary acetylide (step i). Next, we can add either of the two chains. For this example, we will put the ethyl on before the propyl (step ii). We then need to regenerate an acetylide (step iii), which serves as the nucleophile for S_N2 reaction to add the propyl group (step iv).

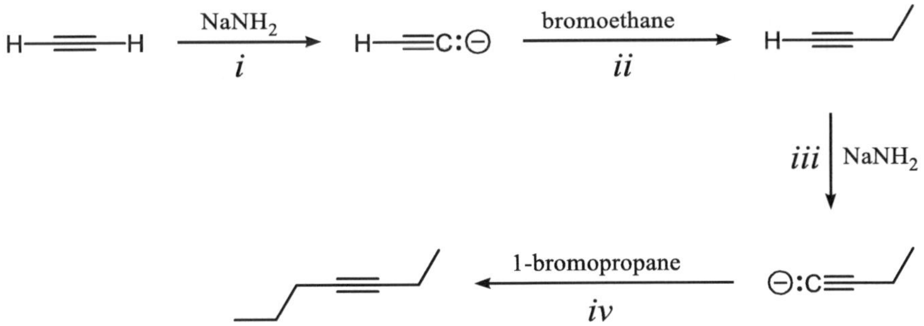

Lesson III.17. Radicals I: Peroxide-Mediated Hydrobromination of Alkenes

Lesson III.17.1. Definition of a Radical and Radical Stability

A radical is a species having an unpaired electron. As with cations, we will limit the current discussion to carbon-centered radicals. Like carbocations, carbon-centered radicals feature an sp^2-hybridized carbon atom, in which the C does not have an octet, so it will have an affinity for electrons:

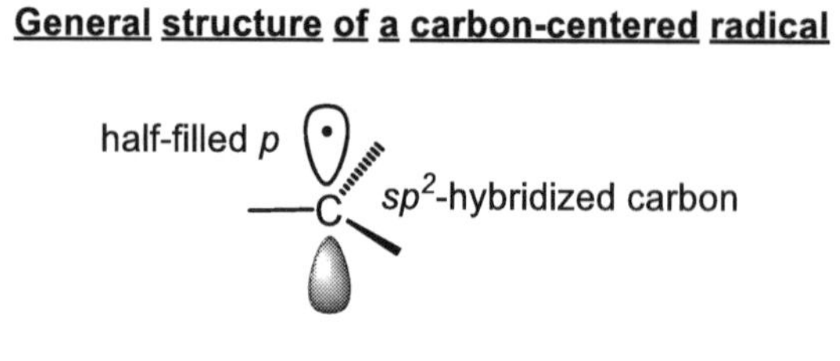

Carbon-centered radicals are similar in electronic structure to carbocations (both lack a complete octet), therefore the trend in radical stability is roughly similar to that for carbocations, but resonance exerts an even higher stabilizing effect in radicals. The stability trends can thus be summarized here:

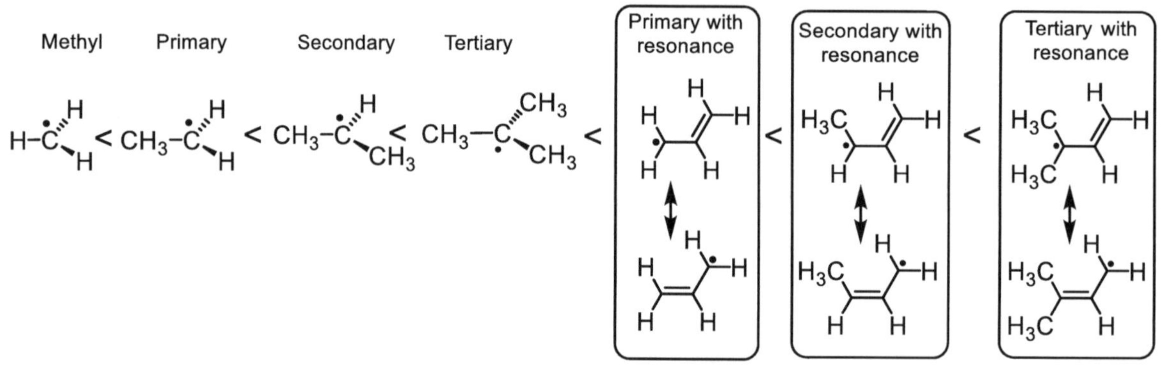

Note that **even a primary radical that has resonance is more stable than a tertiary radical that lacks resonance**.

Example III.17.1

Which is the more stable species within each pair of molecules shown?

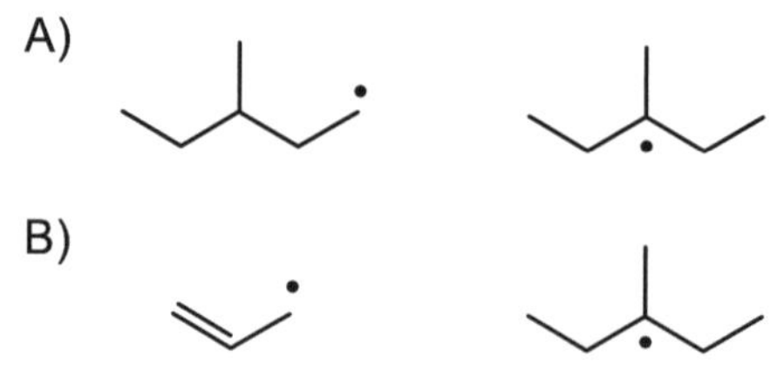

Solution III.17.1

A)  Tertiary is more stable

B) 

Recognize the resonance in this one. The resonance stabilization makes it more stable than the no-resonance tertiary radical

Lesson III.17.2. Radical Initiation

The O–O single bond in a peroxide (RO–OR) is very weak and can easily be broken. Homolysis of the O–O bond can be induced by light or by heat. If this homolysis occurs in the presence of HBr, an alkoxy radical (RO•) will abstract a hydrogen atom (i.e., H•) and generate a bromine radical (Br•), which is the radical propagating species. The peroxide most commonly used for these types of reactions is benzoyl peroxide (BPO, shown below). Generating the first radical to participate in a reaction is called **initiation**.

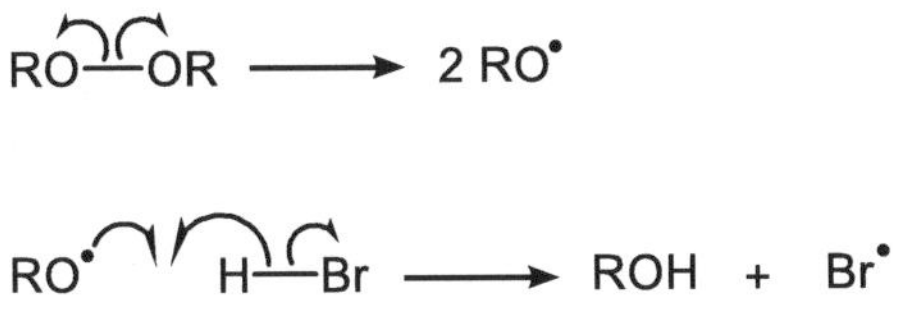

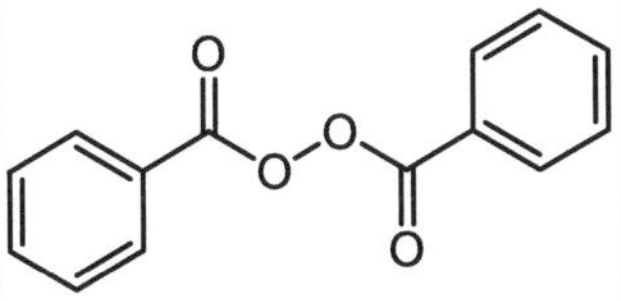

benzoyl peroxide (BPO)

a common ROOR for this reaction

Lesson III.17.3. Radical Propagation

Once generated by initiation, Br• will function as an electrophile because it has fewer than 8 electrons in its valence shell: **Br• needs one more electron to have a full octet**. An alkene will share one electron from its π-bond with Br•, while the other π-electron will stay behind on other carbon and form a carbon radical. The **Br will add to the less substituted alkene carbon**, because the carbon radical will be more stable at the more substituted alkene carbon. This carbon radical can then react with another molecule of HBr, abstracting H• to form a C–H bond and regenerating the propagating species Br•. These two steps, in which a radical reacts with a non-radical to make new bonds towards getting to the product, are called the **propagation** steps:

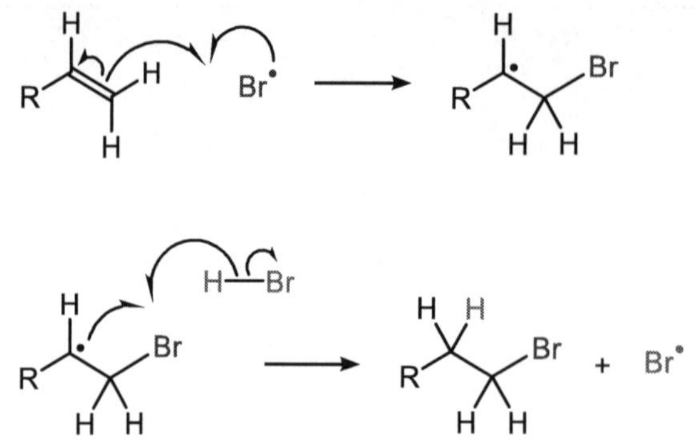

The net result of reacting an alkene with HBr/ROOR is non-Markovnikov addition of H and Br to an alkene with a mix of *syn*/*anti* addition, as summarized in Table III.4.1 (Lesson III.4).

Lesson III.17.4. Radical Termination

A reaction in which a radical reacts with another radical is called **termination**. Some possible termination steps are shown below:

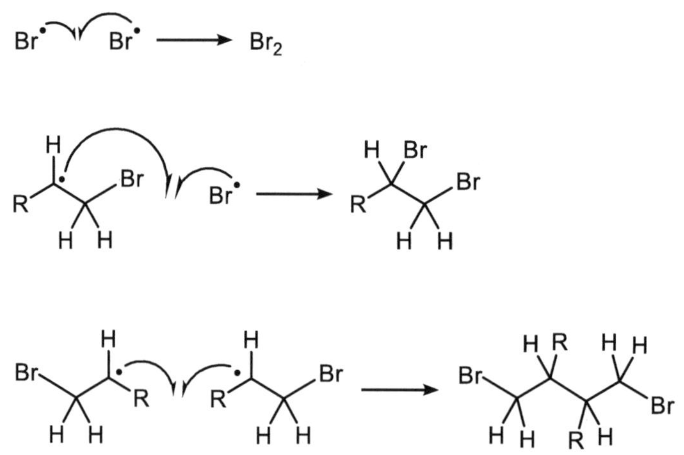

Example III.17.2

Draw the major product(s) for the reaction shown below:

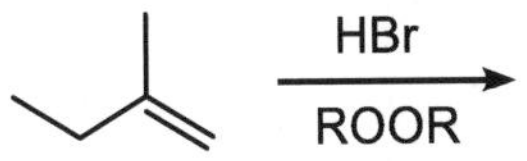

Solution III.17.2

An alkene reacts with hydrobromic acid/ROOR to give non-Markovnikov addition of H and Br. For this alkene, such an addition generates a chiral center, so we will get a 50:50 mixture of enantiomers, a racemic mixture:

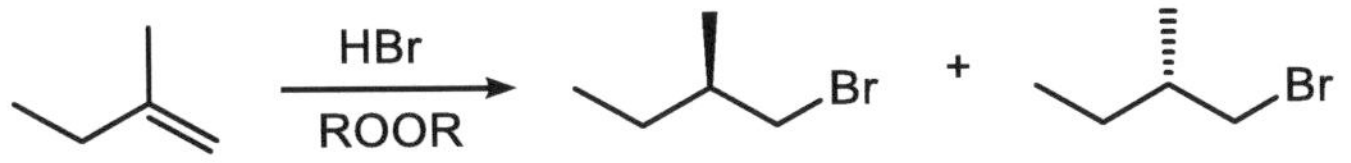

Lesson III.18. Radicals II: Halogenation of Alkanes

Lesson III.18.1. Radical Initiation: Homolysis of the X–X Bond

When an alkane reacts with X_2 (X = Cl or Br) in the presence of light, the net result is substitution of an H for X. Chemists use the symbol "$h\upsilon$" to indicate that light is being shined on a reaction mixture:

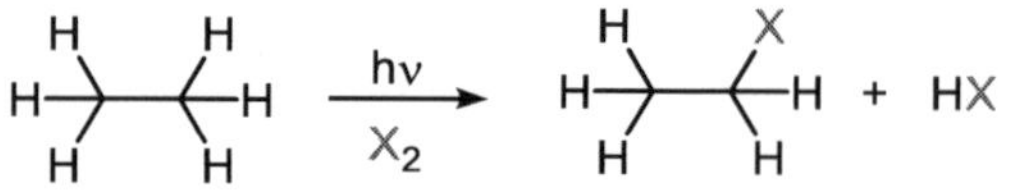

This **substitution of H for Br is called radical halogenation of alkanes**. How does this work? Well, the X–X bond in X_2 is relatively weak and can be broken by the light energy to form $2X^{\bullet}$:

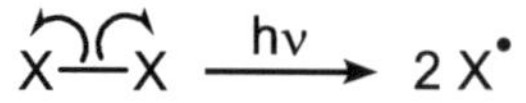

This is the **initiation step** of the radical halogenation of alkanes.

Lesson III.18.2. Radical Propagation: H-atom Abstraction by X•

Once generated, how can $X^{\bullet}$ react with an alkane? Well, $X^{\bullet}$ has only seven electrons in its valence shell, so it needs to pull one more electron from somewhere to get a full octet. In the previous Lesson, $Br^{\bullet}$ reacted with the π-bond of an alkene (remember: π-bonds are weaker than σ-bonds). What happens when $X^{\bullet}$ is generated in the presence of an alkane? The only electrons available for $X^{\bullet}$ to achieve an octet are those in C–H and C–C σ-bonds. In general, C–H σ-bonds are more easily broken than C–C σ-bonds, so when $X^{\bullet}$ encounters an alkane molecule, it will break a C–H by abstracting an $H^{\bullet}$. This process forms HX and a carbon radical. The carbon radical then reacts with a molecule of X_2, breaking the X–X bond homolytically to form a C–X bond and regenerate the radical propagating species $X^{\bullet}$. Because the concentration of $X^{\bullet}$ is usually low, these reaction conditions will typically only afford the monohalogenated product.

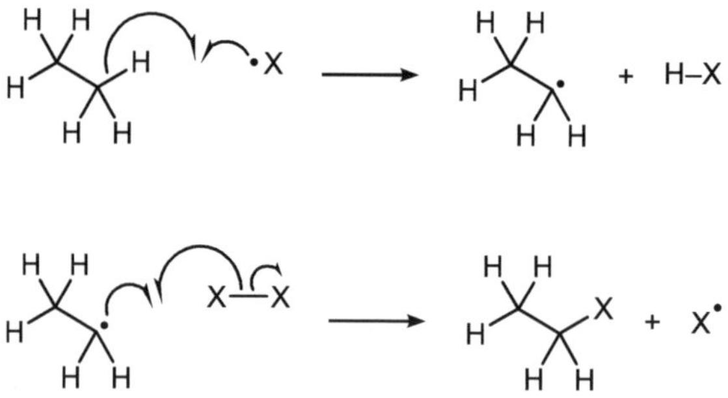

The two steps above form the bonds in the products, and are the **propagation steps**.

Lesson III.18.3. Radical Termination

The radical termination reactions in the free radical halogenation of alkanes are shown below:

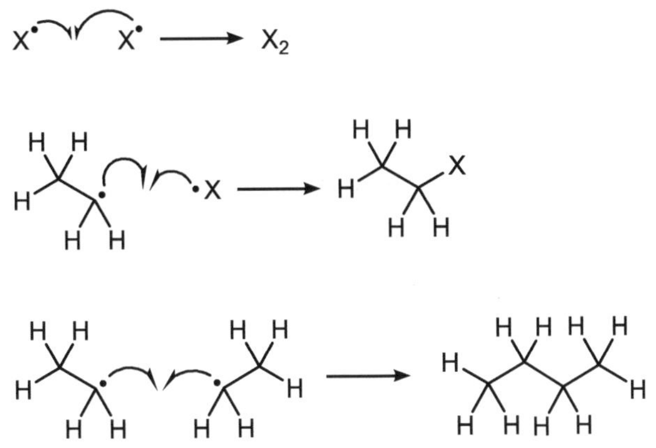

Lesson III.18.4. Regiochemistry and Stereochemistry

Remember that the radical halogenation of an alkane will only replace C–H bonds with C–X bonds, it will never break C–C bonds. So, the easiest way to draw all the possible products of this reaction is to identify all of the chemically unique C–H bonds in a given molecule. If our reactant is 2-methylbutane, there are 4 chemically unique C–H bonds, thus radical halogenation of this molecule will generate 4 different monohalogenated products (which are constitutional isomers of each other):

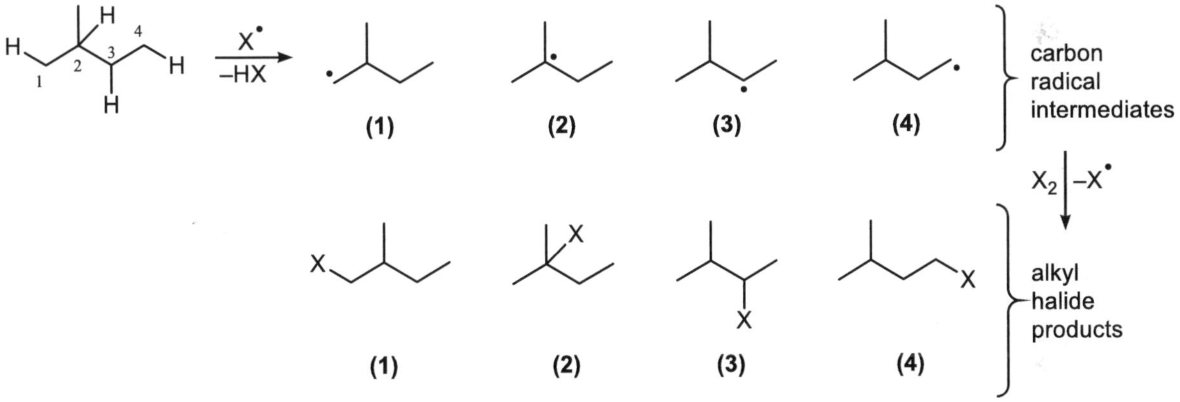

To identify which will be the *major* product, we need to consider the relative stability of the radicals which led to each constitutional isomer. Abstraction of H• from position (1) or (4) will yield a 1° radical. Abstracting the H from position (2) will yield a 3° radical, and abstracting an H from position (3) will yield a 2° radical. Given that the stability of carbon radicals is 3° > 2° > 1°, intermediate (2) is the most stable, and thus we would expect the major product to be derived from that intermediate. **When X = Br, the major product is always formed by replacing the H that is on the most substituted C.** In this case, the major product of radical bromination is product (2):

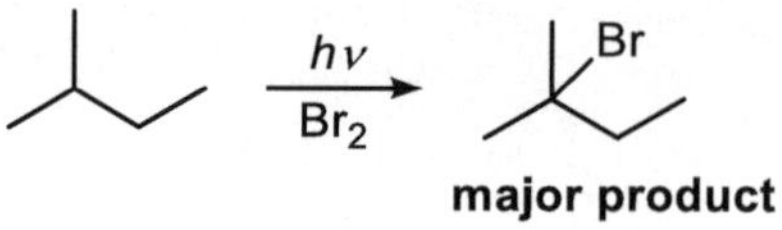

However, when X = Cl, products (1) and (4) are formed in nearly equal amount to product (2)! What is going on here? Well, **chlorine radicals are extremely reactive compared to bromine radicals**. So, whenever a Cl• collides with an alkane near a C–H bond, it rapidly abstracts the H to form HCl and a carbon radical. **Bromine radicals are not nearly as reactive as chlorine radicals, so bromine radicals are more selective.** This means that the bromine radical will preferentially react with the more reactive C–H bonds that would give the most stable carbon radical. The major product of bromination is energetically controlled; the placement of Br where the most stable radical intermediate forms is the major product. The manner in which the different reactivity of Cl and Br radicals leads to different major products leads to a generally applicable concept: **more reactive = less selective**.

Example III.18.1

Draw the monohalogenated products for the reaction shown below. Indicate which is energetically preferred and which is statistically preferred. What is the major product in this case?

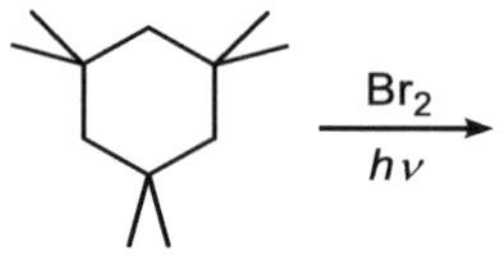

Solution III.18.1

On this reactant, 1,1,3,3,5,5-hexamethylcyclohexane, there are 18 H atoms on the primary sites (the methyl groups), compared to only 6 on the secondary sites. The statistical product is the one derived from replacing a H on a primary C. A secondary radical is more stable than a primary radical, so the energetically preferred product is the one in which the H on a secondary C is replaced by Br:

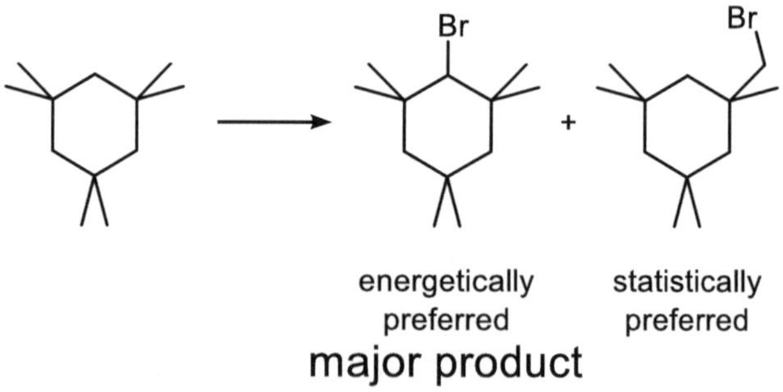

Because Br is less reactive, it is more selective for the energetically preferred product, so the energetically preferred product is the major product.

Lesson III.19. Introduction to Polymers and Radical Polymerization of Alkenes

Lesson III.19.1. Polymers, Monomers and Repeat Units

Many of the materials we encounter in our daily activities have significant mechanical strength, ranging from rigid materials (such as a plastic chair) to flexible materials that can adopt multiple different shapes (such as cloth) to spongy materials that can be compressed and spring back to their original shapes (such as the insoles of shoes). However, nearly all of the organic compounds encountered in introductory organic lab classes are oils, powders, or crystals, and thus lack such mechanical strength. What properties of organic molecules would endow them with mechanical strength, and how can we rationally design materials for specific needs? Many organic materials of high mechanical durability are **polymers**.

A polymer is a very long molecule that is obtained by chemically linking a group of small molecules called **monomers** into a chain. Each polymer is represented as a multiple of its **repeat unit**. To draw a polymer, the repeat unit is drawn out and placed in parentheses. A bond is added that cuts through the parentheses on each side to indicate that the repeat unit attaches to the next repeat unit at that site. For example, poly(vinyl chloride) is prepared by polymerizing the monomer vinyl chloride. The repeat unit is –CH_2–C(H)Cl– (shown in bold below). Similarly, polymerization of styrene yields polystyrene, which comprises –CH_2–C(H)Ph– (shown in bold below) as the repeat unit (Ph is an abbreviation for the word “phenyl”, which is a benzene ring as a substituent):

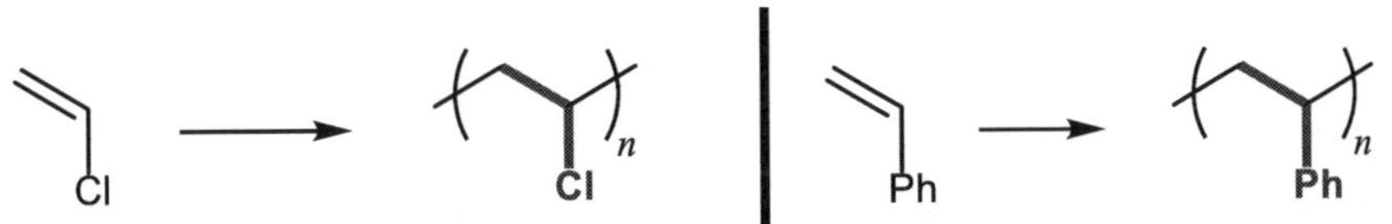

vinyl chloride *monomer* | **poly(vinyl chloride)** *polymer (repeat unit in bold = CH_2CHCl)* | **styrene** *monomer* | **poly(styrene)** *polymer (repeat unit in bold = CH_2CHPh)*

Connecting multiple small molecules with covalent bonds can endow the resulting compound with mechanical strength, but we can tune the chemical and mechanical properties of that polymer by making changes to the monomer, guided by our understanding of chemical principles.

Example III.19.1

Predict which of the following polymers would exhibit a lower melting point and provide an explanation for your answer.

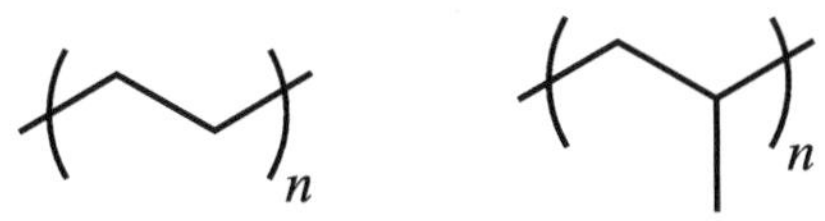

poly(ethylene) **poly(propylene)**

Solution III.19.1

We know from Lesson I.3 that a branched alkane has a lower boiling/melting point than a linear alkene because the branches push the alkane molecules farther apart and diminishes the intermolecular forces holding the molecules together. The same premise applies to polymers. If we draw out a segment of each polymer, we see that the chains are farther apart in poly (propylene). For this reason, polypropylene has a lower melting point.

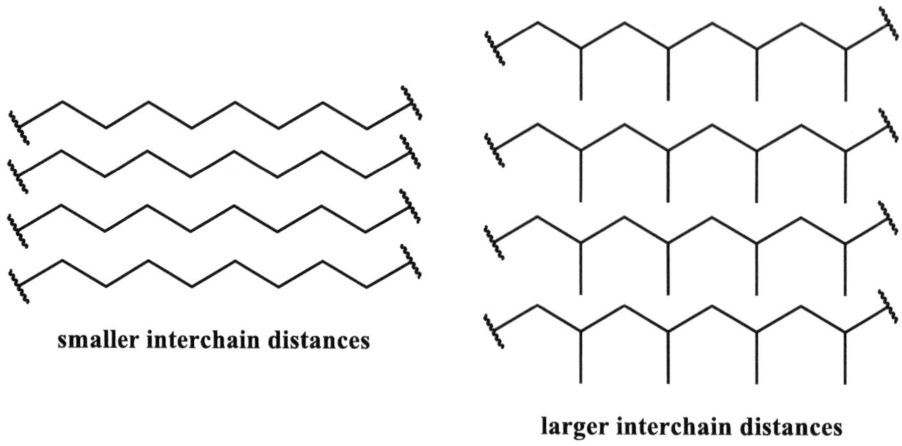

Lesson III.19.2. Radical Polymerization of Alkenes

Growth of a polymer chain can be initiated if we generate a radical in solution with alkenes, leading to **radical polymerization of alkenes**. The most common radical initiators used to get this polymerization started are azobisisobutrylnitrile (AIBN) and benzoyl peroxide (BPO). In Reaction 1, the AIBN or BPO are heated to generate the radical initiator (R•). The initiation step is reaction of the initiator radical with the alkene (Reaction 2). We know from the radical stability rules that this step will favor making the more stable, more substituted radical:

Reaction 1:
(Make initiator, R•)

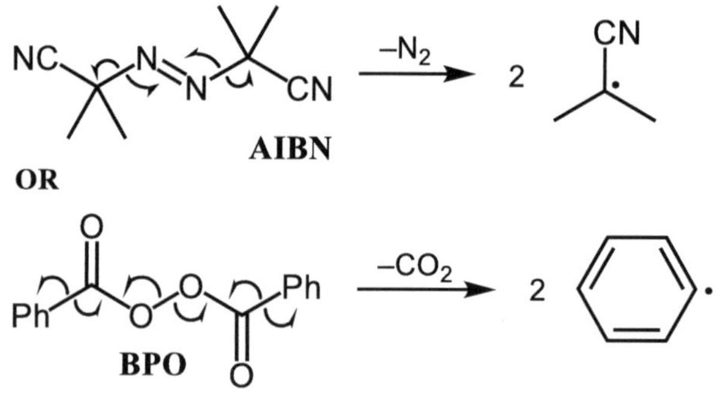

Reaction 2:
(Initiation)

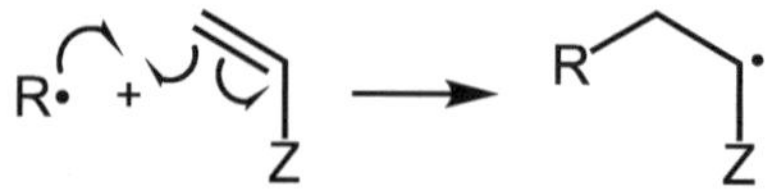

The species generated in Reaction 2 is also a radical, so it can react with another monomer molecule to give another radical (Reaction 3). This new radical species can react with another monomer molecule, and so on (repeating step 4 over and over again), to give an ever-growing polymer chain (represented as **P**· in reaction 5) that can keep adding monomer until a termination step (two radicals combine) stops the reaction as for other radical chain reactions we have learned about. Steps 3-5 are the **propagation steps** of radical polymerization:

Reaction 3:

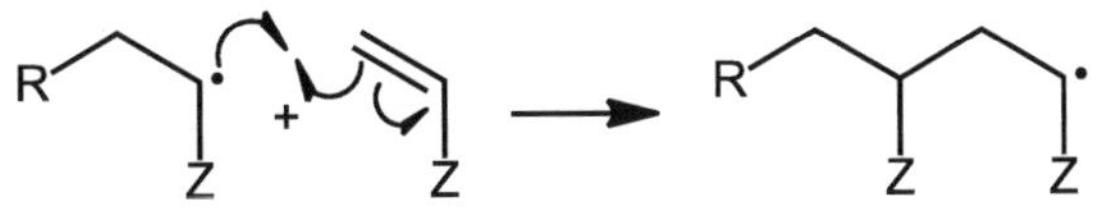

Reaction 4:

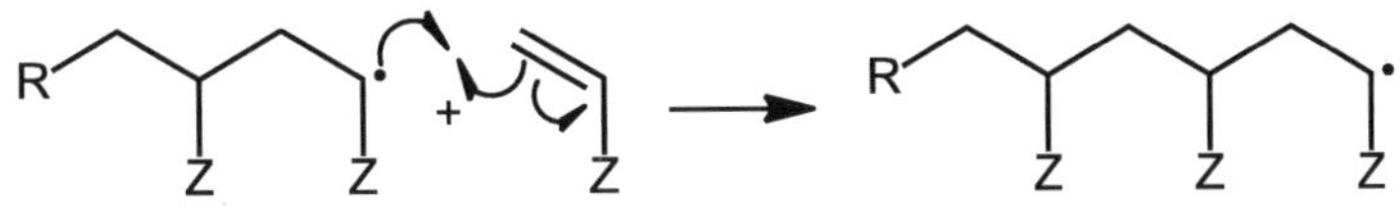

Reaction 5:

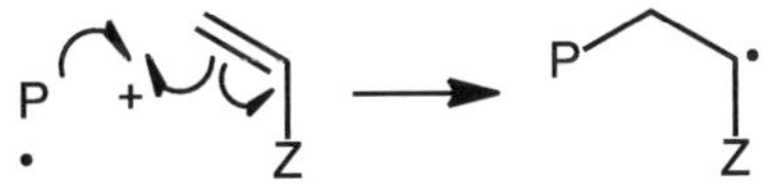

BONUS Material

The remaining Lessons appear in both the "Organic Chemistry 1 Primer" and the "Organic Chemistry 2 Primer" to account for flexibility with respect to what chapters instructors choose to teach in the first and second semester classes at different schools. These bonus Lessons are numbered in the same way that they are in the "Organic Chemistry 2 Primer" so that the companion texts can be used with either Primer independently.

Bonus Lessons from Part IV: Properties and Reactions of Conjugated and Aromatic Molecules

Lesson IV.1. Pi Conjugation Stabilizes a Molecule

Lesson IV.1. π-Bond Conjugation is Stabilizing and π-Bond Cumulation is Destabilizing

In Lessons III.1-10, we considered a variety of reactions typical for individual C=C double bonds in isolated alkene units. Next, we will consider the stability trends for molecules that have *more than one* C=C bond and then examine some reactions of dienes. Remember that π-bonds are made up of carbon *p*-orbitals. In 1,4-pentadiene, the two C=C bonds are separated by an sp^3-hybridized carbon, so the *p*-orbitals cannot overlap, so each of the π bonds is **isolated**:

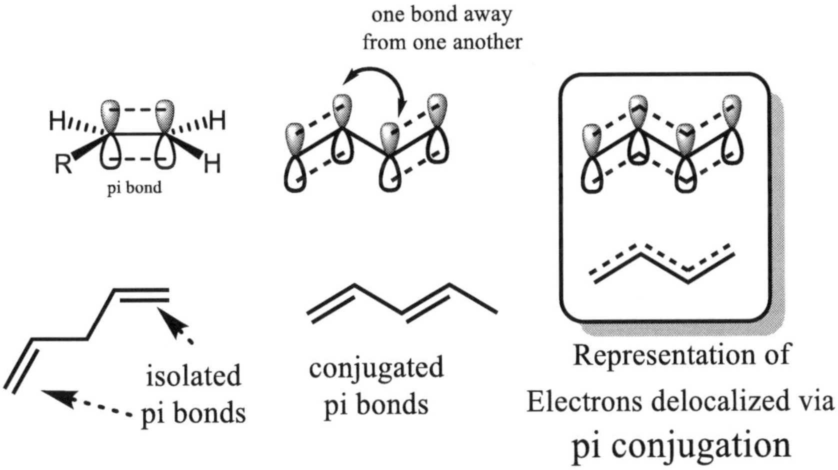

In *trans*-1,3-pentadiene, however, there are no sp^3-hybridized carbons separating the two C=C bonds, so all four of the *p*-orbitals involved in making π bonds can overlap with each other, and we classify this type of diene as **conjugated**. When two C=C bonds are conjugated, each of the 4 π-electrons can spread out over 4 positions, this delocalization is why **conjugation significantly stabilizes π-bonds**. Recall from Lesson III.2 that we measure alkene stabilities using heats of hydrogenation (ΔH_h). For 1-pentene and *trans*-2-pentene, the ΔH_h values are –30.1 and –28.6 kcal/mol, respectively. At first glance, we might expect the heat of hydrogenation for *trans*-1,3-pentadiene to be simply the sum of the values for 1-pentene and *trans*-2-pentene (–58.7 kcal/mol, because it has one monosubstituted alkene and one *trans* alkene). The experimentally observed ΔH_h for *trans*-1,3-pentadiene is –54.1 kcal/mol, which demonstrates that **conjugation of the two C=C bonds provides an extra 4.6 kcal/mol of stabilization:**

		ΔH_h (kcal/mol)
	1-pentene	–30.1
	*trans*-2-pentene	–28.6
	1,4-pentadiene	–60.8
	*trans*-1,3-pentadiene	–54.1
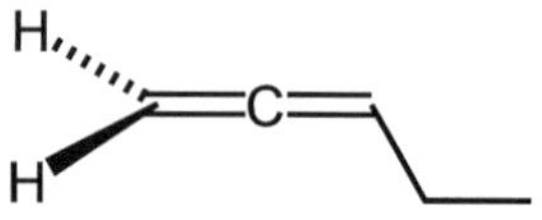	1,2-pentadiene	–69.8

When two C=C bonds begin at the same carbon, we get a C=C=C unit, in which the C=C bonds are classified as **cumulated**. Note that the central carbon is *sp*-hybridized, so the *p*-orbitals used to make the C=C bond to the left carbon is orthogonal to those used to make the C=C bond to the right, and thus the two π-bonds cannot overlap. This is illustrated for $H_2C{=}C{=}CH_2$ here:

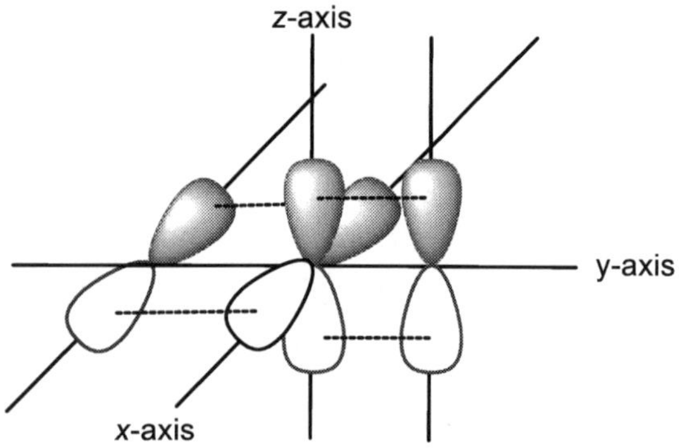

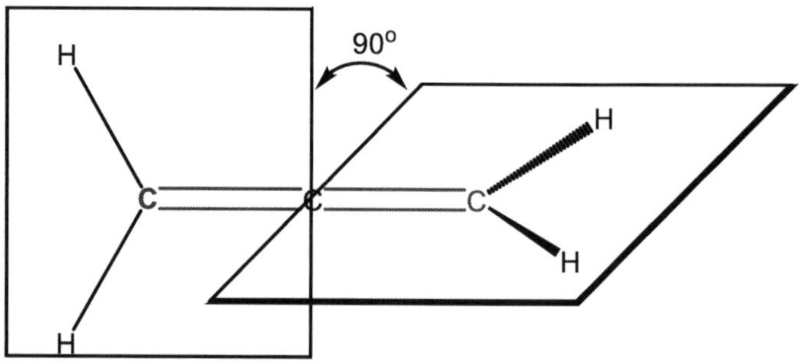

The diagram on the left shows the *p*-orbitals used to make the π-bonds (p_x for the left π bond and p_y for the right π-bond). The diagram on the right shows the geometry in a typical structural drawing.

When we measure the ΔH_h for such a C=C=C unit in 1,2-pentadiene, it is significantly higher than ΔH_h for 1,4-pentadiene (–69.8 vs. –60.8 kcal/mol, respectively), which indicates that **cumulated π-bonds are significantly less stable than isolated π-bonds**. This phenomenon is due to the fact that an *sp*-hybridized carbon is much more electronegative than an sp^2-hybridized carbon, so a C=C bond between an *sp*-hybridized carbon and an sp^2-hybridized carbon is more electron-deficient than a C=C bond between two sp^2-hybridized carbons. Remember that alkene units are stabilized by inductive electron-donating groups like alkyl groups (sp^3 hybridized C atoms).

<u>Example IV.1.1</u>

Which of the following alkenes would have the higher heat of hydrogenation?

<u>Solution IV.1.1</u>

In the molecule on the left (1,4-cyclohexadiene), the two π-bonds are isolated, but in the molecule on the right (1,3-cyclohexadiene), the two π-bonds are conjugated. When π-bonds are isolated from each other by one or more sp^3-hybridized carbons, they cannot interact with each other and thus cannot be stabilized or destabilized by each other. However, when π-bonds are conjugated with each other, the π-electrons are able to delocalize over a greater number of positions, which increases stability. As a result, 1,3-cyclohexadiene will be more stable than 1,4-cyclohexadiene, which means that 1,3-cyclohexadiene will release less energy upon hydrogenation of its C=C bonds. Thus, a higher heat of hydrogenation will be observed with 1,4-cyclohexadiene.

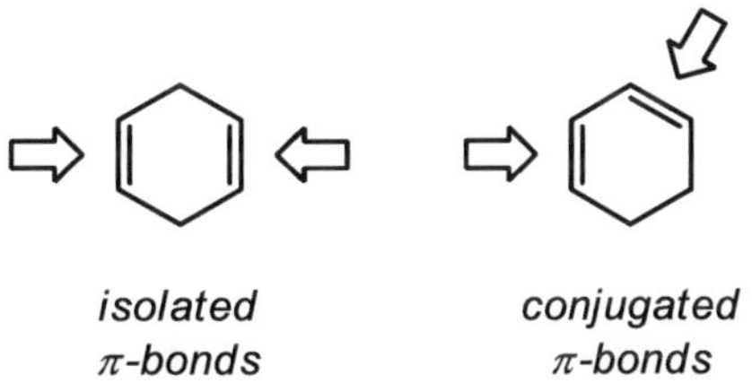

Lesson VI.2.1 Introduction to Addition Reactions of Conjugated Dienes

We saw in Lesson III.5 that reaction of one equivalent of HX (X = Cl, Br or I) to a C=C bond will produce the more substituted alkyl halide as the major product (Markovnikov's Rule). However, two products are possible when one equivalent of HX reacts with a 1,3-diene:

Electrophilic Additions of HX:

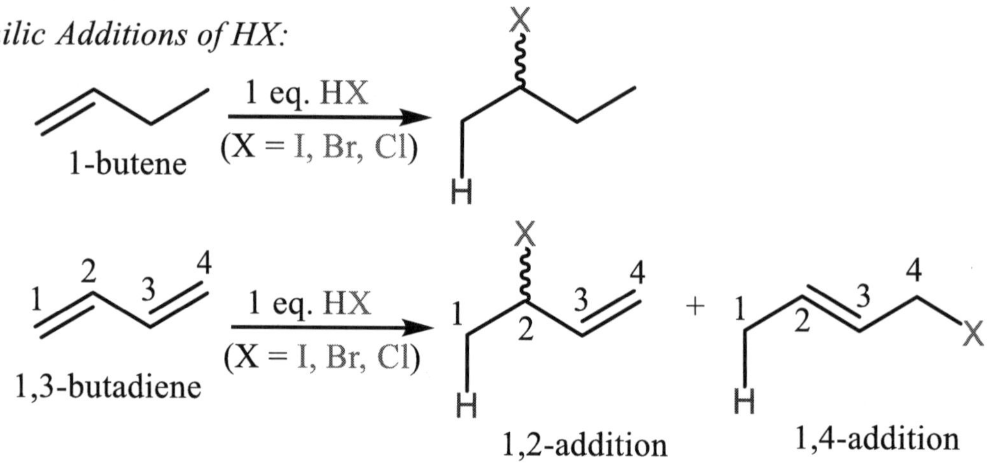

The **1,2-product** (also called the **direct addition** product) and **1,4-product** (also called the **conjugate addition** product) are possible because the carbocation formed by electrophilic addition of the proton to the C1 end of the C=C bond leads to a resonance-stabilized carbocation having δ+ character on both C2 and C4.

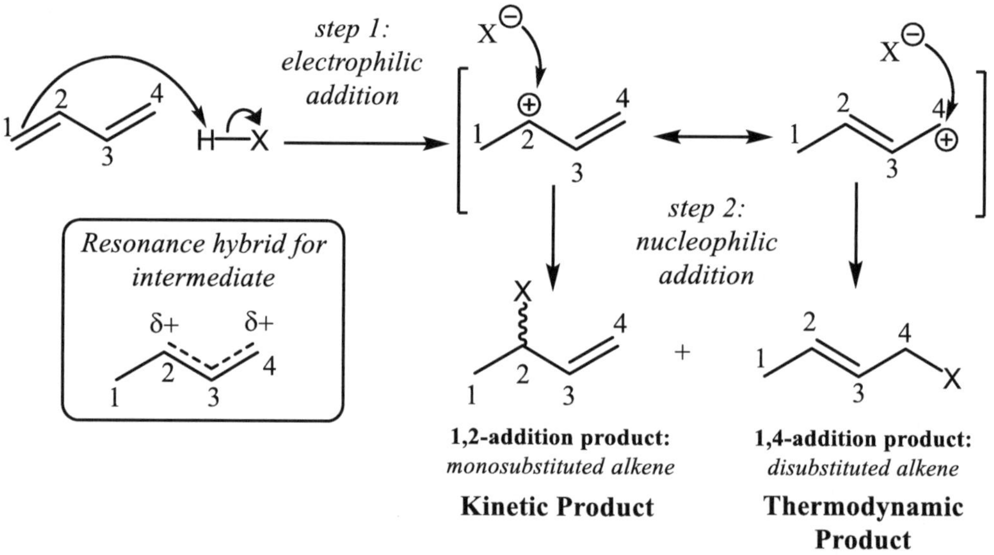

The major product isolated from these reactions depends on the conditions of the reaction.

Both 1,2- and 1,4-addition products have the first mechanistic step: electrophilic addition of the proton to one of the C=C bonds. The second step of the 1,2-addition is coordination of the bromide to C2, whereas the second step of 1,4-addition is coordination of bromide to C4. In the case of addition of HX to 1,3-butadiene, the product of 1,2-addition is a monosubstituted alkene, whereas the product of 1,4-addition is a disubstituted alkene. The 1,4-addition produces the more stable alkene (the **thermodynamic product**). The 1,2-addition is faster (lower energy of activation, $E_{a1,2}$ in the figure below), however, because the bromide is closer to C2 when it is produced in the electrophilic addition step. This is known as a **proximity effect**, and the result of this is that the 1,2-addition product is formed faster – it is the **kinetic product**. The qualitative reaction coordinate diagram below illustrates the energetics of these processes:

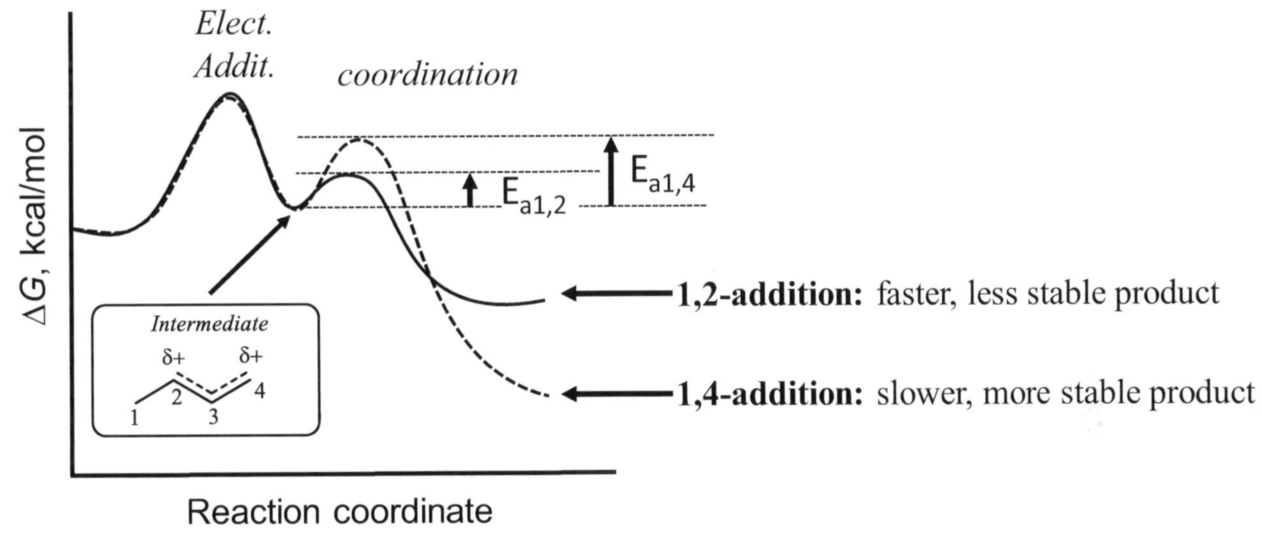

The 1,2-addition is faster, but the 1,4-addition produces the more stable product in this case. So, how to we determine which is the major product? A fundamental property of chemical reactions is that **the major product of irreversible reactions is the kinetic product** (the one formed at a higher rate). For a reaction **at equilibrium (reversible), the major product is the thermodynamic product** (the more stable product). We can control which product is the major product by controlling the reaction conditions. **At low temperature (below ~20 °C for these reactions) the reaction is irreversible** because there is not enough energy to overcome the larger energy barrier for the reverse reaction, so the kinetic product is the major product, and the reaction is said to be under **kinetic control**.

At higher temperature (above ~50 °C for these reactions), the reaction is reversible because there is enough energy to overcome the energy barrier to the reverse reaction, so the thermodynamic product is the major product, and the reaction is said to be under **thermodynamic control**.

The 1,2-product is always the kinetic product due to the proximity effect. The thermodynamic product is always the more stable alkene, which may be either the 1,2- or 1,4-addition product depending on the substitution pattern. For example, consider addition of HBr to 2,5-dimethyl-2,4-hexadiene:

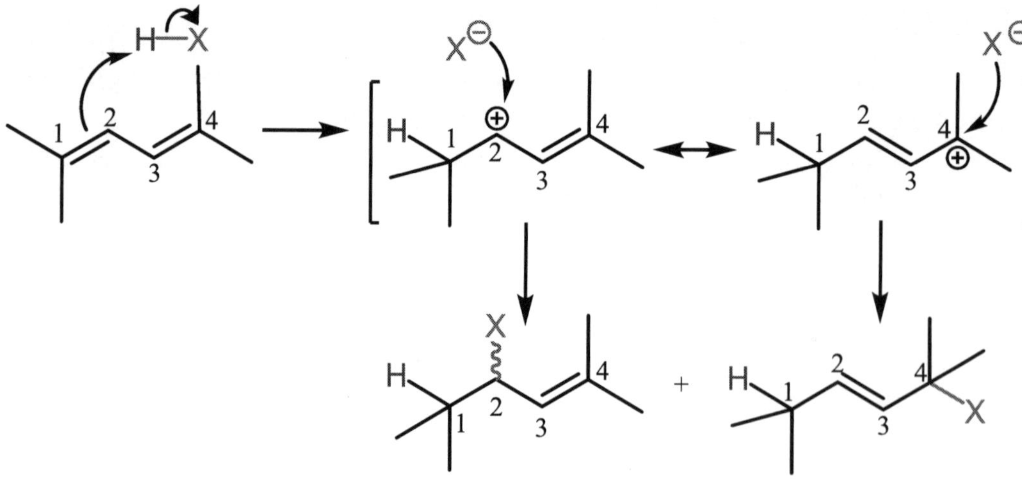

1,2-addition product:
trisubstituted alkene
Kinetic product
<u>and</u>
Thermodynamic product

1,4-addition product:
disubstituted alkene

In this case, the 1,2-addition product is *both* the kinetic product *and* the thermodynamic product.

<u>To approach these problems effectively, follow this procedure:</u>

1. Number the π-conjugated part 1–4.
2. Add H^+ to carbon #1 of the conjugated segment
3. Draw both resonance contributors for the cation you get
4. Place the X^- on each cation to get your final products
5. The 1,2-product is the kinetic product
6. The most substituted alkene is the thermodynamic product

<u>Example IV.2.1</u>

Provide the major product for each reaction. Label each reaction as favoring either the kinetic or thermodynamic product.

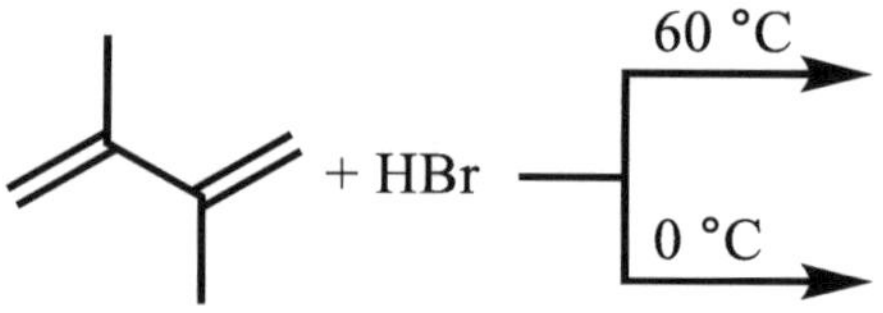

Solution IV.2.1

First, we consider what the 1,2- and 1,4-addition products would be:

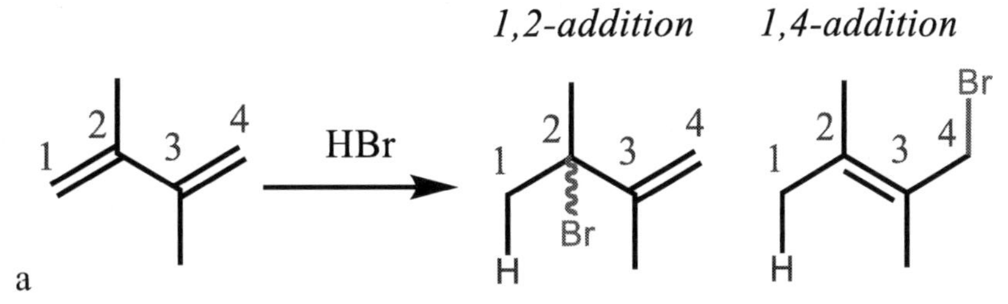

At 60 °C, the thermodynamic product is favored. This will be the most stable alkene. In this case, that is the 1,4-addition product, a tetrasubstituted alkene.

At 0 °C, the kinetic product is favored. This will always be the 1,2-addition product, due to the proximity effect.

IV.3.1. Introduction to the Diels-Alder Reaction

The Diels-Alder reaction is a [4 + 2] **cycloaddition reaction** of a diene and an alkene to form a cyclohexene ring. The Diels-Alder reaction is one of the most widely-used reactions in synthesis. The reaction takes place between a diene, which acts as a nucleophile and an alkene which acts as an electrophile (often referred to in this reaction as a **dienophile**). For the most effective Diels-Alder reactions, the diene is electron-rich and the dienophile is electron poor. The scheme below shows the general mechanism of the reaction:

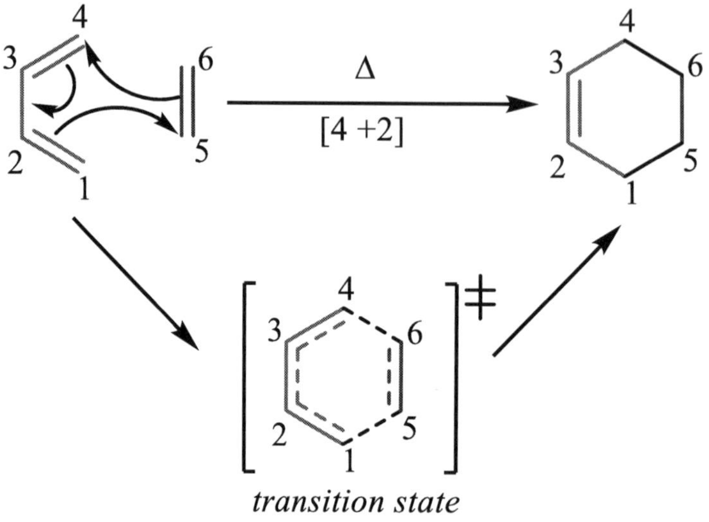

The Diels-Alder reaction is an example of a **pericyclic reaction**: a concerted, cyclic flow of π electrons within a cyclic transition state structure. The π bond between carbons 1 and 2 breaks and these electrons form a new σ bond between carbons 1 and 5, causing the π bond between carbons 5 and 6 to flow into a σ-bonding orbital between carbons 6 and 4, which in turn breaks the π bond between carbons 4 and 3, pushing those electrons to make a π bond between carbons 3 and 2.

The diene of the Diels-Alder reaction has to be a s-cis configuration for the cycloaddition to take place in a concerted mechanism. Even if the diene in the starting material is in the thermodynamically more stable configuration (s-trans), the added heat to the reaction will promote the rotation of the C—C σ bond to form the less stable configuration for the reaction to take place as the figure below shows.

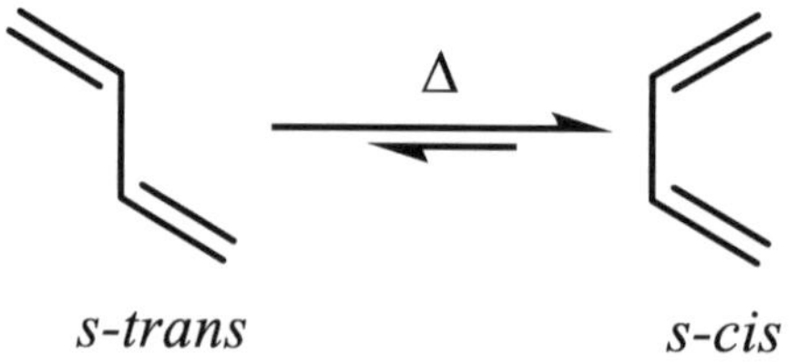

The Diels-Alder reaction proceeds best when the dienophile has an electron-withdrawing group (an electronegative atom, or an atom with a δ+ on it) attached to carbon 6. The presence of an electron-withdrawing group such as carbonyl groups (C=O), nitro groups (–NO_2) or nitrile groups (–C≡N) creates a dipole through resonance, placing +δ and –δ charges on carbons 5 and 6 respectively:

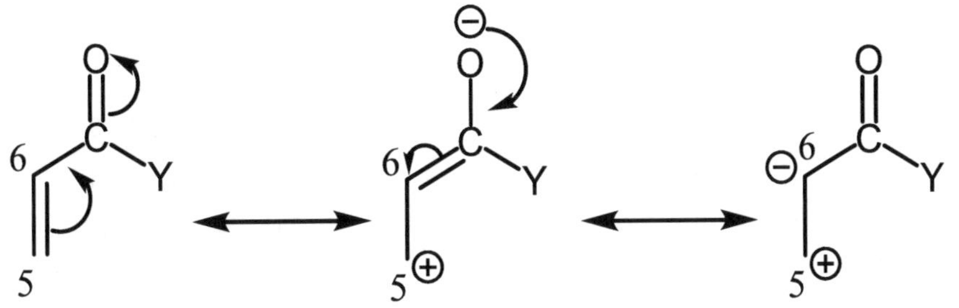

The Diels-Alder reaction is a stereoselective reaction: a cis- alkene (dienophile) will produce a cis-cyclohexene and a trans-alkene will produce a trans-cyclohexene as the following figure shows.

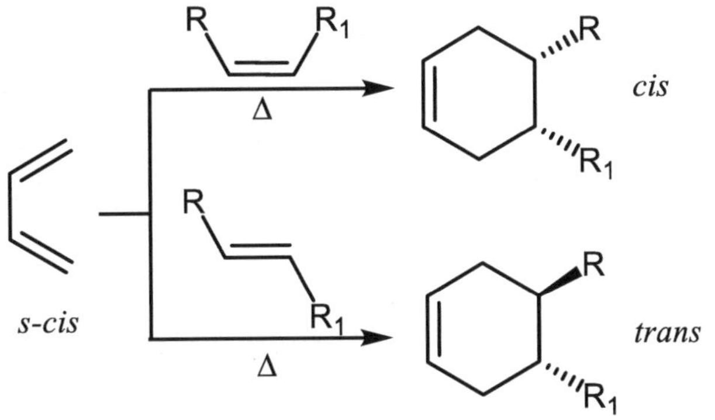

Example VI.3.1

Give the final product of the following Diels-Alder cycloaddition

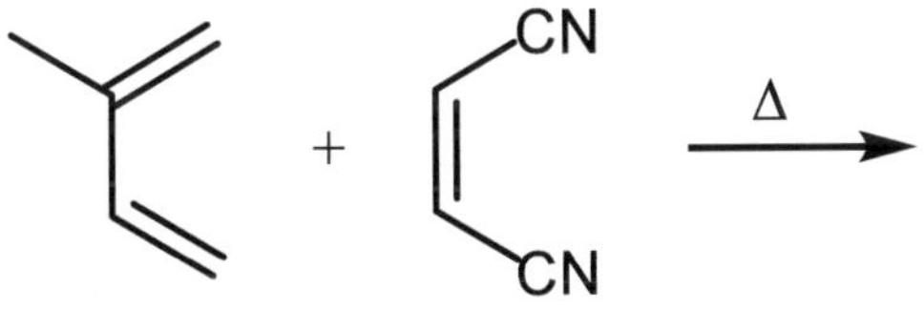

Solution VI.3.1

To give the product of this reaction, we number the reactants in a sequence of 1 to 4 on the diene and 5 and 6 on the dienophile to figure out where the new bonds are formed and the bonds are broken. Then we give the cyclohexene, the double bond is always formed between carbons 2 and 3. Then we add the substituents, the methyl group on carbon 3 and two –CN groups on carbons 5 and 6, since the alkene is a *cis* alkene then two nitrile groups are going to be *cis* in configuration. As we can see the product has two chiral centers (marked by asterisk), this will result in a racemic mixture of two enantiomers:

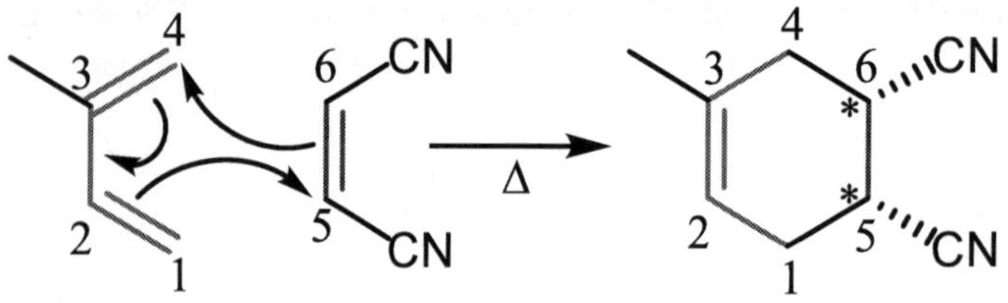

IV.3.2. The Endo- and Exo- Products

When you have a cyclic diene, you sometimes have a choice between two products that differ only in the position of the new π bond in the product relative to the substituent that came from the dienophile. In the endo product, the substituent points towards the same direction as the pi bond. In the exo product, the substituent points away from the π bond. The rule is that if the substituent R has a π bond in it itself, the endo product is the major product.

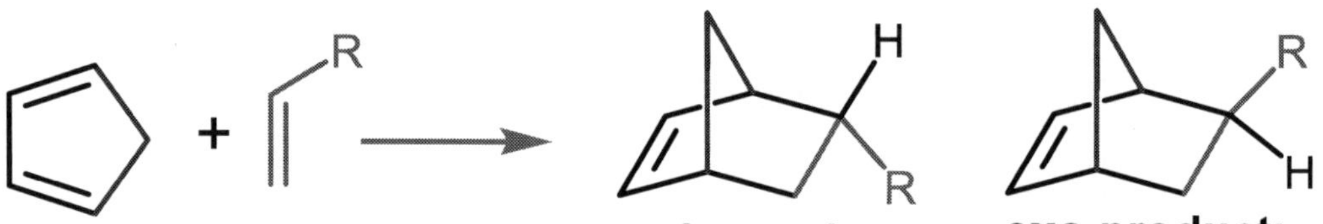

endo product: (major product) R is down and the pi bond in the product is down

exo product: R is up and the pi bond in the product is down

Bonus Lessons from Part VII: Methods for Determining the Structure of Organic Compounds

Lesson VII.1. Interaction of Ultraviolet and Visible Light with Molecules

Lesson VII.2. UV–Visible Spectroscopy

Lesson VII.3. Interaction of Infrared Light with Molecules

Lesson VII.4. Infrared Spectroscopy

Lesson VII.5. Introduction to Nuclear Magnetic Resonance

Lesson VII.6. Carbon-13 Nuclear Magnetic Resonance Spectrometry

Lesson VII.7. Proton Nuclear Magnetic Resonance Spectrometry

Lesson VII.1. Interaction of Ultraviolet and Visible Light with Molecules

Lesson VII.1.1 The Electromagnetic Spectrum

The electromagnetic spectrum stretches from small wavelength gamma rays to long wavelength radio waves. Different wavelengths of light interact with matter in different ways. The region of the electromagnetic spectrum with which we are most familiar is the visible region because light in this range is detectable by our unaided eyes. At longer wavelengths than visible light is **infrared (IR) light**, which we can feel as heat. At shorter wavelengths than visible light is **ultraviolet (UV) light**, which we (unintentionally) detect as sunburned skin and other damage to our tissues. In the next few lessons, we will learn how we can gain structural information about molecules by studying how they interact with electromagnetism in the UV, visible, IR, and radio frequency ranges of the electromagnetic spectrum. These regions of the electromagnetic spectrum are shown in terms of relative energies below:

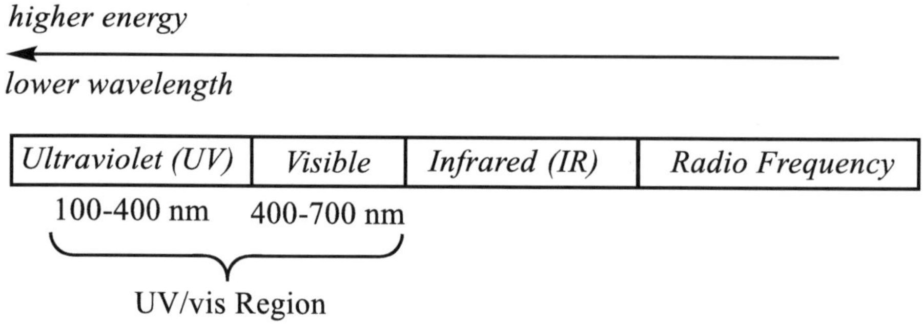

In this lesson, we will focus on how UV and visible light, collectively abbreviated **UV–visible (UV/vis) light**, interact with organic molecules. The UV/vis part of the spectrum we will consider spans a wavelength range from ~100–700 nm.

Lesson VII.1.2 UV and Visible Light Cause Electronic Transitions in Molecules

Each covalent bond in an organic compound consists of two electrons shared between the nuclei joined by the bond. When a molecule absorbs UV/vis radiation of an appropriate energy, it causes one of the electrons to undergo an **electronic transition** to a higher-energy orbital that. In terms of energy, the **highest occupied molecular orbital** (abbreviated HOMO) is commonly the orbital holding the electron that is promoted to higher energy upon absorption of light energy. The **lowest unoccupied molecular orbital** is abbreviated LUMO. Here, "occupied" and "unoccupied" refer to whether or not the orbitals contain electrons in them before energy absorption. Energy absorption (in the form of a photon) can promote an electron from the HOMO to the LUMO. The orbital that contains a pair of bonding electrons is called a **bonding orbital**. If the bonding orbital holds electrons in a σ-bond, the orbital is given the symbol σ, whereas if the bonding orbital holds electrons in a π-bond, it is given the symbol π. Upon absorption of an appropriate energy photon of UV/vis light, a σ-bonding electron will generally be promoted to a σ-**antibonding orbital**, given the symbol σ*:

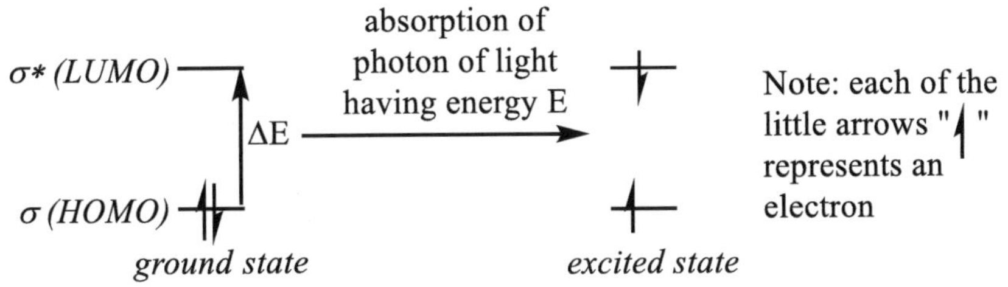

This type of transition is called a **σ→σ* transition** (read "sigma to sigma star transition"). Similarly, a π-bonding electron generally gets promoted to a π-antibonding orbital, given the symbol π*:

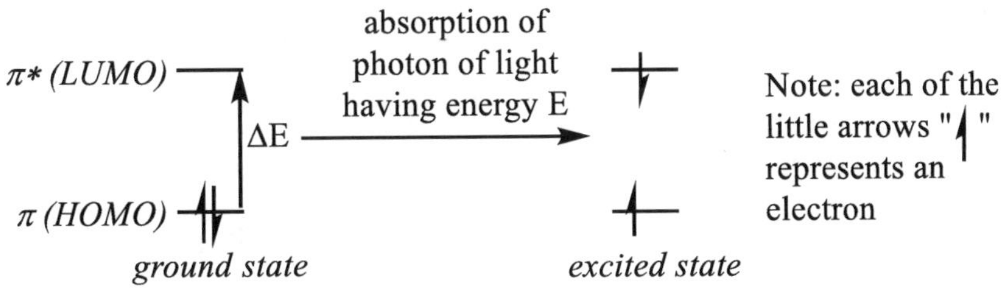

This type of transition is called a **π→π* transition** (read "pi to pi star transition").

Regardless of it is a σ→σ* transition or a π→π* transition, the energy of the photon must be identical to the energy gap between the electronic energy levels (ΔE in the figures above) for the photon to be absorbed and cause the electronic transition to occur. Longer wavelength light – towards the visible end of the UV/vis range – can only cause transitions requiring less energy. The lower energy transitions in organic molecules are usually π→π* transitions. By measuring the wavelength of UV/vis light that is absorbed by the sample, some knowledge about the bonding in the sample molecule can be inferred. The types of information we can gain by measuring absorption of UV-vis light are detailed in Lesson VII.2.

Example VII.1.1

Which of these molecules will have a lower-energy electronic transition? What type of transition would the HOMO–LUMO transition represent in each?

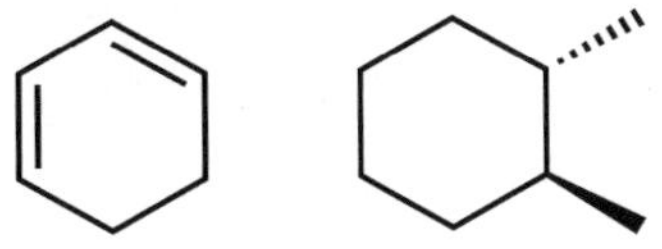

I **II**

Solution VII.1.1

The lower-energy transitions in organic molecules tend to be π–π* transitions, so molecule **I**, which has a π-bond, would be expected to have a lower energy (longer wavelength) transition than would molecule **II**. The transition observed in the UV-vis spectrum of molecule **I** will be a π–π* transition, while the transition observed for molecule **II** would be a σ–σ* transition.

Lesson VII.1.3 Comparing σ→σ to π→π* Transitions, and the Effect of π-Conjugation*

Because σ-bonding electrons lie nearer the nuclei than π-bonding electrons, they are held more tightly by Coulombic attraction to the nucleus than are π-bonding electrons. It therefore takes more energy (ΔE_2 in the figure below) to pull an electron out of a σ orbital to promote it to a σ*-orbital than it does to promote a π-bonding electron to a π*-orbital (requiring ΔE_1 in the figure below). A qualitative diagram showing both the σ→σ* and the π→π* transitions in a molecule that has both types of bonds will consequently look like this:

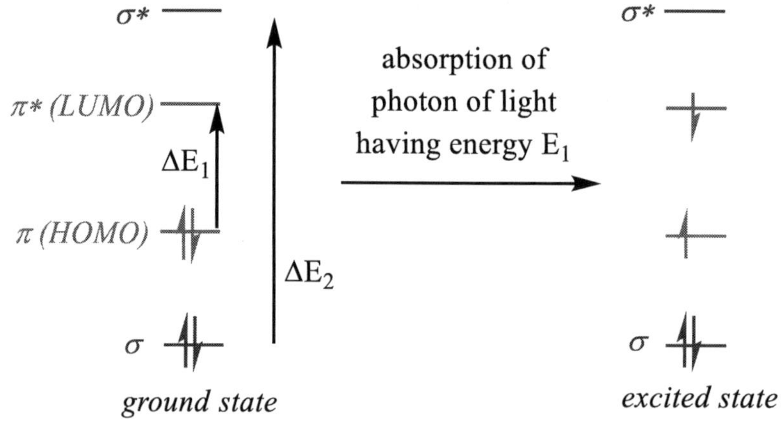

What if a molecule has more than one π-bond? The answer depends on whether the π-bonds in the molecule are isolated or conjugated (see Lesson IV.1). If the two π-bonds are isolated, they will not interact with one another, so the π-orbitals lie at similar energies. This results in an electronic transition at about the same energy as if there were only one π-bond. However, two photons will be needed per molecule to cause both to promote both the π→π* transitions (i.e., a molecule with two isolated π-bonds will absorb twice as much light as a molecule with only one π bond). The amount of light absorbed per mole of a sample is called the **molar absorptivity** or **molar extinction coefficient**.

We also know that conjugated π-bonds are more stable than isolated π-bonds. It takes less energy to cause an electronic transition of an electron in a conjugated π-system than it does to cause the transition in an isolated π-bond. One reason for this is that the single electron that is left in the bonding orbital – which we can think of as a radical – has resonance stabilization, so it takes less energy to form than it would take to form the radical resulting from an isolated π-bond that lacks resonance stabilization. In fact, **the longer the π conjugated system, the lower the energy of the photon needed to promote the π→π* transition.** The energy of UV-vis light absorbed and the number of photons absorbed at that energy allow us to assess the relative number of π-bonds in a compound and the extent to which these π-bonds are conjugated.

Example VII.1.2

Which of these molecules will have the lowest-energy electronic transition? Which would have the highest molar absorptivity?

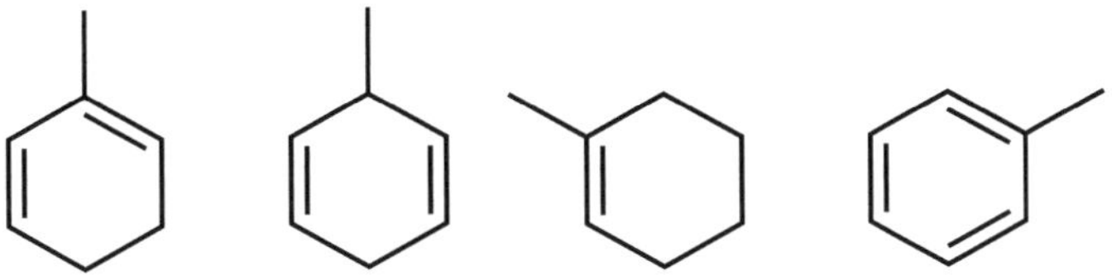

I **II** **III** **IV**

Solution VII.1.2

The lowest-energy transition will occur for the molecule with the most-extended π-conjugated system. In molecules **II** and **III** have only isolated alkenes. Molecule **I** has a π-conjugated system of two π-bonds, whereas molecule **IV** features a π-conjugated system of three π bonds. This analysis suggests that molecule **IV** will have the lowest-energy (highest wavelength) band in its UV-vis spectrum. Molecule **IV** is also expected to have a higher molar absorptivity because it has the greatest number of π bonds per molecule as well.

Lesson VII.2. UV–Visible Spectroscopy

Lesson VII.2.1 The UV/vis Spectrum

In Lesson VII.1, we learned that an electronic transition occurs when a molecule absorbs UV-vis light of an appropriate energy. In the current lesson, we will learn how scientists have developed a technique, called **UV/vis spectroscopy**, that correlates the absorption of light at a given wavelength with molecular properties, intermolecular forces, and chemical reactions. The operating principle of a typical UV-vis spectrometer is illustrated as follows:

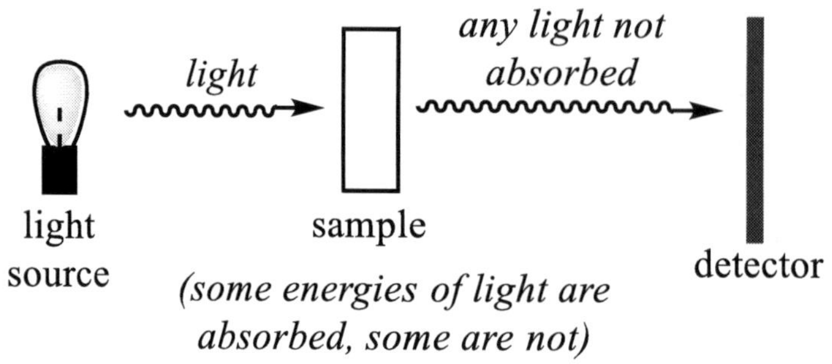

The light source generates a beam of light, which is directed through the sample. The source provides one specific wavelength at a time, scanning over the range of wavelengths that is set by the experimentalist. The detector, located on the opposite side of the sample relative to the source, detects the intensity of the light that passes through the sample relative to the amount of light emitted by the source. If the sample absorbs light at a particular wavelength, the detector will detect a decrease in light coming through at that wavelength. The computer system then provides the user with a plot of how much light was absorbed versus the wavelength at which it was absorbed. A sample that has electrons in bonds that undergo an electronic transition at 220 nm, for example, might produce a UV-vis spectrum like this:

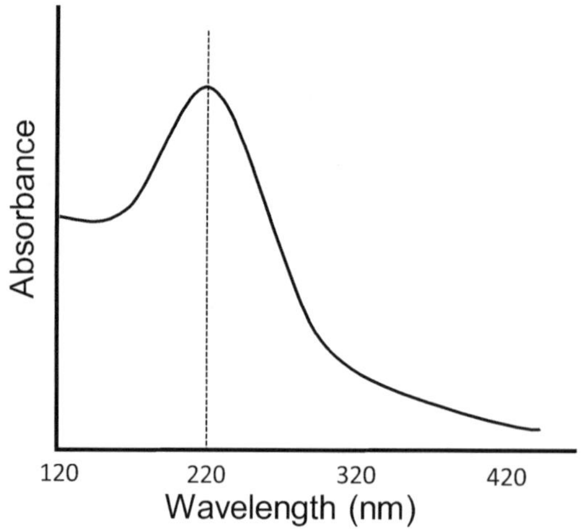

A few things should be noted at this point. If we increase the **concentration** (c) of the compound in the sample cell that is absorbing light at 220 nm, then more light will be absorbed. If we increase the size of the sample cell (the **pathlength**, b), then then more light will be absorbed. For a certain concentration and a certain pathlength, the amount of light absorbed will be a constant for a given electronic transition. This constant that relates the absorbance to the concentration and the pathlength is called the **absorptivity**, or the **extinction coefficient**, and it is given the symbol ε. The **Beer-Lambert Law** provides an equation relating the absorbance (A), pathlength (b), concentration (c) and extinction coefficient (ε):

$$A = \varepsilon bc$$

The UV-vis spectra shown below illustrate the Beer-Lambert Law:

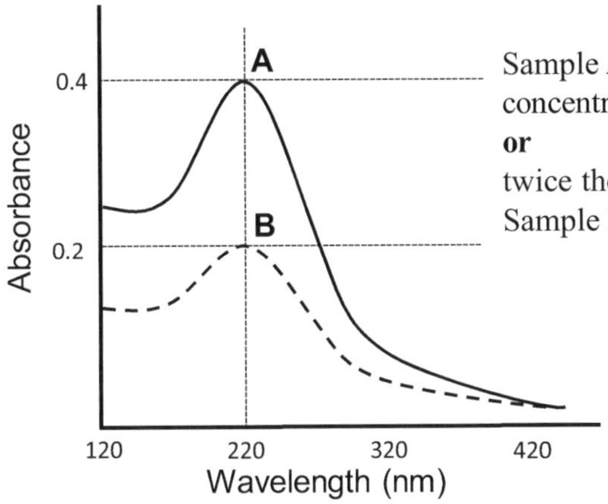

Sample A has twice the concentration
or
twice the pathlength of Sample B

Usually, a sample cell with a pathlength of 1 cm is used, which simplifies the relationship between absorbance and concentration. The well-defined way that absorbance relates to concentration for a given pathlength allows us to measure the concentration of a molecule as a function of time with very high accuracy and sensitivity. As a result, **UV-vis spectroscopy is an excellent way to quantify reaction rates**. Let us consider, as an example, a sample that contains 1 mole each of reactants **A** and **B**, where **B** is the only species that absorbs at 500 nm, and you experimentally measure an absorbance of 0.6. After 1 h has elapsed, you measure the absorbance at 500 nm and now obtain a value of 0.3, which allow you to deduce that the reaction between **A** and **B** has proceeded at a rate of 0.5 moles h^{-1} during this time period.

Lesson VII.3. Interaction of Infrared Light with Molecules

Lesson VII.3.1 Infrared Radiation Causes Vibrational Transitions in Molecules

In Lessons VII.1–2, we saw how UV-vis light interacted with organic molecules to cause electronic transitions. In this lesson, we will learn what happens when lower-energy IR light interacts with organic molecules. Unlike UV-vis light, infrared light does not have enough energy to cause an electron to be promoted to a higher electronic level in the majority of organic molecules. Instead, molecules undergo **vibration** upon interaction with IR light. The wavelength of light that is most commonly used to study the vibrational modes of organic molecule runs from about 2500-20,000 nm. There are several modes of molecular vibration that result when a molecule absorbs infrared energy. Molecular vibrational modes can be broadly divided into **stretching** and **bending** modes. Stretching modes involve changes in the bond lengths. Bending modes involve changes in bond angles. Bond stretching and bond bending in molecules occurs in well-defined combinations at specific energies. Let us begin by looking at the stretching modes. Consider how the C–H bonds in a CH_2 unit might undergo stretching in a molecule:

Symmetric stretch: bonds lengthen and shorten at the same time

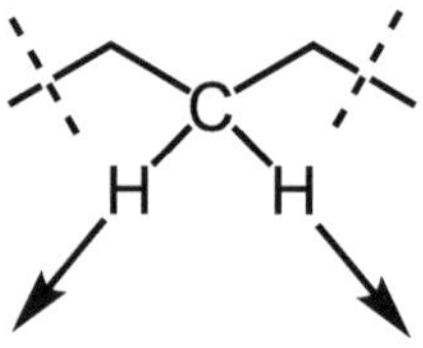

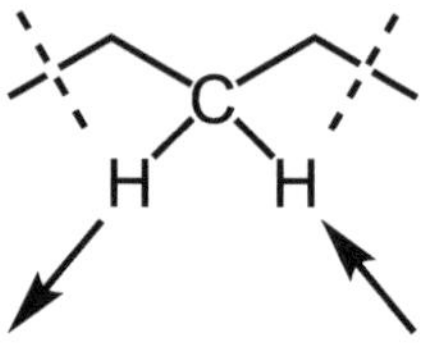

Asymmetric stretch: One bond lengthens as the other shortens

There are more possibilities for bending modes. Not only may a given bending mode be symmetric or asymmetric, but the different bending modes may involve the atoms remaining coplanar as they bend, or they may bend in a way that they are no longer coplanar:

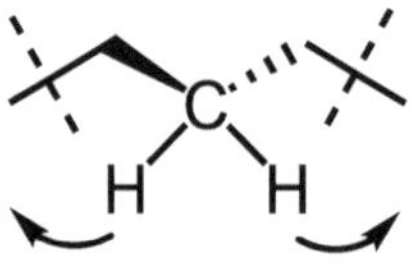

Symmetric in-plane bending: Both angles increase or decrease at the same time by bending in the plane of the page

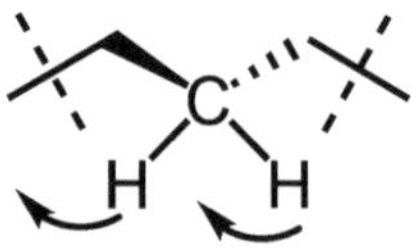

Asymmetric in-plane bending: One angle increases as the other decreases by bending in the plane of the page

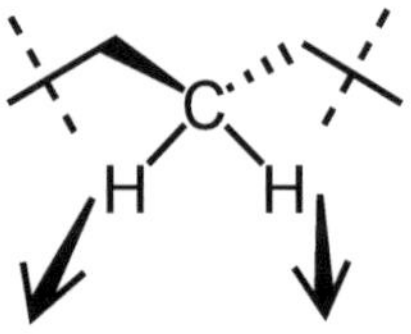

Symmetric out-of-plane bending: Both angles increase or decrease at the same time by bending out of the plane of page (here, both shown bending towards the viewer)

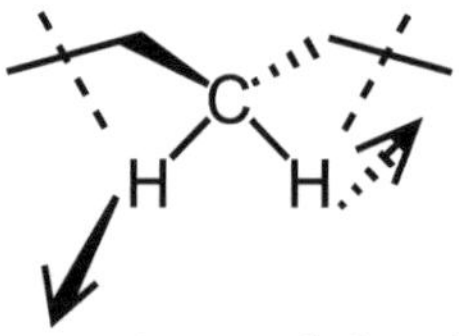

Asymmetric out-of-plane bending: One angle increases as the other decreases by bending out of the plane of page (here, one bending towards, one away from the viewer)

Lesson VII.3.2 Different Types of Bonds Absorb IR Radiation at Different Energies

Bonds can be thought of as springs. The strength of a spring is reflected in how much energy it takes to stretch and compress it. Likewise, the energy at which a bond stretches depends on the strength of the bond. A stronger bond requires more energy to stretch than does a weaker bond. This means that a set of strong bonds absorbs a higher energy of IR light than does a set of weaker bonds. This means that a vibrational mode involving a set of strong bonds will absorb higher energy light than a set of weaker bonds. This phenomenon is invaluable for determining the chemical identity of bonds present in a given sample based on what energies of IR light the sample absorbs.

In addition to the energy of the IR light absorbed, the *amount* of light absorbed per molecule can also provide us with information about the bonds present. Due to some considerations that are outside the scope of this course, one result of how light and charged species interact is that the more polar bonds in a sample absorb more of the IR energy to which the sample is exposed than do less polar bonds. The next lesson will illustrate how the interaction of a molecule with IR light can be used to infer structural information about that molecule.

Lesson VII.4.1 The IR Spectrum

In Lesson VII.3, we learned that a molecule absorbs IR light of an appropriate energy to excite vibrational modes for specific sets of bonds in that molecule. In the current lesson, we will learn how scientists have developed a technique that allows us to use this property of matter to gain a wide range of knowledge about molecules and reactions. The infrared spectrometer instrument setup is very similar to the UV-vis spectrometer we saw in Lesson VII.2, but using IR radiation in place of UV-vis wavelength light:

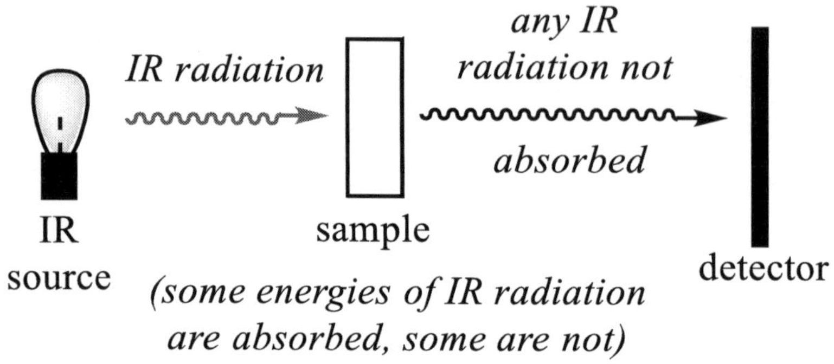

A typical IR spectrum provided by the IR spectrophotometer is a plot of transmittance (the % of light that is **not absorbed** by the sample) versus the energy of light (in units of **wavenumbers**, cm^{-1}). A higher wavenumber value corresponds to higher energy. If a sample absorbs all of the IR radiation from the IR source, the transmittance would be zero and a downward peak would extend all the way to the bottom of the spectrum:

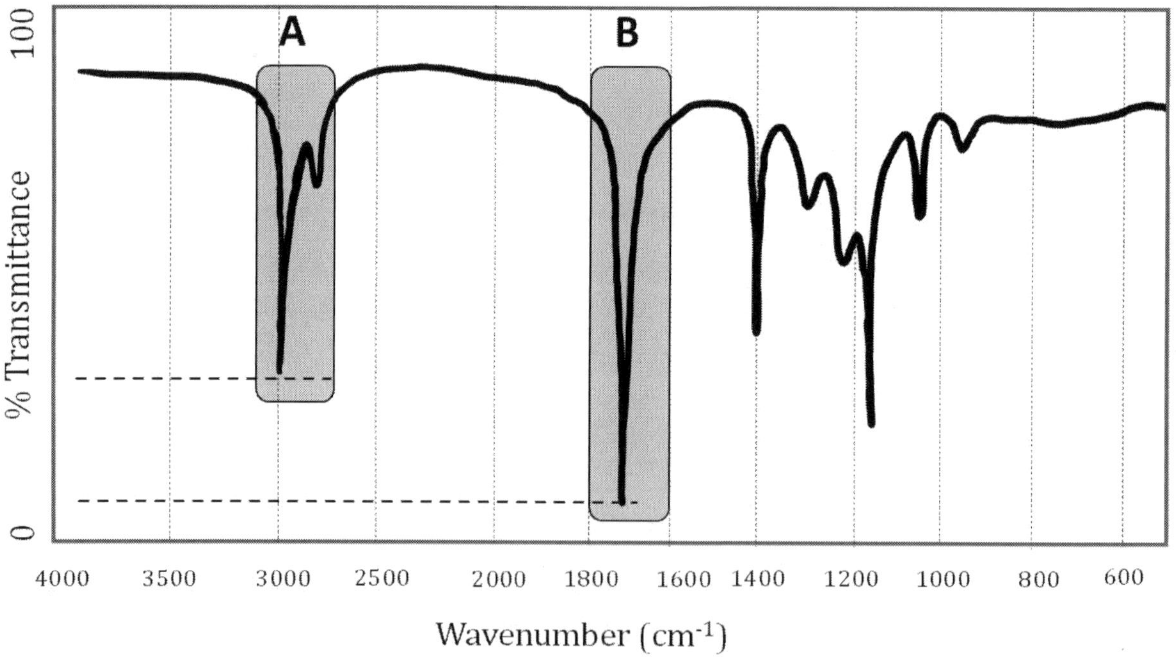

The above spectrum shows the typical *x*-axis range of 4000–500 cm^{-1} (high energy to low energy). Two bands are highlighted. Band **A** corresponds to a set of bonds that absorbs ~75% of the IR light at 3000 cm^{-1} emitted by the source (i.e., 25% transmittance), whereas band **B** corresponds to a set of bonds that absorbs nearly 95% of the IR light at 1750 cm^{-1} (i.e., 5% transmittance).

Lesson VII.4.2 Information Provided by the IR Spectrum

The infrared spectrum of a compound can provide information about the types of bonds, and therefore the functional groups, that are present in a given compound. IR spectra are usually less cluttered at the higher-energy region of the IR spectrum (4000–1500 cm^{-1}), which is referred to as the **functional group region**. The lower-energy region of the IR spectrum of the spectrum (1500–500 cm^{-1}) is referred to as the **fingerprint region**, because the peaks there are characteristic to each given molecule. However, the fingerprint region is significantly more cluttered and is thus more difficult to assign specific transitions. The characteristic peaks for the common functional groups encountered in an undergraduate organic chemistry course are provided below:

Bond	Energy (cm^{-1})	Intensity
N≡C	2255-2220	m-s
C≡C	2260-2100	w-m
C=C	1675-1660	m
N=C	1650-1550	m
	1600 **AND** 1500-1425	w-s
C=O	1775-1650	s
C—O	1250-1000	s
C—N	1230-1000	m
O—H	3650-3200	s (br)
O—H	3300-2500	s (br)
N—H	3500-3300	m (br)
C—H	3300-2725	m

C-H Bond (Stretch)	Energy (cm^{-1})
C≡C—H	3300-ish
C=C—H	3100-3000
C—C—H	2950-2850
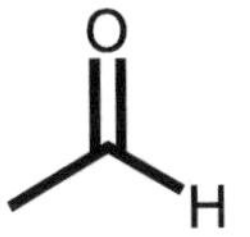	2820-ish and 2720-ish

C-H Bond (Bending)

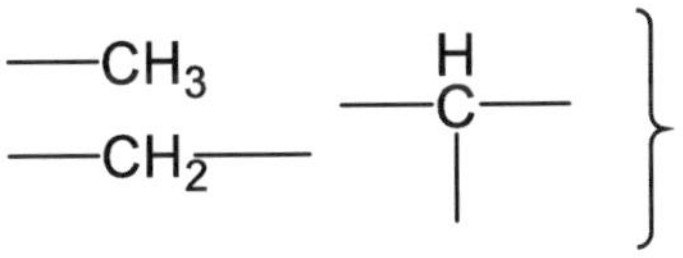

1450-1400

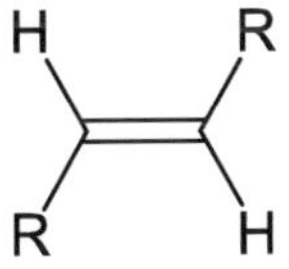

980-960
trans

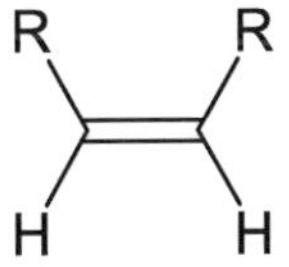

730-670
cis

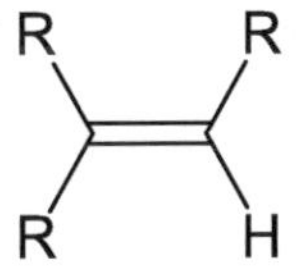

840-800
trisubstituted

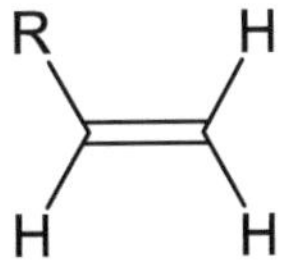

990 and 910
monosubstituted

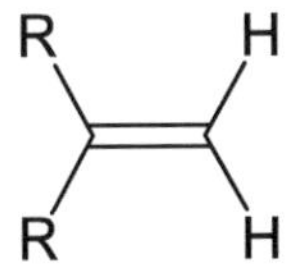

890
disubstituted terminal

Lesson VII.4.3 Interpreting IR Spectra for Simple Molecules

Knowing the energy at which each of the common organic functional groups absorbs in the IR spectrum, we are now able to look at an IR spectrum and identify what specific functional groups are present or absent. The following examples demonstrate the method by which one deduces the possible structure of a compound from its IR spectrum. Numerous examples to illustrate how IR spectra can be used to gain structural information are provided in Part VIII.

Lesson VII.4.4 Influence of Resonance and Pi-Conjugation on IR Band Energies

We know that the resonance hybrid structure is a better representation of the "real" structure of a molecule. The resonance hybrid reflects the fact that the "real" bond order in the hybrid structure is different from that in any of the individual resonance contributors. Consider, for example, the resonance hybrid for a carboxamide that takes into account the two contributors:

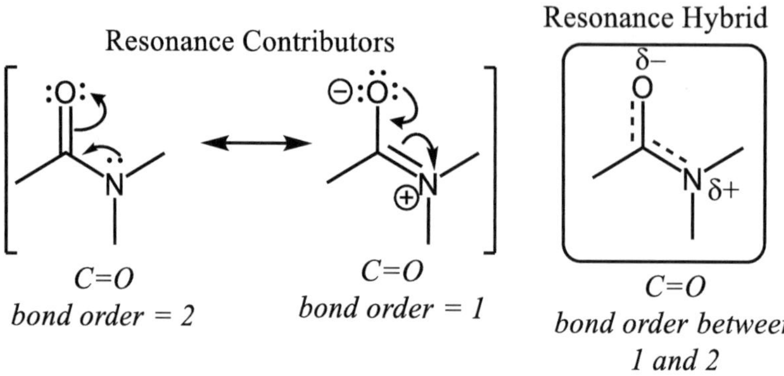

One result of the resonance is that the actual C–O bond order is < 2. For this reason, the C=O bond in a carboxamide is weaker than a C=O bond that does not participate in resonance delocalization, and will thus have an IR band at a lower wavenumber (~1680 cm^{-1}). As a rule of thumb, a C=O bond that can engage in resonance delocalization with an adjacent C=C bond will have a peak at ~20 cm^{-1} lower in energy than a C=O bond that does not participate in resonance. Resonance effects can also be observed in the IR absorptions by C=C bonds. Notice in the table of IR absorptions in Lesson VII.4.2 that a C=C bond in a typical alkene absorbs in the 1660–1675 cm^{-1} range, whereas a C=C bond in an arene absorbs at energies below 1600 cm^{-1}. This reflects the resonance delocalization of π-bonds in arene rings, whose highest C–C bond order is 1.5, rather than the 2 for a typical alkene.

Lesson VII.5. Introduction to Nuclear Magnetic Resonance

Lesson VII.5.1 Behavior of Nuclei in a Magnetic Field

Although chemistry primarily focuses on electrons, the behavior of some nuclei can provide useful information about the bonding within a molecule. This information can be obtained due to the fact that some nuclei have *magnetic* dipoles (not to be confused with dipoles in polar molecules). Not all elements and not all isotopes of a single element have a nuclear magnetic dipole, but fortunately many of the elements present in common organic molecules do have isotopes whose nuclei have magnetic dipoles: hydrogen-1 (^{1}H), carbon-13 (^{13}C), phosphorus-31 (^{31}P), and fluorine-19 (^{19}F). Nuclei such as these that have magnetic dipoles are called **NMR active** nuclei and can be observed by **Nuclear Magnetic Resonance (NMR) spectrometry**. Under typical conditions, the magnetic dipole vectors for the nuclei in a sample will be oriented in random directions. If an external magnetic field is applied, however, these magnetic vectors will align such that they are pointing in either the same direction (parallel) or opposite direction (antiparallel) of the applied field:

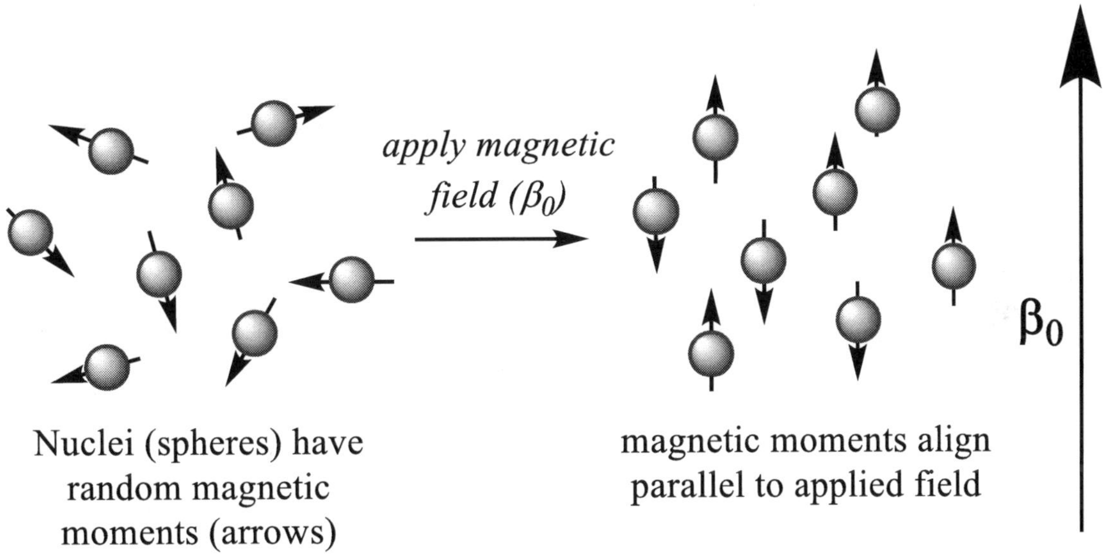

Nuclei that are aligned parallel with a magnetic field will have a lower energy than nuclei aligned antiparallel. The stronger the applied field, the greater the energy difference (ΔE) between these two spin states. In the magnetic field strength used in NMR spectroscopy, the energy difference between these two spin states is comparable to the energy of a photon in the radio frequency range (usually 60–1000 MHz). When **radio frequency (RF) radiation** is transmitted into the sample at an energy that matches ΔE, these photons will be absorbed and will promote nuclei from the lower-energy parallel spin state to the higher-energy antiparallel spin state:

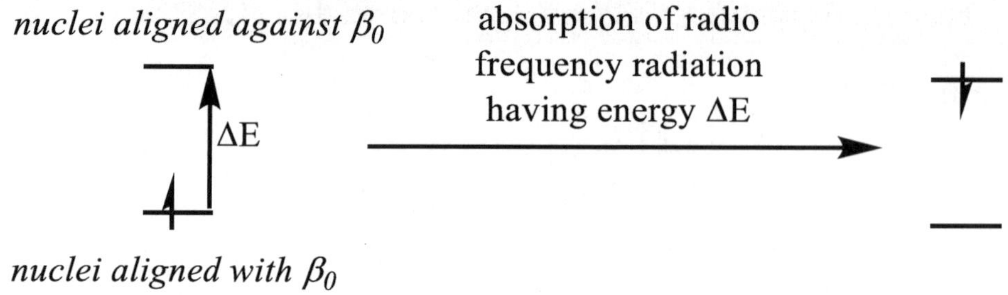

The detector in an NMR spectrometer measures the absorption of the RF energy and can thus report the energy between states for each nucleus in a sample.

Lesson VII.5.2 The NMR Spectrum

An NMR spectrometer subjects a sample to a strong magnetic field. A coil around the sample detects absorption of RF photons applied by the instrument. Each instance of the absorption of RF radiation is detected and reported to a computer, whereby this energy absorption is represented as an upwards-pointing peak (also known as a **signal**). The particular magnetic field and radio frequency range required depend on the specific type of nucleus that is being detected. A ^{1}H NMR spectrum will have an appearance like this:

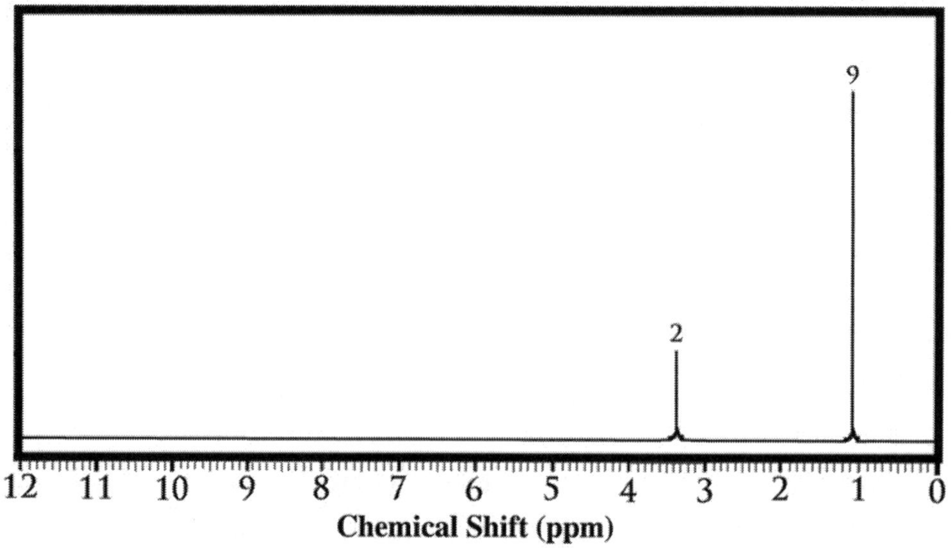

This example features two signals: one at 1.1 ppm and one at 3.3 ppm on the *x*-axis. The units of the *x*-axis are **parts per million (ppm)** and correspond to **chemical shift**, which correlates with energy. In the rest of this Lesson, we will discuss how the chemical shifts of the different nuclei are influenced by (1) the other nuclei and (2) the electrons in the sample. In the few lessons that follow this one, we will study how NMR spectrometry (^{1}H and ^{13}C NMR in particular) can be used to elucidate organic molecule structure.

Lesson VII.5.3 Factors Influencing the Energy Between Spin States

Not all nuclei in a molecule exhibit the same ΔE between spin states, because (1) the electrons in a molecule affect how the various nuclei "experience" β_0 and (2) the electronic environment varies for each nucleus according to its chemical environment. For example, there will be more electron density near a nucleus at the δ^- end of a polar bond than at the δ^+ end. How does this influence the effective magnetic field strength at a given nucleus? Well, not only does β_0 applied by the NMR spectrometer interact with the magnetic spin of nuclei, it also interacts with the magnetic spin of electrons. As a result, the electrons generate their own localized magnetic field that opposes β_0. Thus, as the electron density that surrounds a nucleus increases, the opposing magnetic field those electrons create also increases, which decreases the β_0 that particular nucleus experiences, and consequently it takes less energy to promote that nucleus to a higher spin state. As a result, **electrons shield the nuclei from the applied magnetic field.**

The NMR spectrometry experiment operates on a timescale slower than the rate of σ-bond rotation and slower than the rate of molecular diffusion or rotation. This means that all nuclei that interconvert by free bond rotation or that can be interconverted by molecular rotation are **magnetically equivalent** and will be promoted at identical energies. On the other hand, nuclei that cannot be interconverted by σ-bond or molecular rotation will have slightly different electron densities around them and will thus require a slightly different energy to be promoted. **Each set of magnetically-inequivalent nuclei will give a signal at a different energy in the NMR spectrum**.

Lesson VII.5.4 Effect of Neighboring Nuclei on NMR Energies

As we learned in Lesson VII.5.1, each nucleus has a magnetic dipole that aligns either parallel or antiparallel with the applied magnetic field of an NMR spectrometer. However, just as the magnetic spin of the electrons surrounding a nucleus can affect the β_0 it "experiences", so too can the magnetic spin a neighboring nucleus. This means that, depending on the direction of nucleus A's magnetic field, nucleus A may shield or reinforce the magnetic field experienced by its neighboring nucleus B:

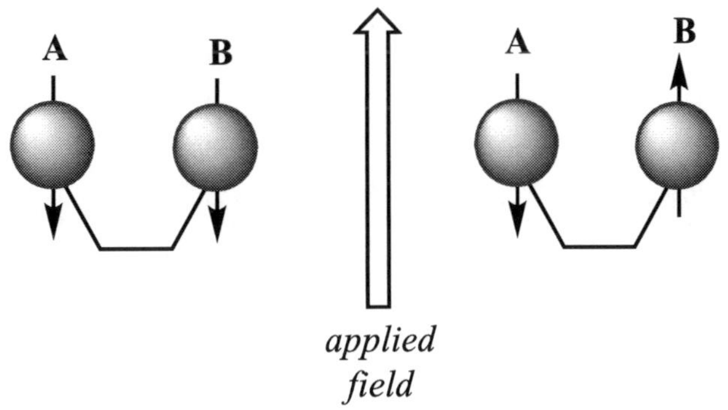

For a nucleus having one nearby nucleus, then, there are two slightly different energies needed to promote the nucleus. The difference in energies is very slight in cases we will encounter in this book, so the result is that the peak for the nucleus is "split" into two peaks. The peak looks like this:

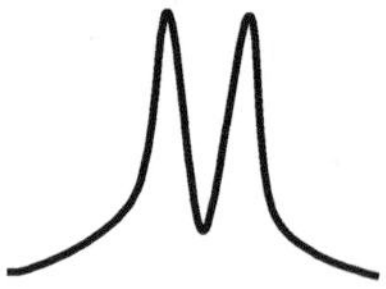

This peak shape is called a doublet. As more and more nuclei are present on neighboring sites, the number of smaller signals into which a signal splits increases, as does the complexity of the peak shape. The number of peaks into which a signal splits is called the **multiplicity** (*m*). For the nuclei studied in organic chemistry, $m = n+1$, where n is the number of nuclei (only counting those capable of causing splitting) on adjacent C atoms. In common organic molecules, a signal that is well-defined will generally not have a multiplicity greater than seven (six splitting atoms on adjacent C atoms). Those signals having a higher multiplicity generally are too broad to readily see the splitting on the strength of spectrometers in common use. The following schematic shows the typical shape of simple **multiplets**:

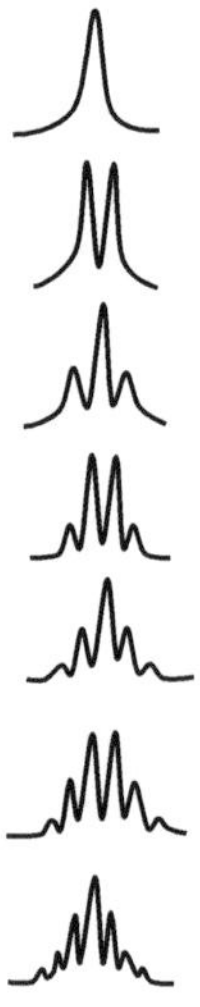

Type of peak	Ratio of heights
singlet	1
doublet	1:1
triplet	1:2:1
quartet	1:3:3:1
quintet	1:4:6:4:1
sextet	1:5:10:10:5:1
septet	1:6:15:20:15:6:1

Lesson VII.6. Carbon-13 Nuclear Magnetic Resonance Spectrometry

Lesson VII.6.1 The NMR Spectrum

In the current Lesson, we will learn how a ^{13}C NMR spectrum can reveal structural information about a molecule. In the previous Lesson, we learned about nuclear magnetic resonance and that the energy at which a nucleus absorbs energy is influenced by its electronic environment. This means that the chemical shifts at which peaks appear in a ^{13}C NMR spectrum reveals the types of carbons in the molecule. The diagram below is a representative sample of a ^{13}C NMR spectrum:

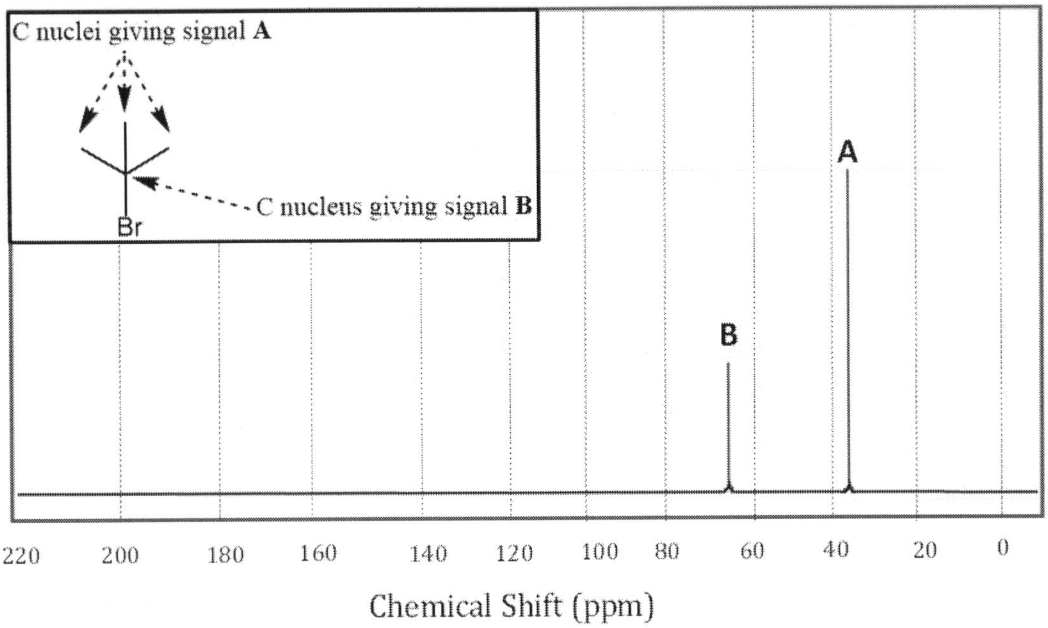

In this spectrum, each ^{13}C nucleus (or set of magnetically inequivalent ^{13}C nuclei) has a resonance that produces an upward-pointing peak.

Lesson VII.6.2 Avoiding Splitting in ^{13}C NMR spectra

We saw in Lesson VII.5 that the interaction of one nuclear magnetic dipole with another can cause an NMR signal to split into a multiplet. This phenomenon can make spectra substantially more difficult to interpret than if each peak was just a singlet. Most ^{13}C NMR spectra typically contain only singlets. Why is this? The first reason is that, although a carbon-13 nucleus is NMR-active, 99% of all carbon atoms in the universe are the carbon-12 isotope which is not NMR-active. It is therefore statistically very unlikely for one ^{13}C atom to be bonded directly to another ^{13}C atom (i.e., 0.01% probability), so C–C splitting is rarely observed. Most organic molecules feature multiple H atoms attached to C. Protons (^{1}H) are NMR-active nuclei, so the fact that each ^{13}C atom has one or more ^{1}H atom would produce a lot of splitting in a ^{13}C NMR spectrum and render interpretation nearly impossible. To prevent the splitting of ^{13}C NMR signals by ^{1}H nuclei, the ^{13}C NMR spectrometer is typically run in **proton-**

decoupled mode. In this mode, the spectrometer suppresses the ^{1}H splitting effect so that each ^{13}C nucleus in a molecule appears as a single, unsplit peak in the ^{13}C NMR spectrum.

Lesson VII.6.3 Carbon-13 NMR Spectra Provide Information on Types of C in a Sample

We know that the electronic environment around an atom differs based on its chemical environment (number and types of bonds). These differences produce predictable shifts in ^{13}C NMR resonances depending on what substituents the C atoms carry. Chemical shifts for ^{13}C nuclei in the bonding environments most frequently encountered in organic molecules are listed below:

Carbon (Shown)	Chemical Shift	Carbon (Shown)	Chemical Shift
$Si(CH_3)_4$	0	O—C	50-80
—CH_3	10-35	N—C	40-60
—CH_2—	15-50	C—X	X = I 0-35 X = Br 20-65 X = Cl 35-80
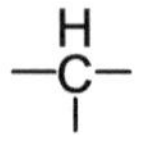	20-60		
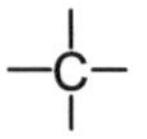	30-40		
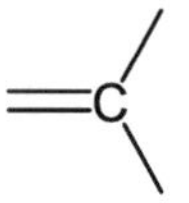	100-150	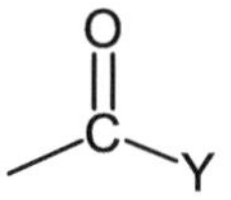	Y = H 190-200 R 200-220 OR 160-180 OH 175-185 NR_2 165-175
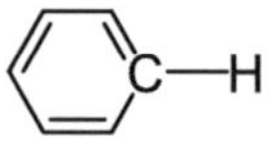	110-175		
≡C—	60-85		

Examining the peaks in a ^{13}C NMR spectrum and comparing the chemical shifts therein to this table can provide significant insight into the structure of the molecule. In the ^{13}C NMR spectrum of 2-bromo-2-methylpropane on the previous page, the three magnetically-equivalent methyl carbons produce peak **A** at 35 ppm, which falls within in the range (10–35 ppm) suggested in the table above. The quaternary C attached to the bromine produces a peak at 65 ppm, again within the typical range for carbons bearing Br substituents in the table (20–65 ppm). Even if we were not provided with the structure and only had the spectrum, we could easily conclude that the sample molecule contains no alkenes, carbonyls, alkynes, or aromatics because no ^{13}C resonances characteristic for these functional groups appear in the spectrum.

Lesson VII.7. Proton Nuclear Magnetic Resonance Spectrometry

Lesson VII.7.1 Chemical Shifts in ^1H NMR Spectra

In Lesson VII.6, we saw that ^{13}C NMR spectra can provide significant insight into the structures of organic molecules. We can gain even more information from a ^1H NMR spectrum, which will look like the following:

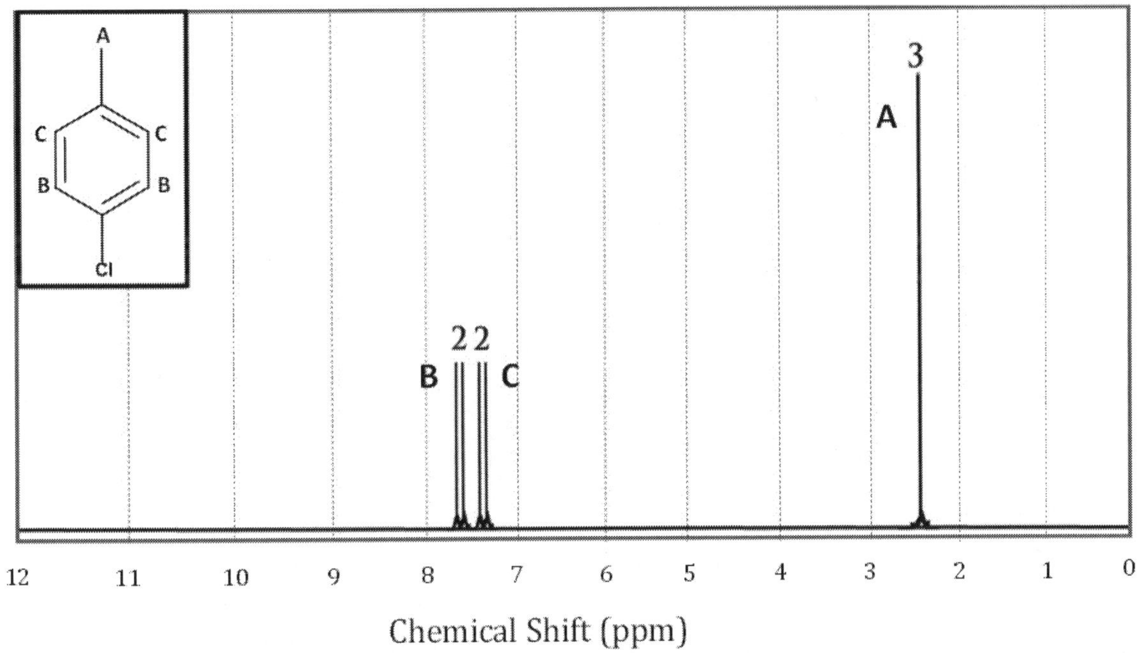

One notable difference between ^{13}C NMR spectra and ^1H NMR spectra is that the chemical shift scale for ^1H signals in typical organic samples fall in the range of about 0–12 ppm (cf. 0–220 ppm for ^{13}C NMR). Another difference is that ^1H NMR spectra often contain multiplets, whereas ^{13}C NMR spectra (by experimental design) only contain singlets.

Although not present in the example above, many ^1H NMR samples include tetramethylsilane (TMS) in addition to the molecule of interest, because the TMS speak is set to be 0 ppm as a common reference point for all ^1H NMR chemical shifts. Another common reference peak is $CHCl_3$ (7.28 ppm) because chloroform is one of the most commonly used solvents for NMR samples.

Chemical shifts for ^1H nuclei in the bonding environments most frequently encountered in organic molecules are listed below:

Protons (Shown)	**Chemical Shift**	**Protons (Shown)**	**Chemical Shift**
Si$(CH_3)_4$	0	R—OCH_3	3.3
—CH_3	0.9	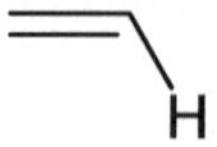	4.5-5.5
—CH_2—	1.2	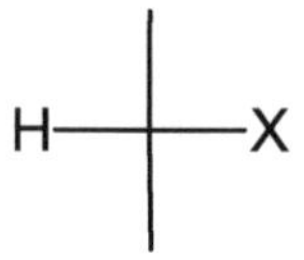	X = I 2.5-4 X = Br 2.5-4 X = Cl 3-4 X = F 4-4.5
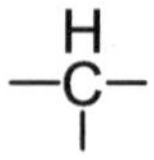	1.4	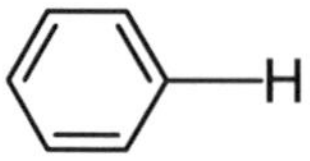	6.5-8.0
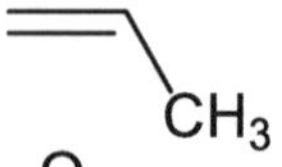	1.7	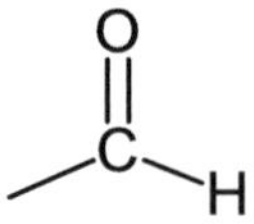	10
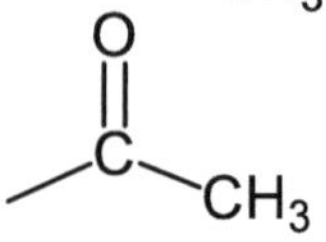	2.1		
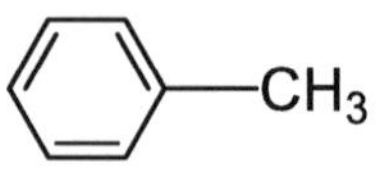	2.3		
	2.4		

Lesson VII.7.2 Integration of Proton Nuclear Magnetic Resonance Signals

In a ^1H NMR spectrum, the area under each peak (the **integration or integral**) is proportional to the number of protons that absorb at that energy (the number of protons in that set of magnetically equivalent protons in that group in the sample). The same is not true for a proton-decoupled ^{13}C NMR spectrum. There are several ways to indicate the integration for a given peak in a ^1H NMR spectrum. In this book, the integration will be provided as a number printed above the peak. For the spectrum of 4-chlorotoluene (shown in Lesson VII.7.1), the integration of peak **A** is 3, the integration of peak **B** is 2, as is the integration of **C**. The integrations can be very useful in figuring out what type of units are present in a molecule based on NMR data.

In practical settings, the spectrometer does not have a way to know how many protons give a signal, it can only measure the relative integration of one peak with respect to another. As a result, the spectrometer might automatically and arbitrarily provide integrations of 0.60, 0.40 and 0.40 for peaks **A**, **B** and **C** in the spectrum of 4-chlorotoluene. The ratio, 223, must then be calculated by the user (or software used with the spectrometer) to get the values displayed in the spectrum as shown in this book.

Lesson VII.7.3 Splitting in ^1H NMR Spectra

In Lesson VII.5, we saw that the magnetic spin of one nucleus can split the energy of absorption at nearby nuclei. In ^1H NMR spectra, the signal for a proton on carbon "X" is notably split into $n+1$ subpeaks, where n is the total number of H atoms on the carbons directly adjacent to carbon "X". If the directly adjacent atoms are not carbon (i.e., oxygen or nitrogen), the H atoms on these directly adjacent atoms typically do not produce observable splitting. This property allows us to tell **how many H are on C atoms directly adjacent to the carbon bearing the ^{1}H nucleus (or set of ^{1}H nuclei) that produces a particular signal** in a ^{1}H NMR spectrum. If a particular signal is a triplet ($m = 3$), for example, we know that the protons giving rise to this signal are attached to a carbon which has carbons directly adjacent to it carrying a total of 2 protons among them. Numerous examples to illustrate how the chemical shift, integration and multiplicity of resonances in a ^{1}H NMR spectrum can be used to gain structural information are provided in Part VIII.

Bonus Lessons from Part VIII. Spectroscopy Practice with Solutions

Lesson VIII.1. Charts for IR and NMR Spectral Analysis

Lesson VIII.2. Infrared Spectroscopy Practice Problems

Lesson VIII.3. Answers to Infrared Spectroscopy Practice IR Spectroscopy Problems

Lesson VIII.4. Proton NMR Practice Problems

Lesson VIII.5. Answers to Proton NMR Practice Problems

Lesson VIII.6. Carbon-13 NMR Practice Problems with Solutions

Lesson VIII.1. Charts for IR and NMR Spectral Analysis

Lesson VIII.1.1 Typical Wavenumbers in IR Spectra

This table provides some useful general ranges for where certain types of bonds are likely to show up in IR spectra.

Bond	Energy (cm^{-1})	Intensity
N≡C	2255-2220	m-s
C≡C	2260-2100	w-m
C=C	1675-1660	m
N=C	1650-1550	m
	1600 **AND** 1500-1425	w-s
C=O	1775-1650	s
C—O	1250-1000	s
C—N	1230-1000	m
O—H	3650-3200	s (br)
O—H	3300-2500	s (br)
N—H	3500-3300	m (br)
C—H	3300-2725	m

C-H Bond (Stretch)	Energy (cm^{-1})
C≡C—H	3300-ish
C=C—H	3100-3000
C—C—H	2950-2850
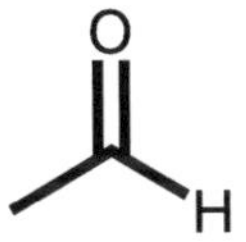	2820-ish and 2720-ish

C-H Bond (Bending)

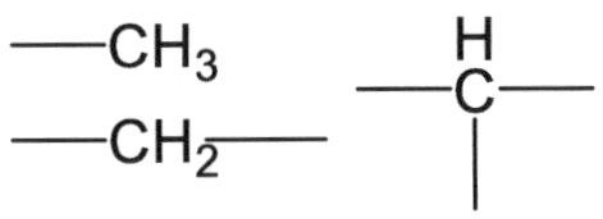

1450-1400

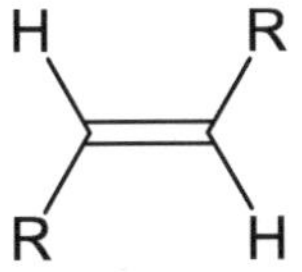

980-960
trans

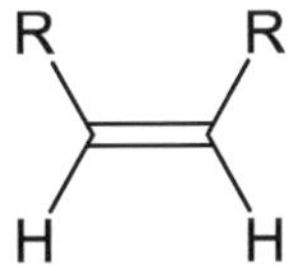

730-670
cis

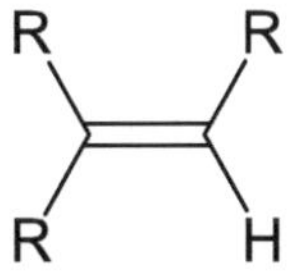

840-800
trisubstituted

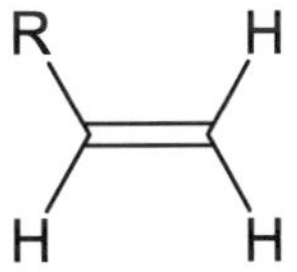

990 and 910
monosubstituted

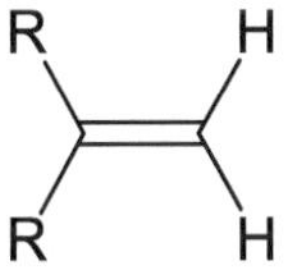

890
disubstituted terminal

If only one functional group is present in a particular compound, the following simplified flowchart is a good starting point for determining which functional group it is:

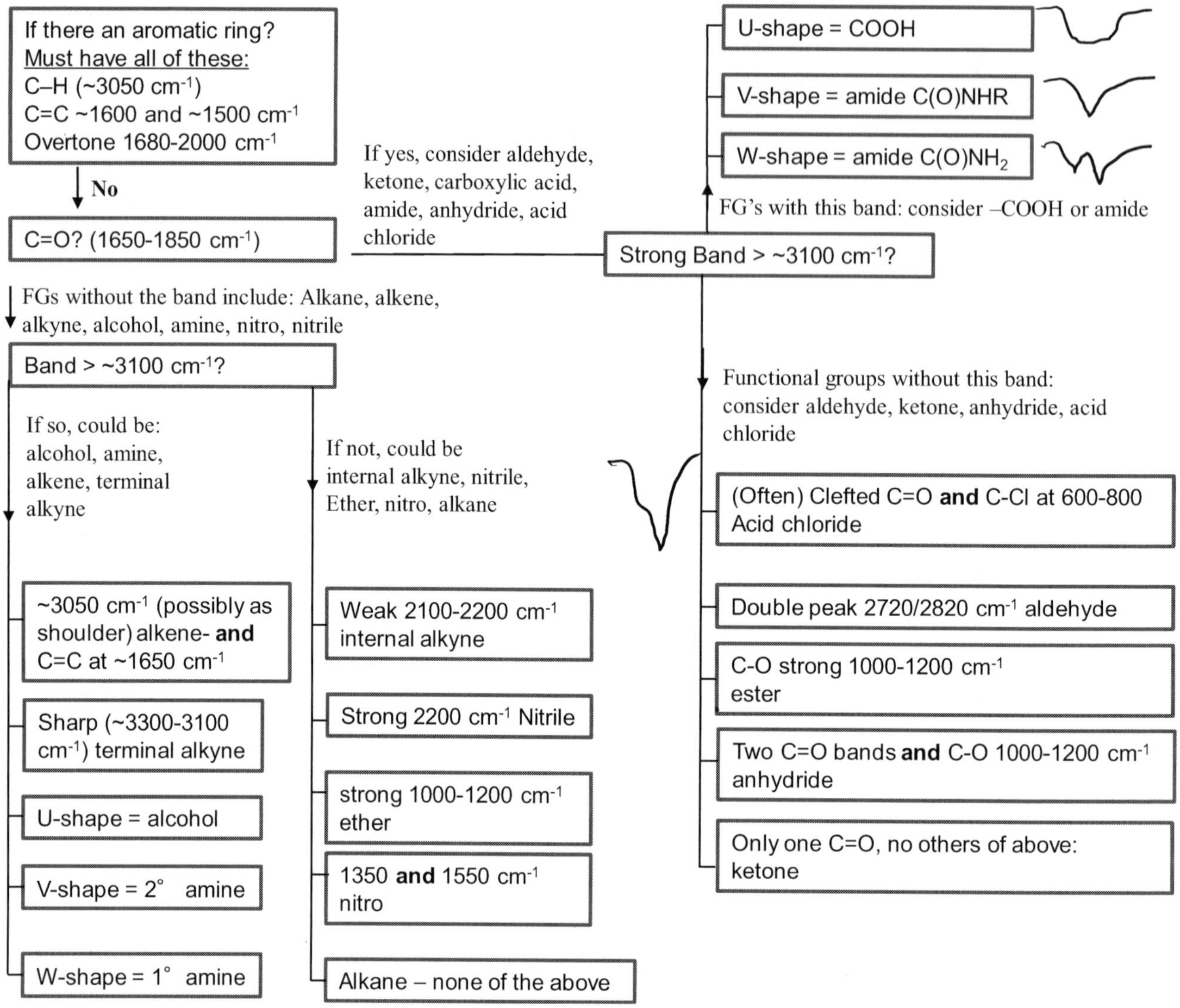

Lesson VIII.1.3 Typical ^{13}C NMR Chemical Shifts

This diagram gives you a good idea for determining the general functional groups present in a sample from the observed ^{13}C NMR chemical shifts. These are very rough estimates that divide the spectrum into four areas:

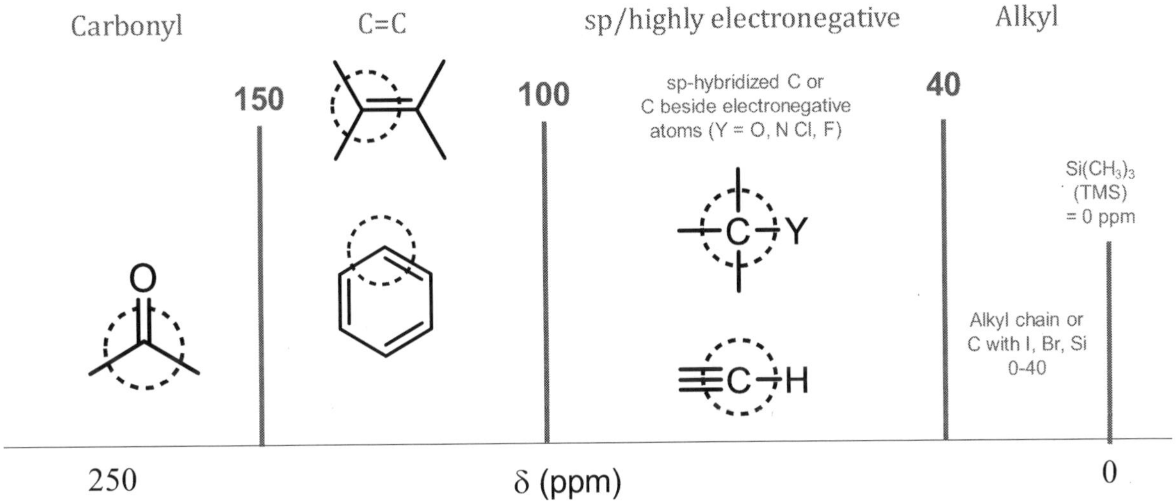

For more fine-tuned determinations than are possible with the above diagram, this table provides useful information on the commonly-observed ^{13}C NMR chemical shifts for different types of carbons:

Carbon (Shown)	Chemical Shift
$Si(CH_3)_4$	0
—CH_3	10-35
—CH_2—	15-50
H–C (tertiary CH)	20-60
C (quaternary)	30-40
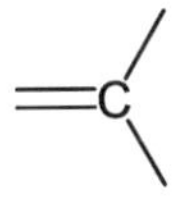	100-150
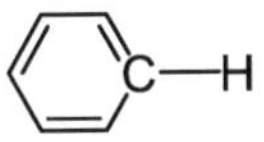	110-175
≡C—	60-85

Carbon (Shown)	Chemical Shift
O—C	50-80
N—C	40-60
C—X	X = I 0-35 X = Br 20-65 X = Cl 35-80
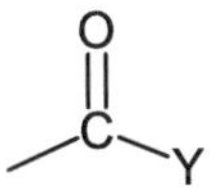	Y = H 190-200 R 200-220 OR 160-180 OH 175-185 NR_2 165-175

Additional note:

Many common ^{13}C NMR spectra also have a peak at 0.00 ppm due to the presence of tetramethylsilane, $Si(CH_3)_4$, a standard, as well as a usually intense triplet at 77 ppm due to the presence of deuterated chloroform, $CDCl_3$, a solvent often used for NMR spectra.

Lesson VIII.1.4 Typical ¹H NMR Chemical Shifts

This diagram gives you a good idea for determining the general functional groups present in a sample from the observed ¹H NMR chemical shifts:

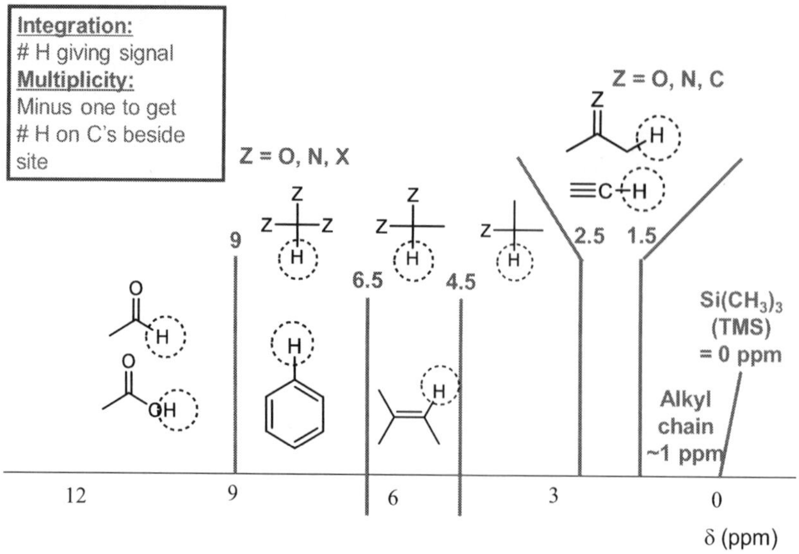

For more fine-tuned determinations than are possible with the above diagram, this table provides useful information on the commonly-observed ¹H NMR chemical shifts for different types of protons:

Protons (Shown)	**Chemical Shift**	**Protons (Shown)**	**Chemical Shift**
$Si(CH_3)_4$	0	R—OCH_3	3.3
—CH_3	0.9	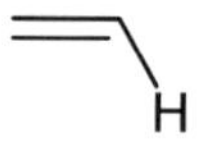	4.5-5.5
—CH_2—	1.2	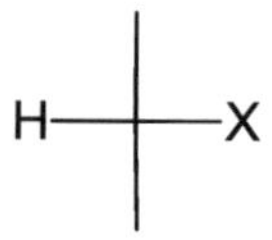	X = I 2.5-4 X = Br 2.5-4 X = Cl 3-4 X = F 4-4.5
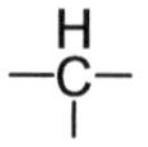	1.4	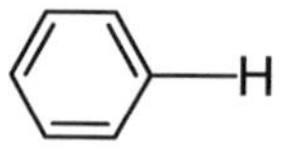	6.5-8.0
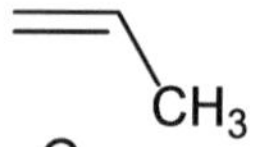	1.7	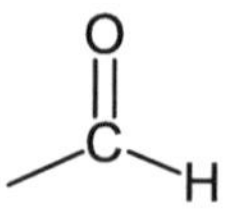	10
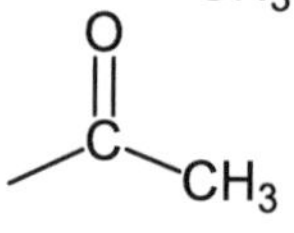	2.1		
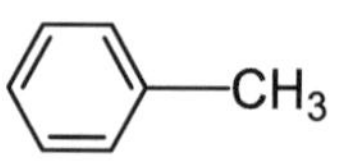	2.3		
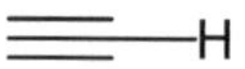	2.4		

Additional note:

Many common ¹H NMR spectra also have a peak at 0.00 ppm due to the presence of tetramethylsilane, $Si(CH_3)_4$, a standard, as well as at 7.28 ppm due to the presence of chloroform, $CHCl_3$, present in the deuterated chloroform that is a solvent often used for NMR spectra.

Lesson VIII.2. Infrared Spectroscopy Practice Problems

For each of the following, determine which of the potential structures is most likely to have produced the provided IR spectrum.

IR Spectroscopy Problem 1

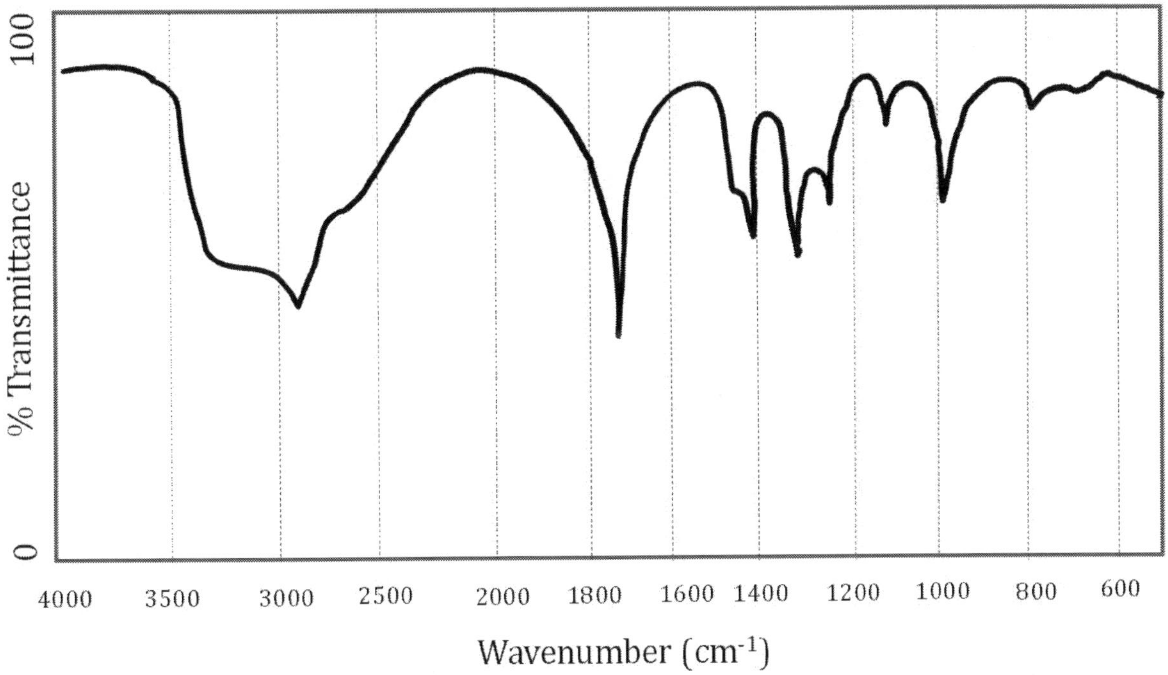

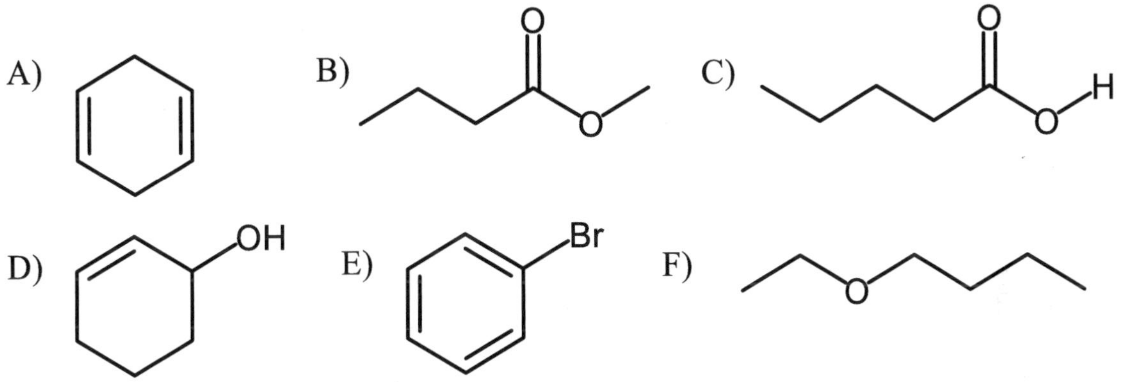

IR Spectroscopy Problem 2

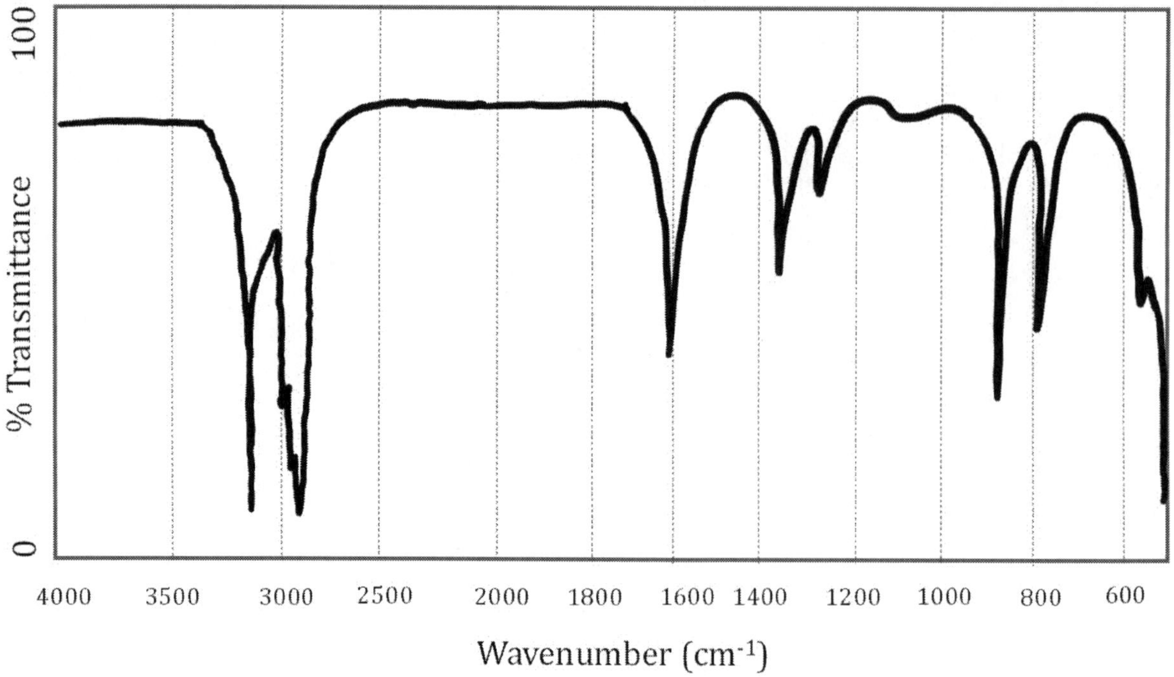

A) 

B) 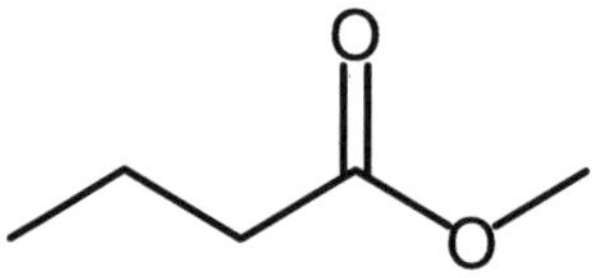

C) 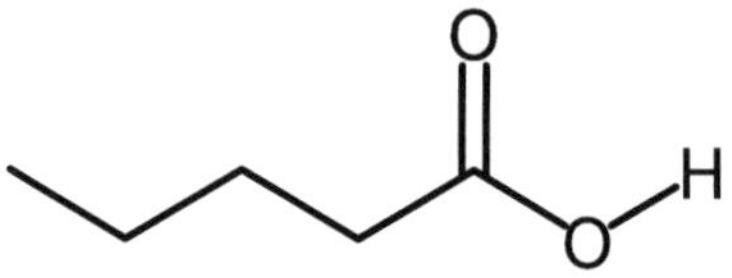

D) 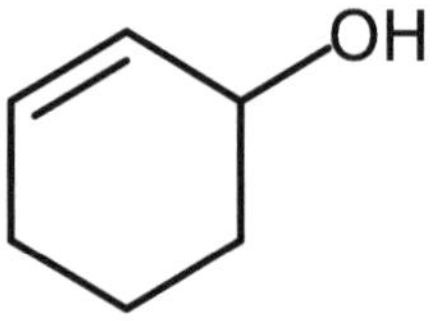

E) 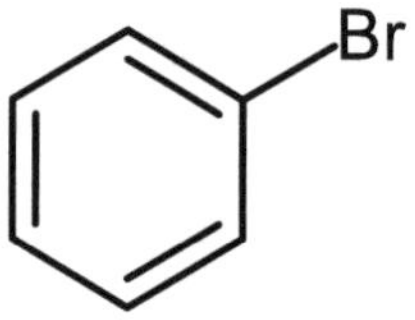

F) 

IR Spectroscopy Problem 3

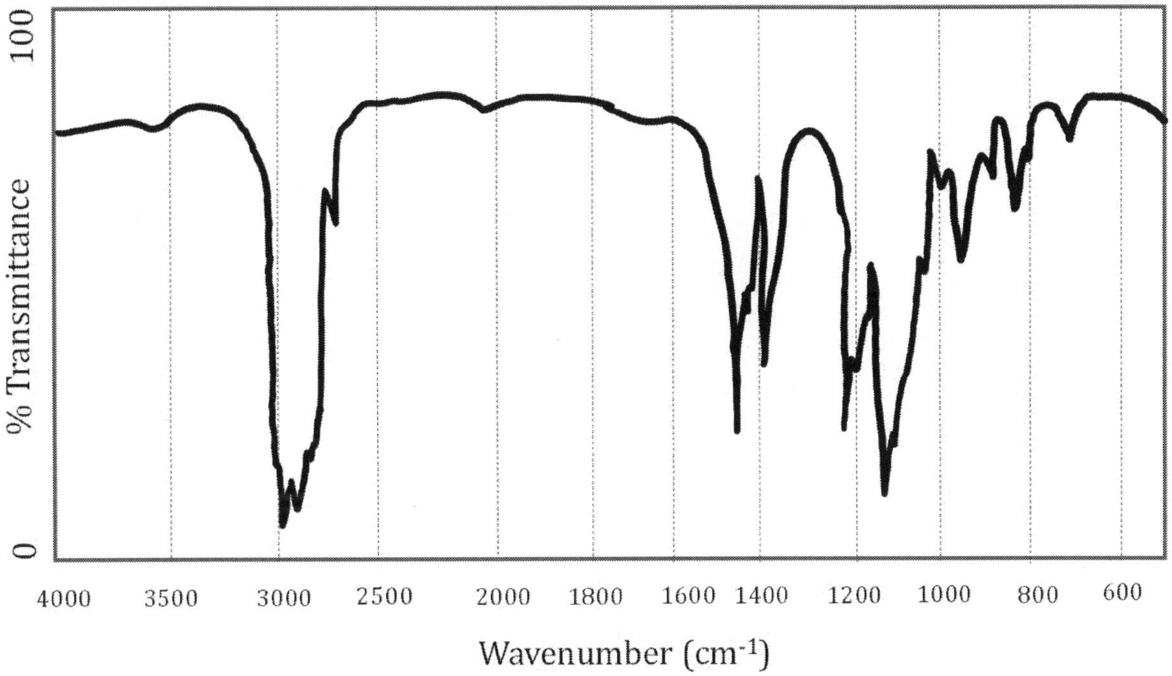

A) 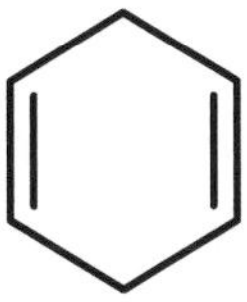

B) 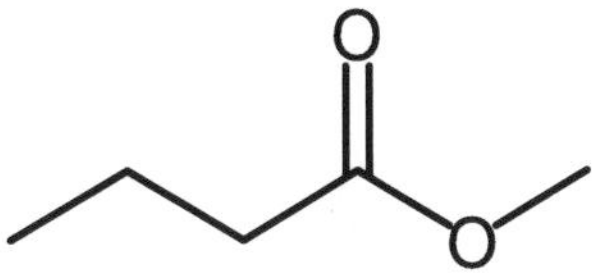

C) 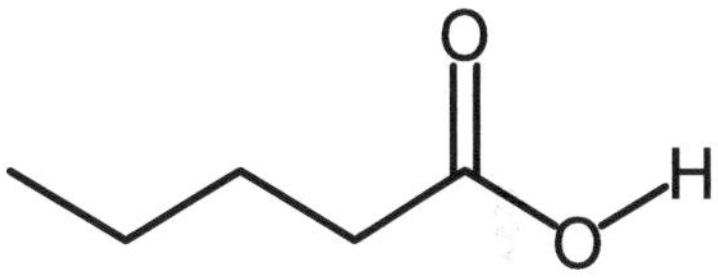

D) 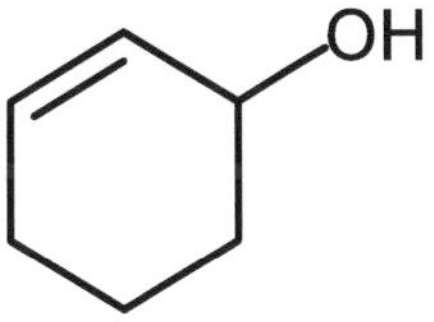

E) 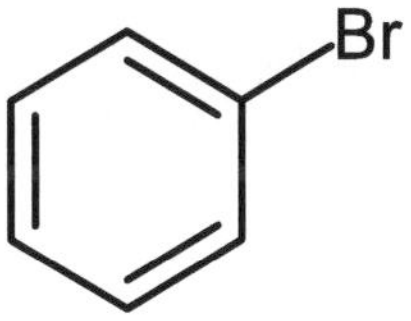

F) 

IR Spectroscopy Problem 4

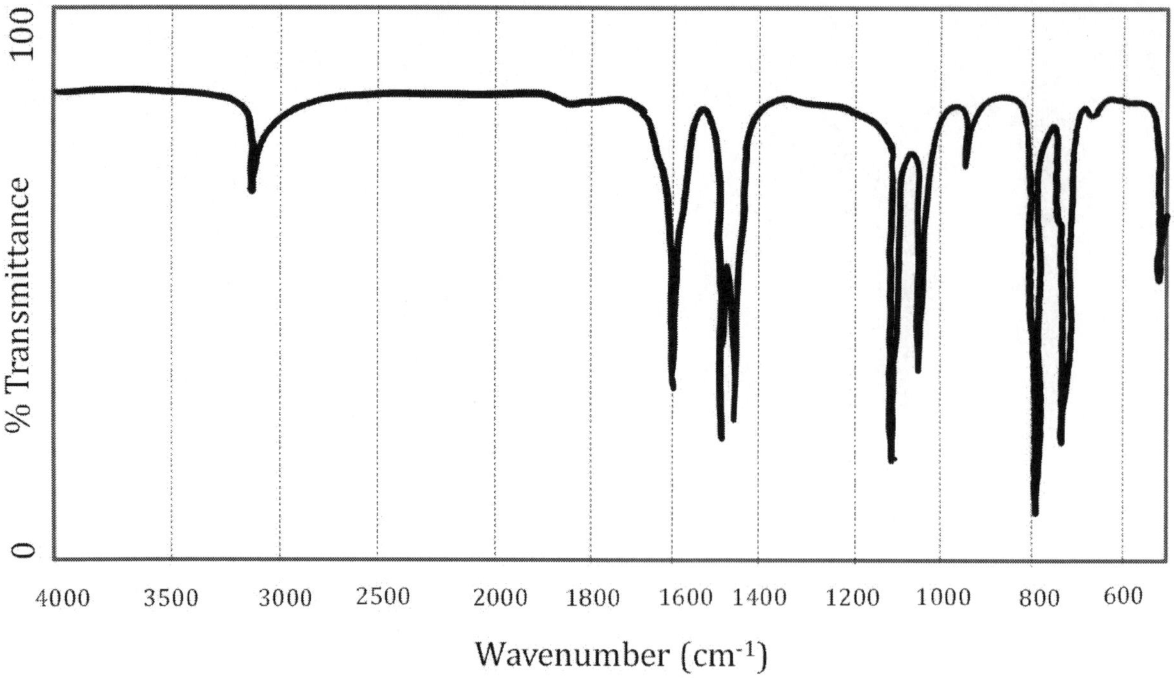

A) 

B) 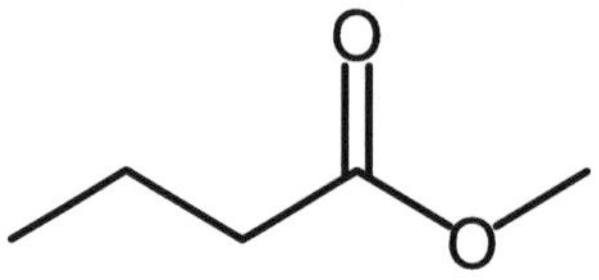

C) 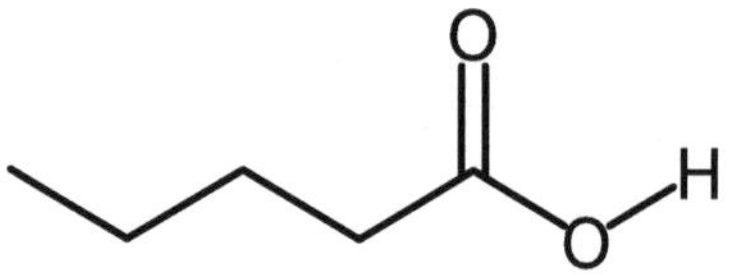

D) 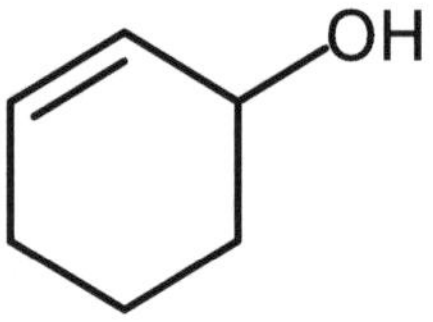

E) 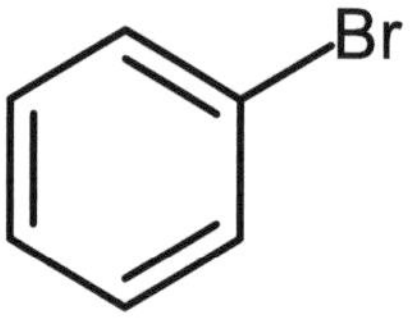

F) 

IR Spectroscopy Problem 5

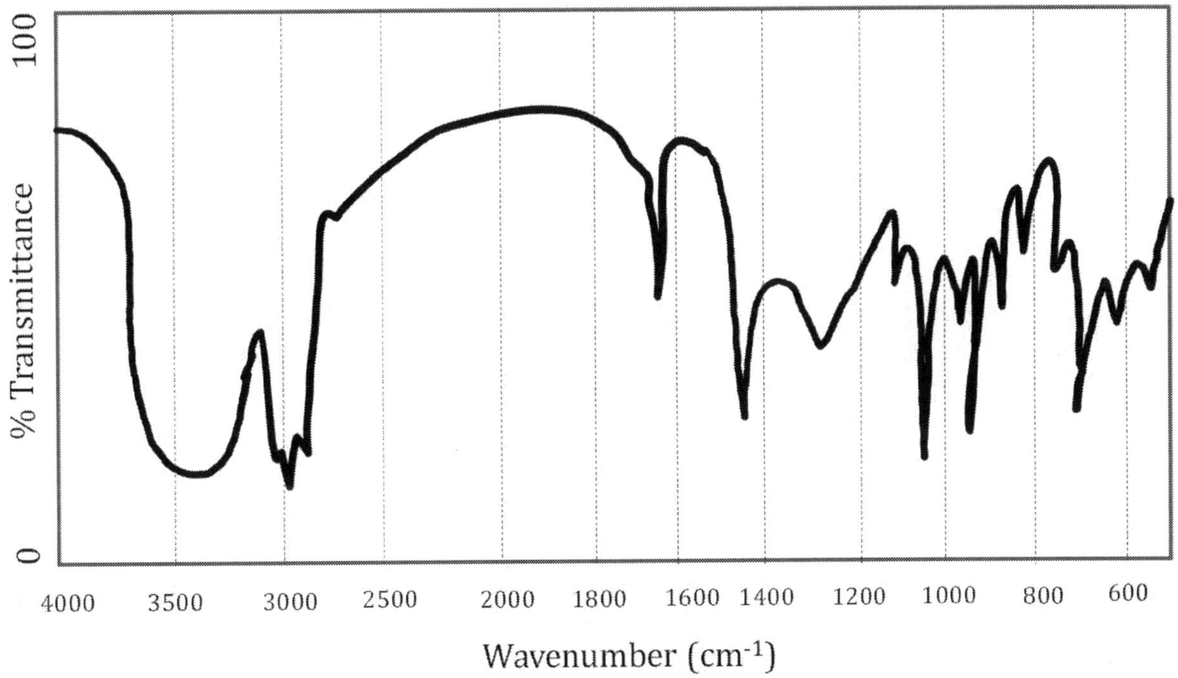

A) 

B) 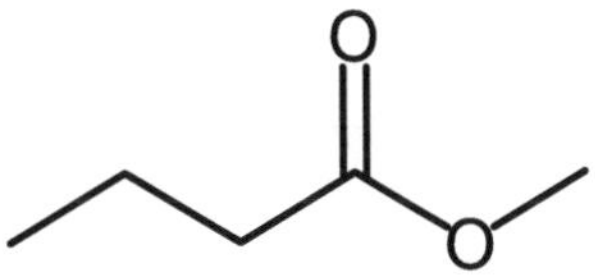

C) 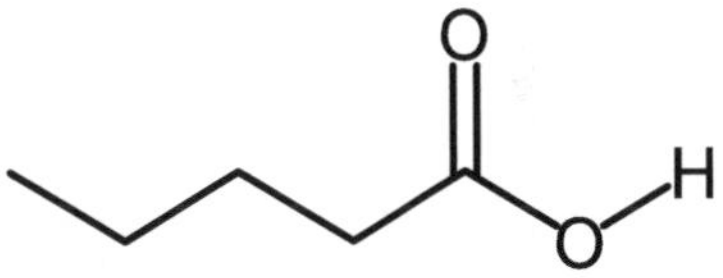

D) 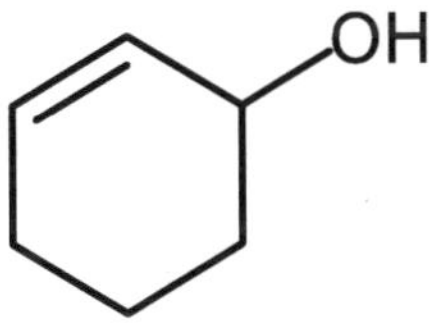

E) 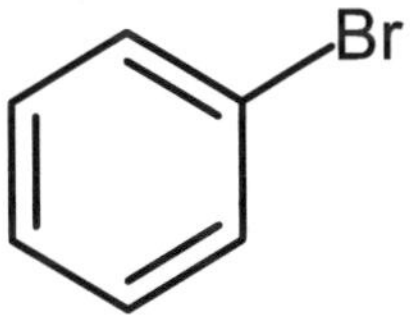

F) 

IR Spectroscopy Problem 6

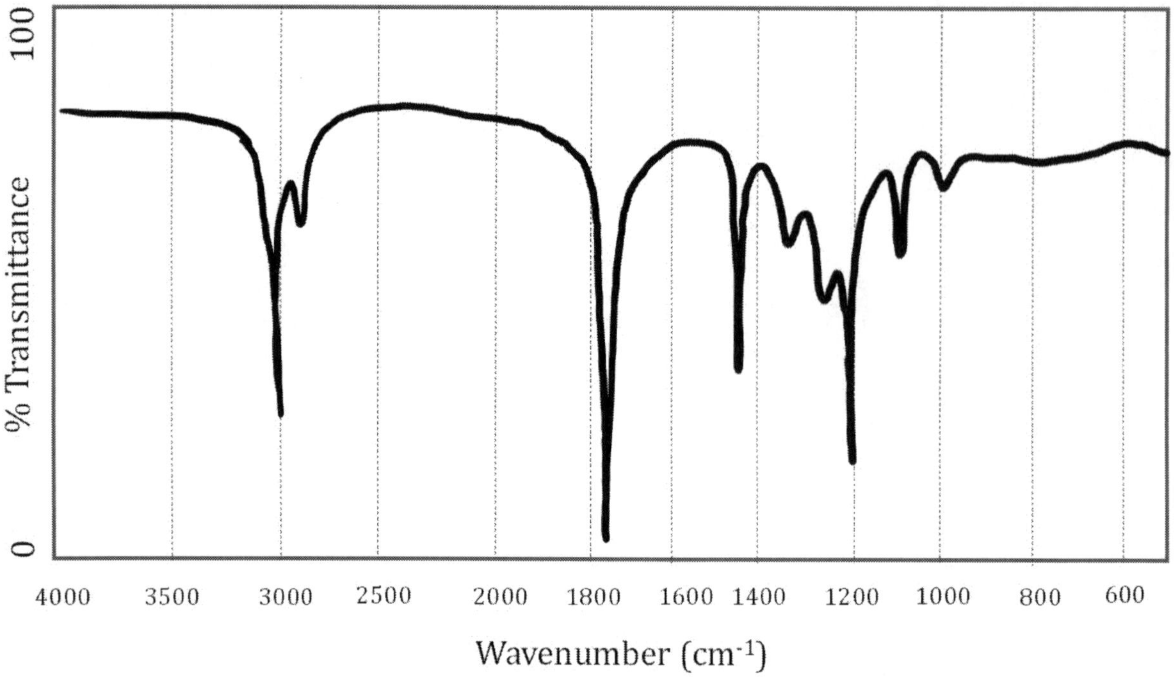

A) 

B) 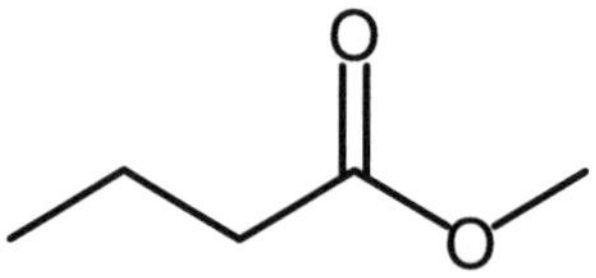

C) 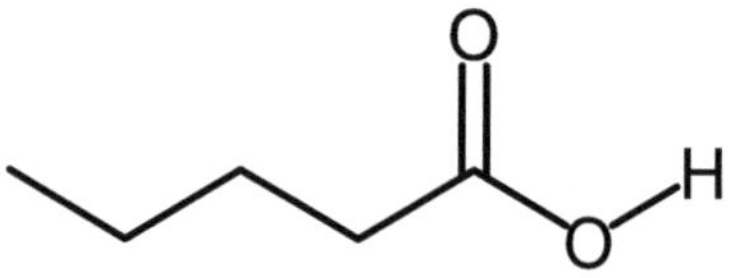

D) 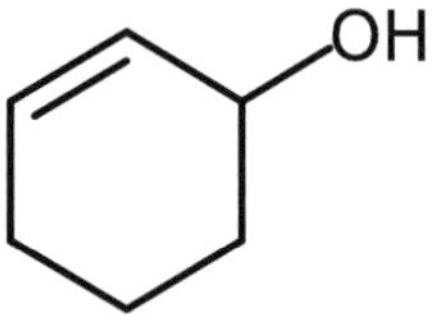

E) 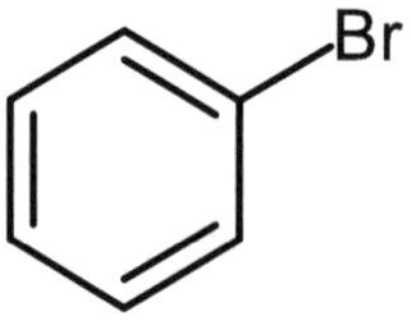

F) 

IR Spectroscopy Problem 7

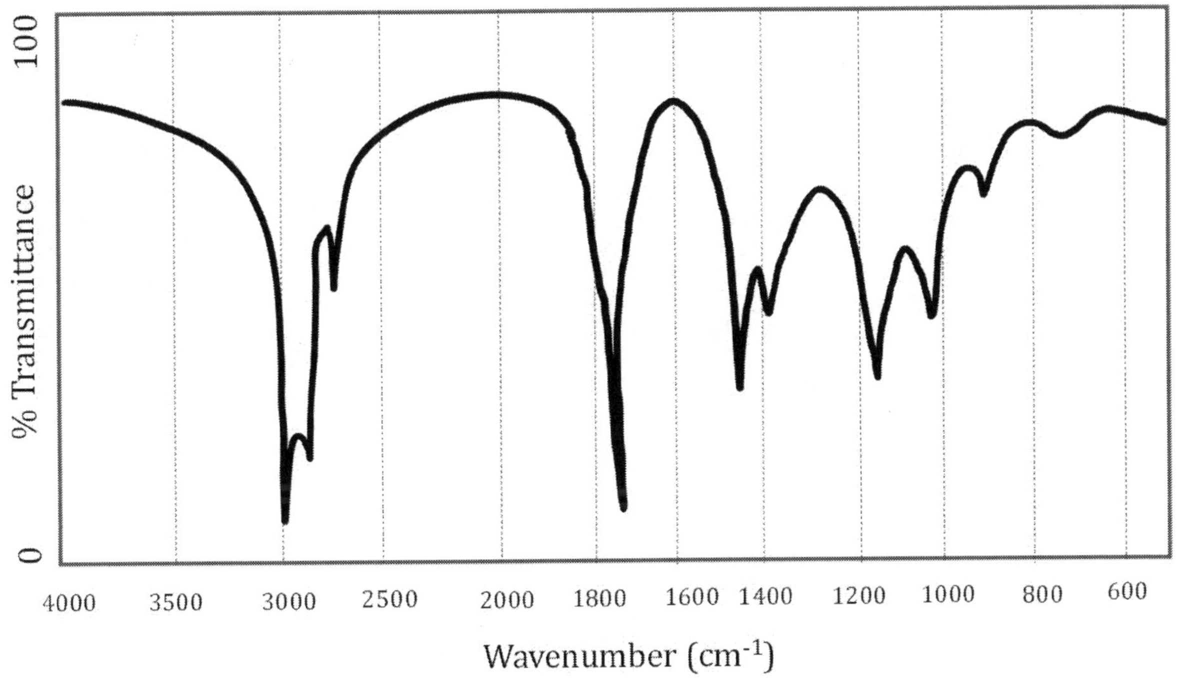

A)

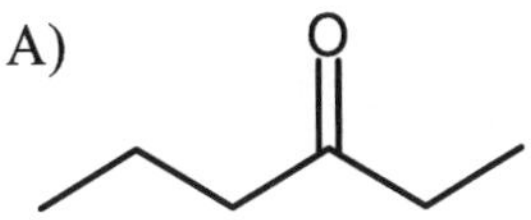

B)

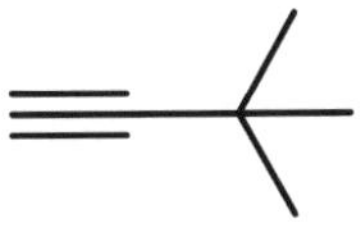

C)

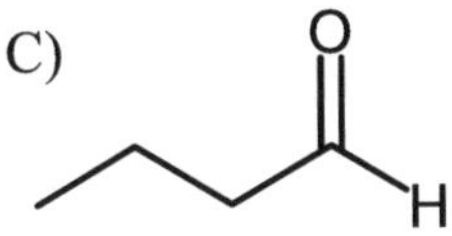

D)

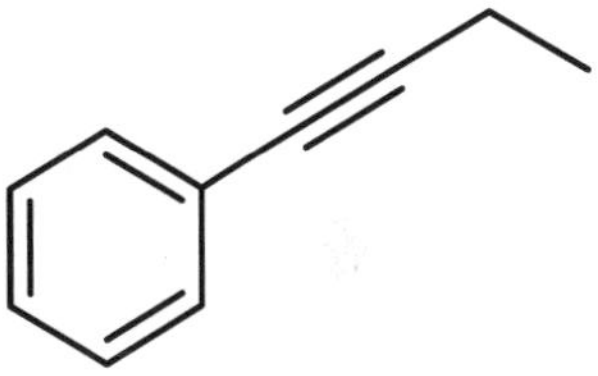

IR Spectroscopy Problem 8:

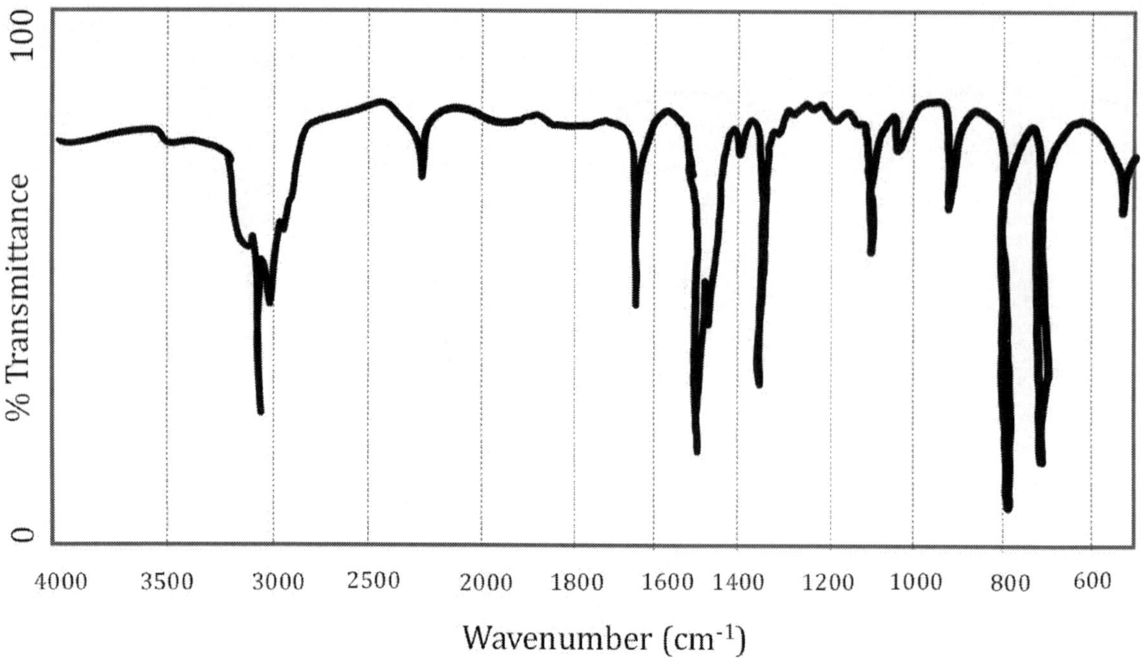

A)

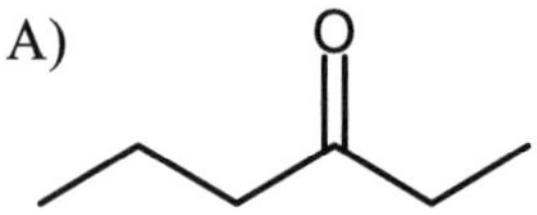

B)

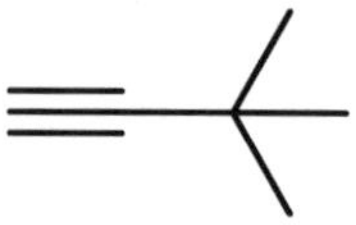

C)

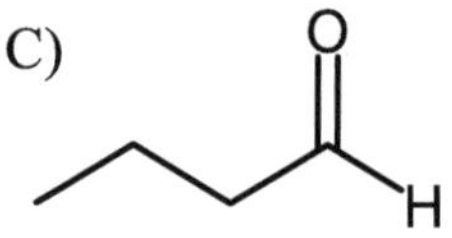

D)

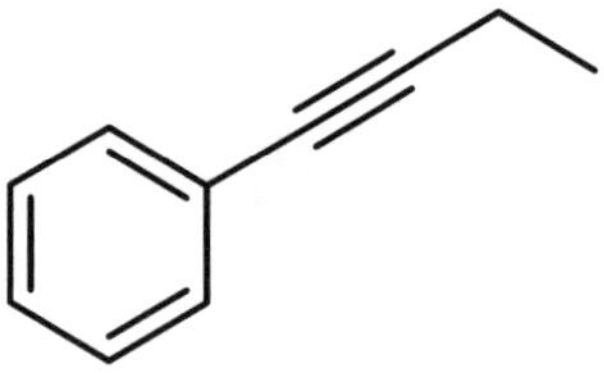

IR Spectroscopy Problem 9:

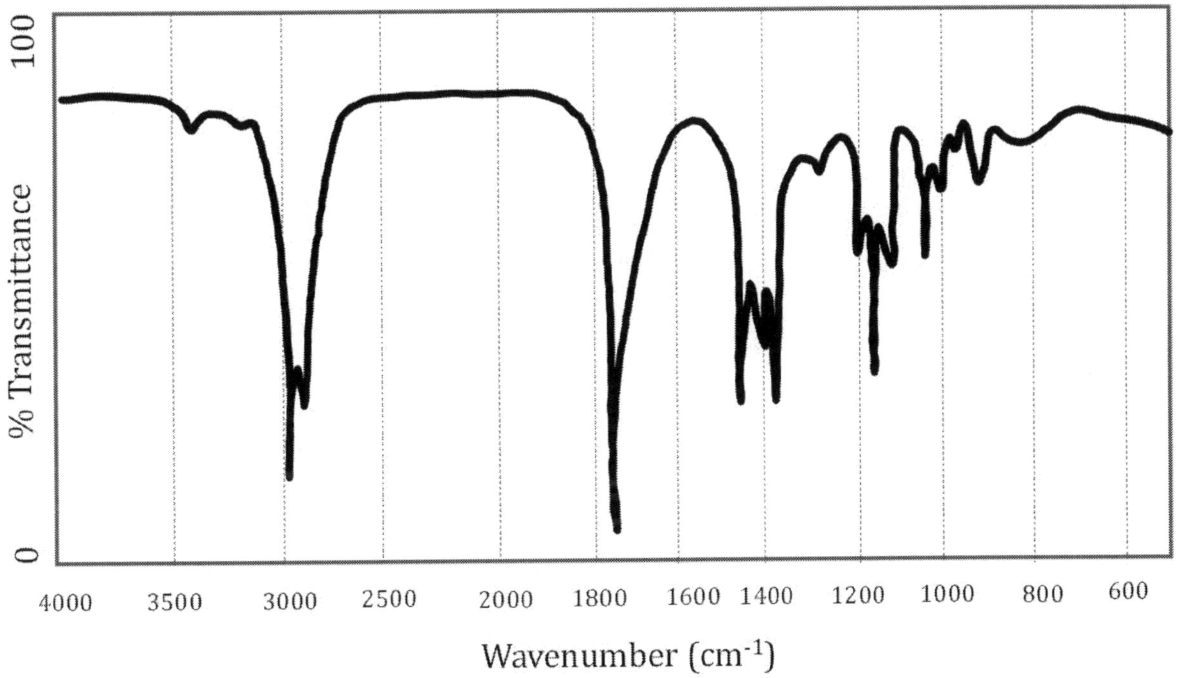

A)

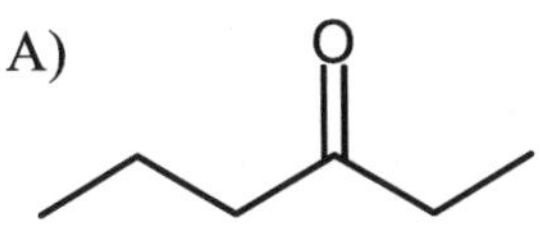

B)

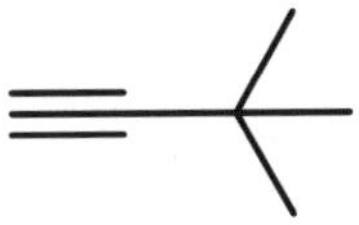

C)

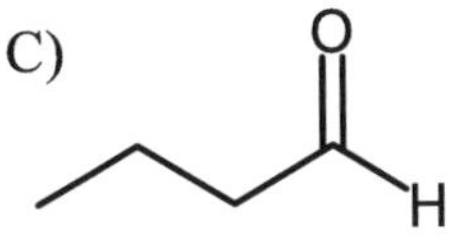

D)

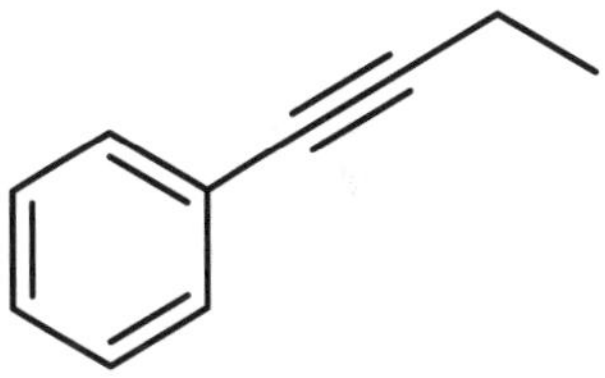

IR Spectroscopy Problem 10:

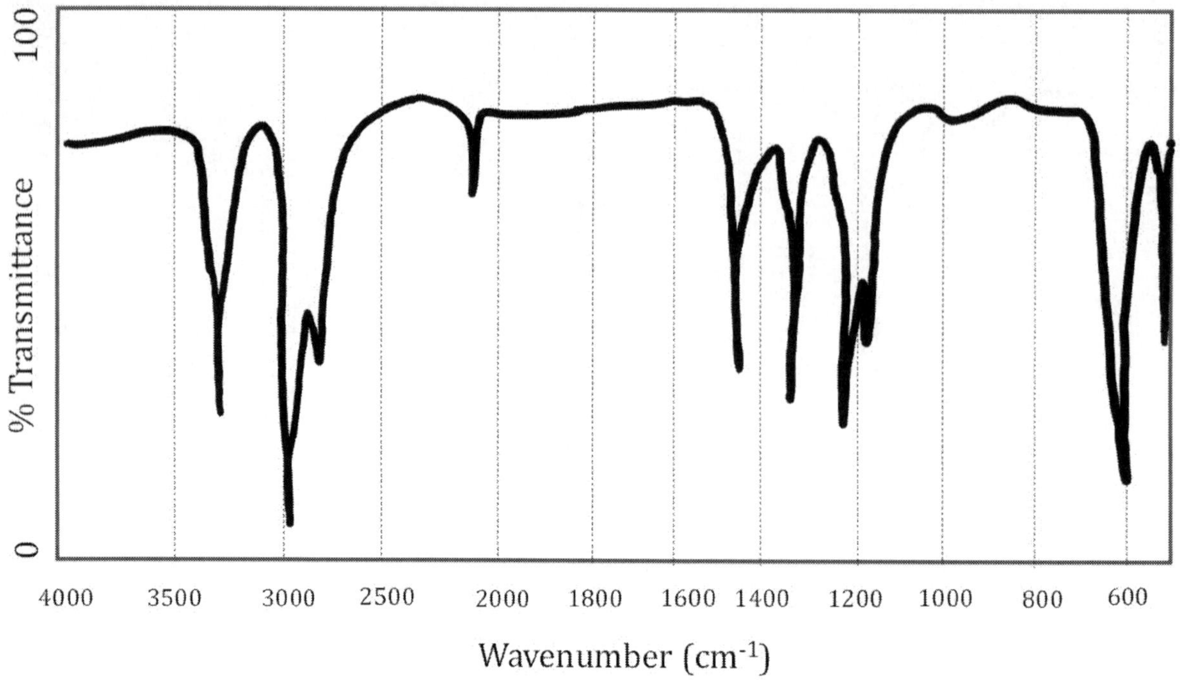

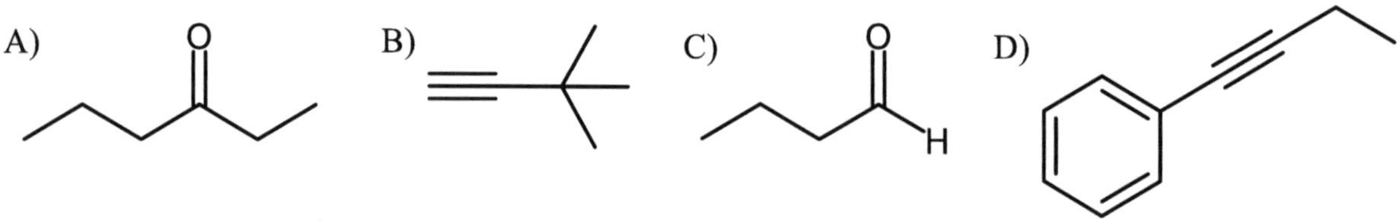

IR Spectroscopy Problem 11

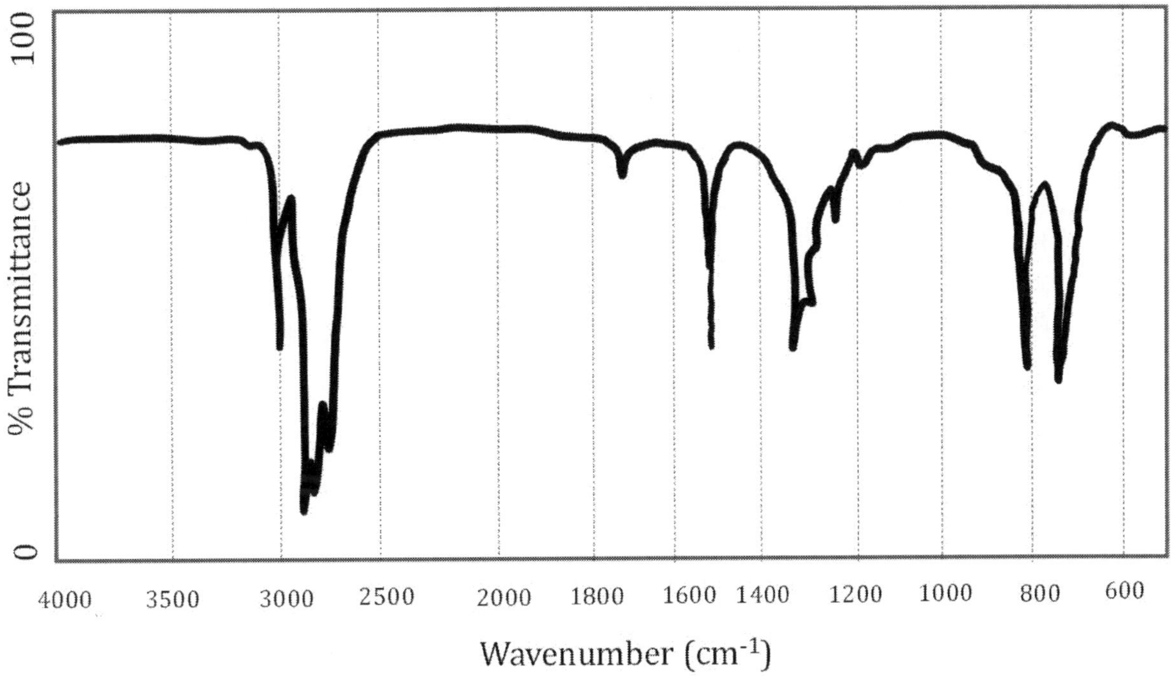

A) 

B) 

C) 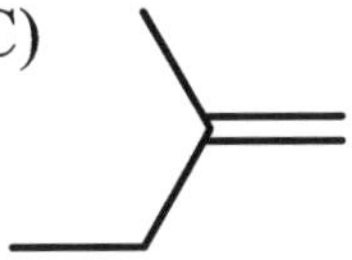

D) 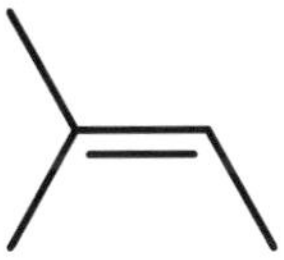

E) 

IR Spectroscopy Problem 12

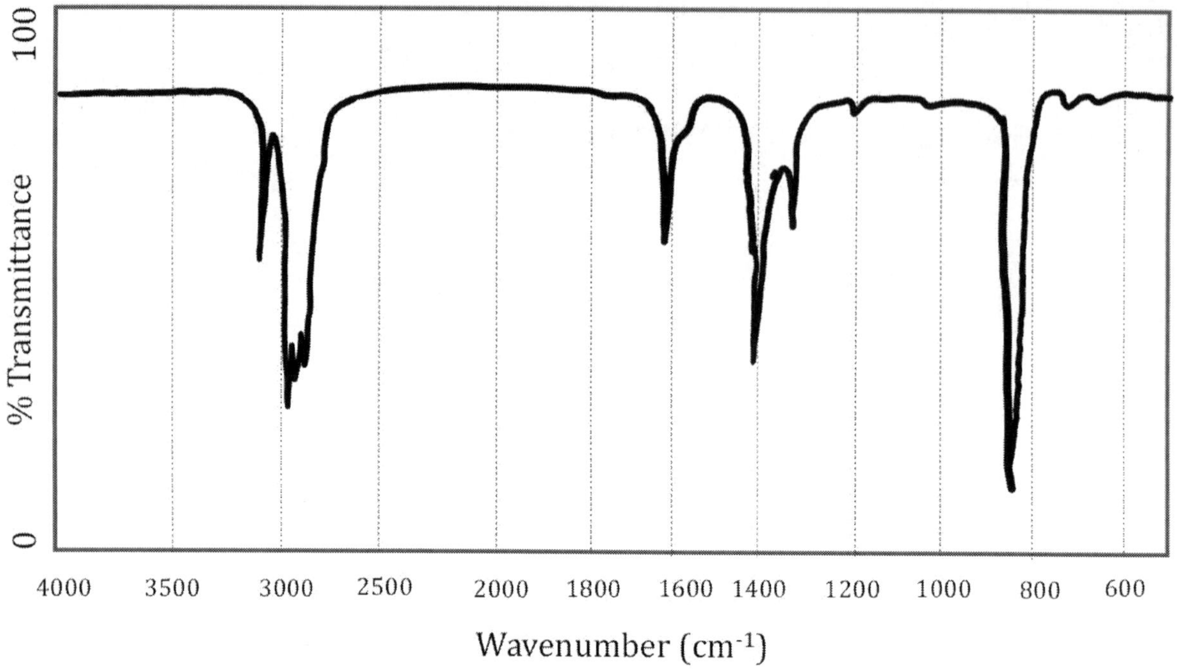

A)

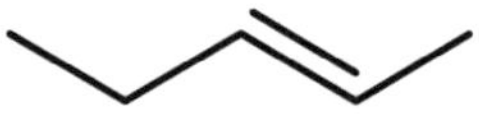

B)

C)

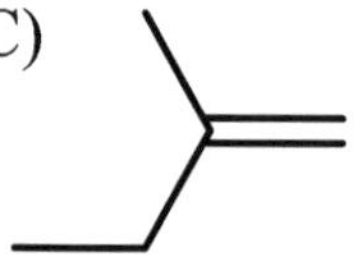

D)

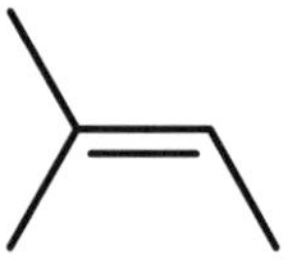

E)

IR Spectroscopy Problem 13

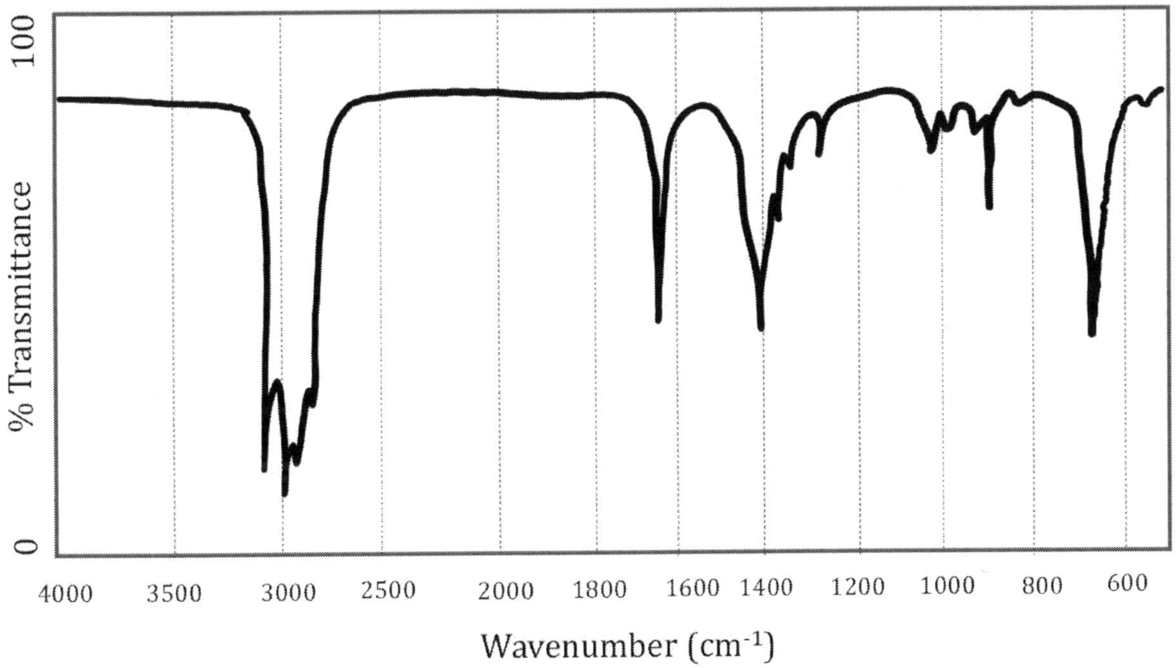

A) 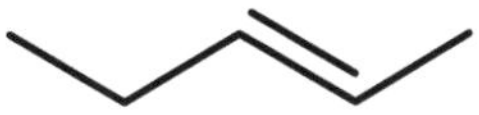

B) 

C) 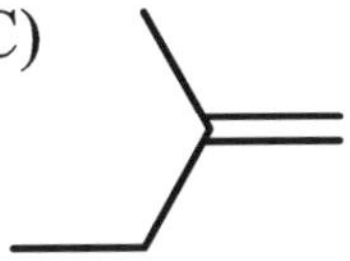

D) 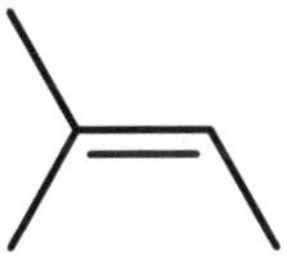

E) 

IR Spectroscopy Problem 14

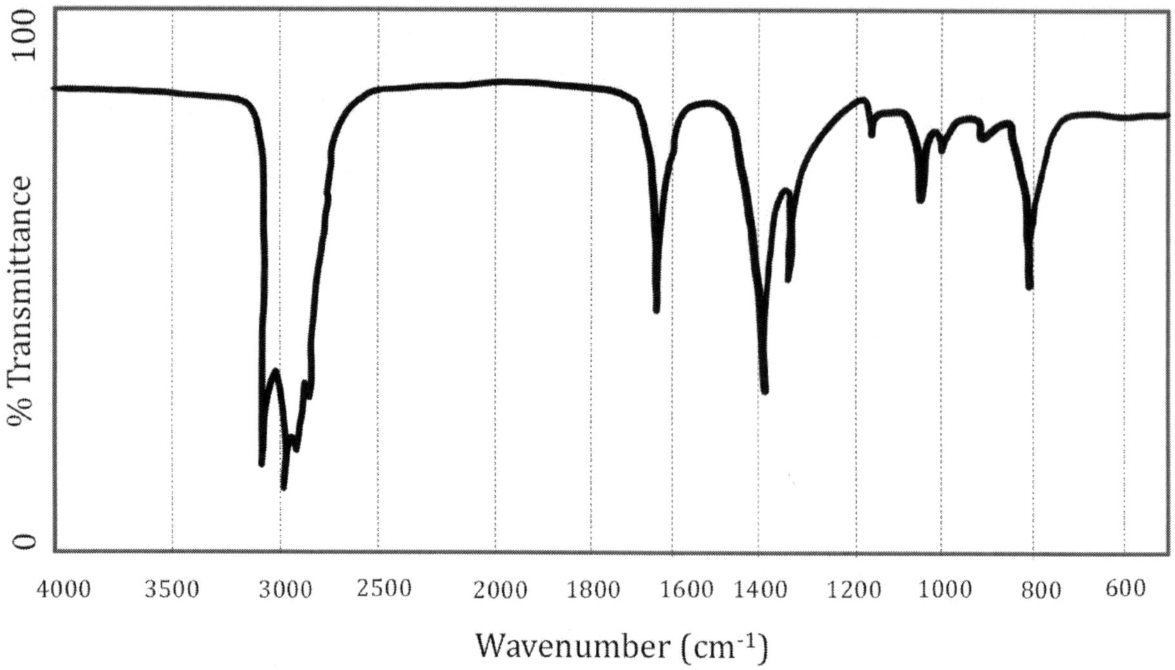

A) 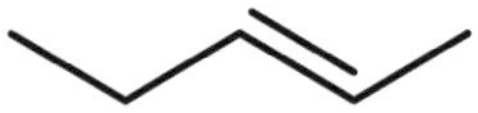

B) 

C) 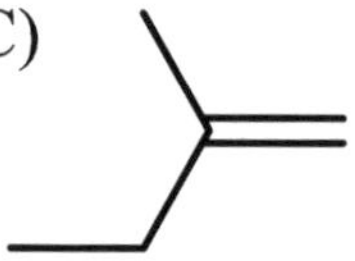

D) 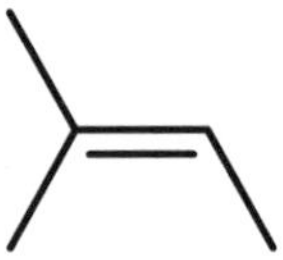

E) 

IR Spectroscopy Problem 15

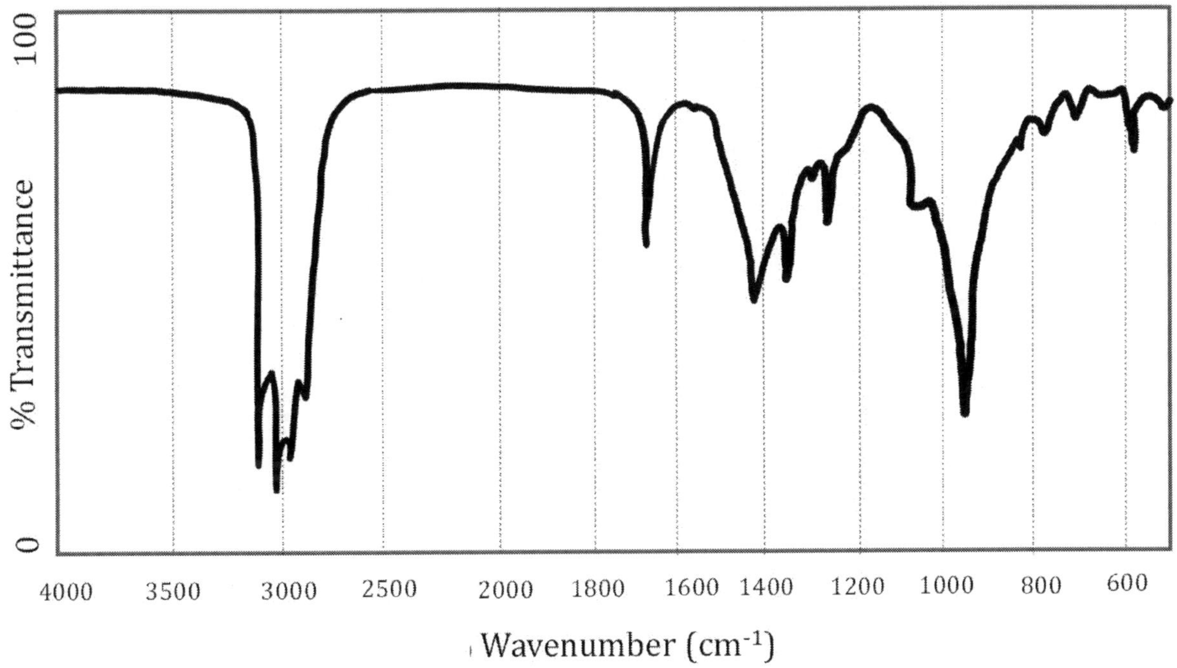

A)

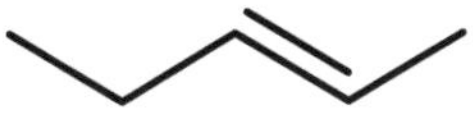

B)

C)

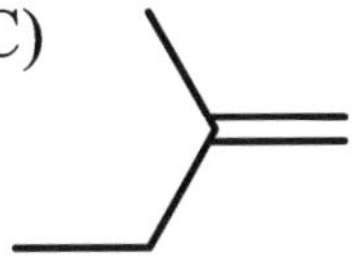

D)

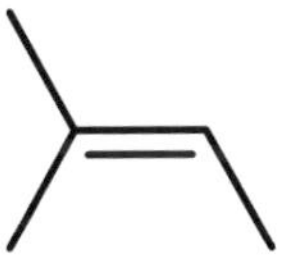

E)

Lesson VIII.3. Answers to Infrared Spectroscopy Practice IR Spectroscopy Problems

Answer to IR Spectroscopy Problem 1

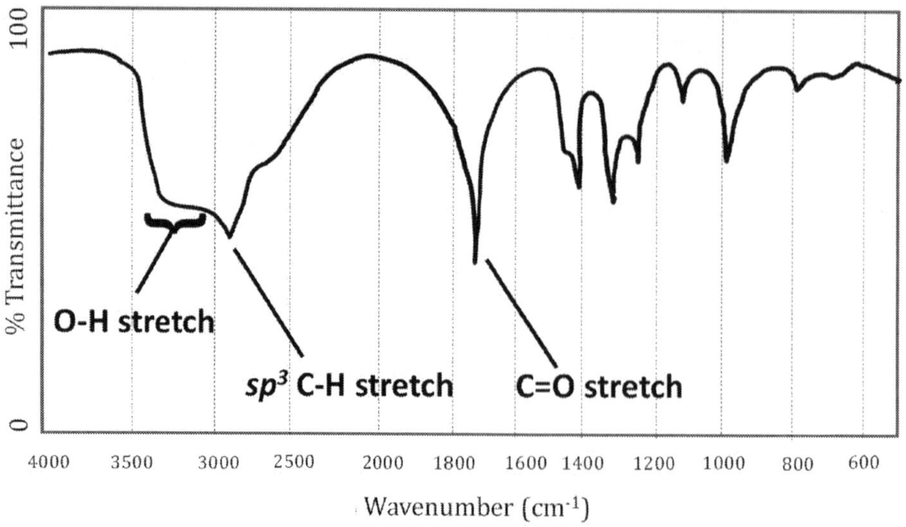

C) 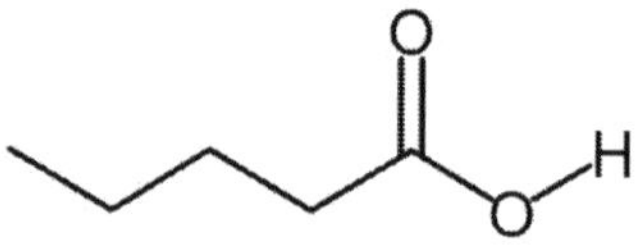

Answer to IR Spectroscopy Problem 2

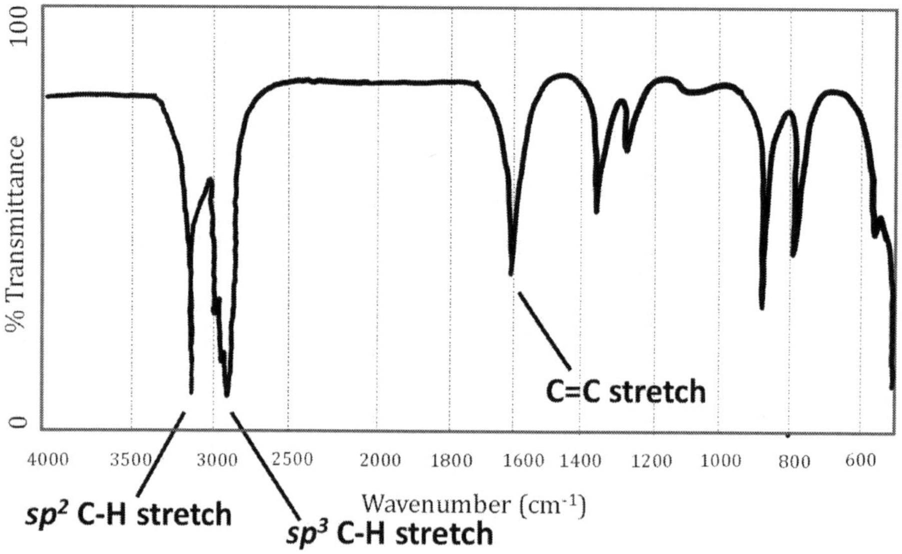

A) 

Answer to IR Spectroscopy Problem 3

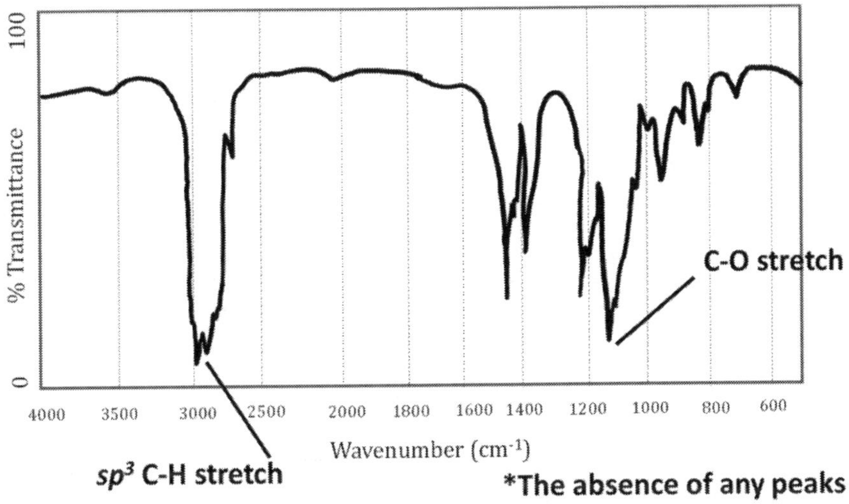

F) 

***The absence of any peaks attributable to a multiple bond or O-H stretch rules out other choices**

Answer to IR Spectroscopy Problem 4

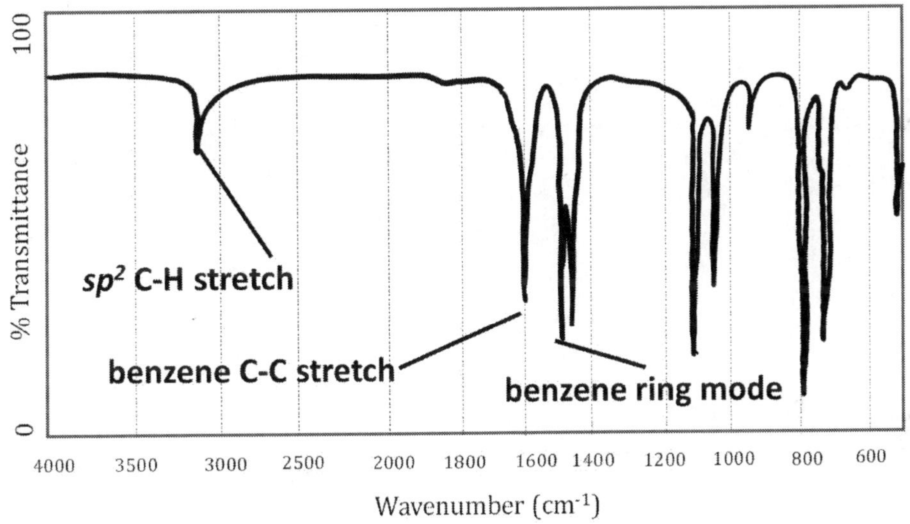

E) 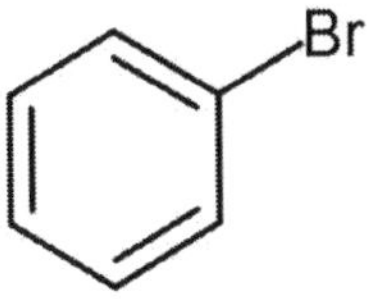

Answer to IR Spectroscopy Problem 5

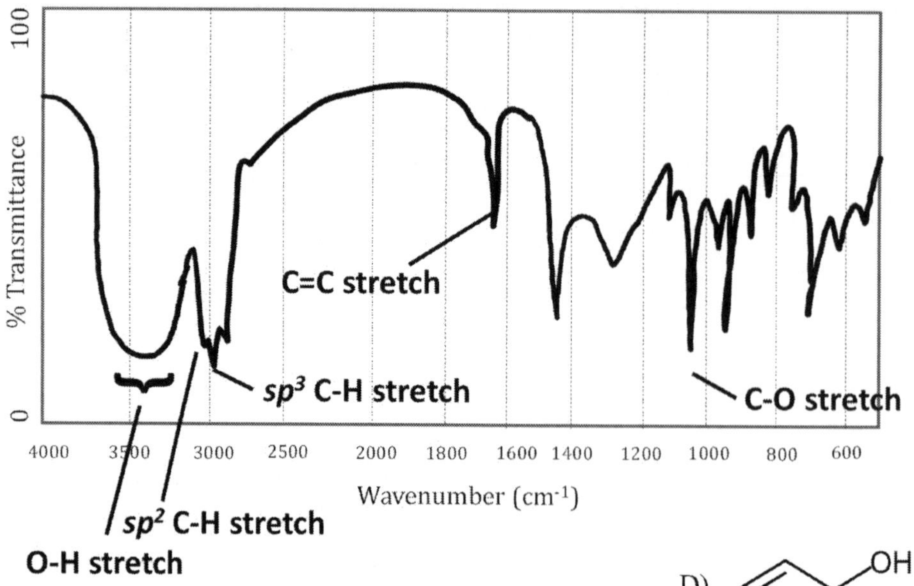

D) 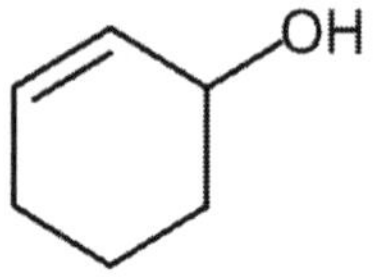

Answer to IR Spectroscopy Problem 6

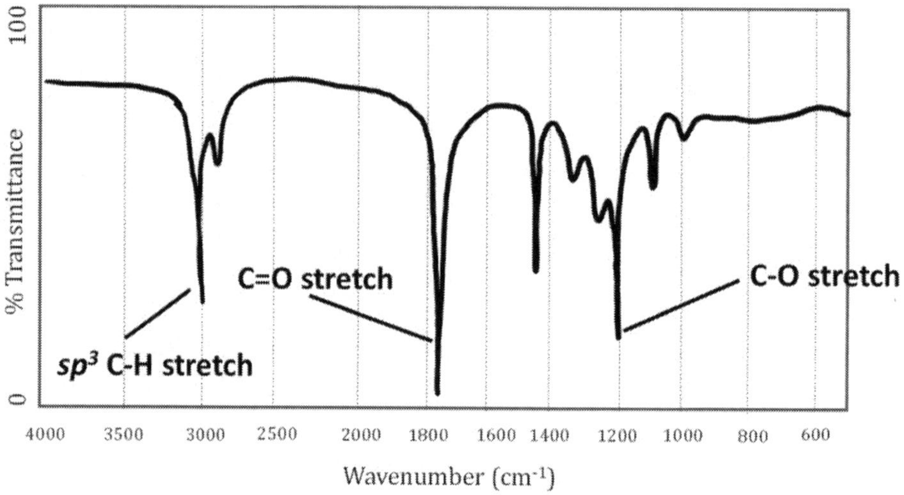

B) 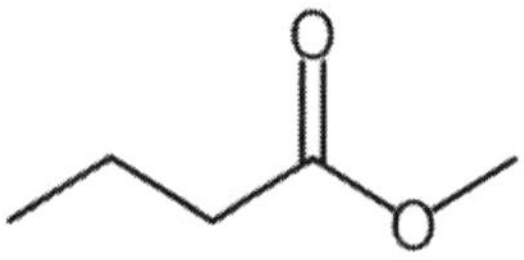

Answer to IR Spectroscopy Problem 7

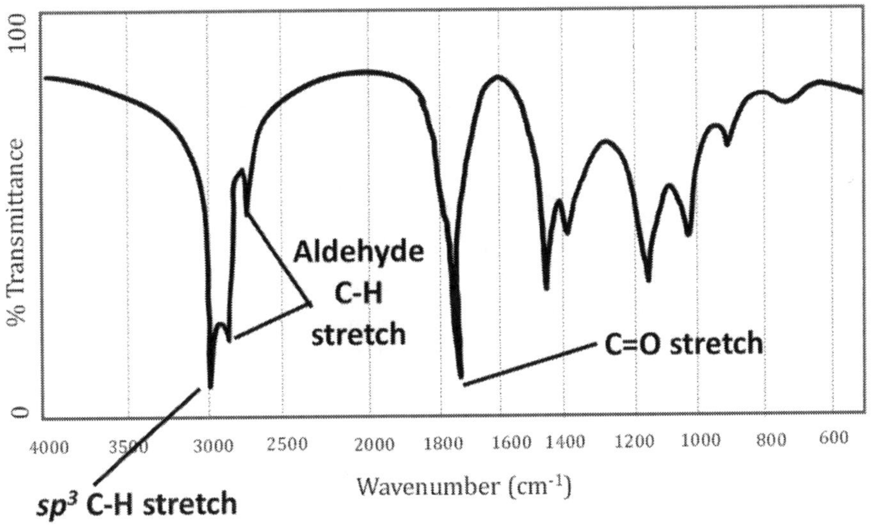

C)

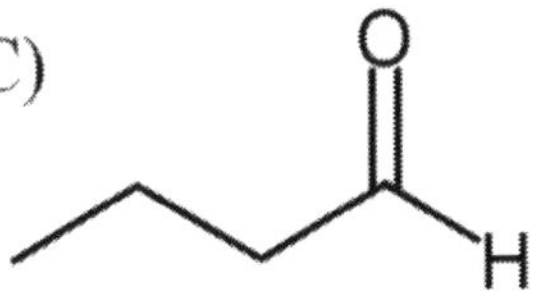

Answer to IR Spectroscopy Problem 8:

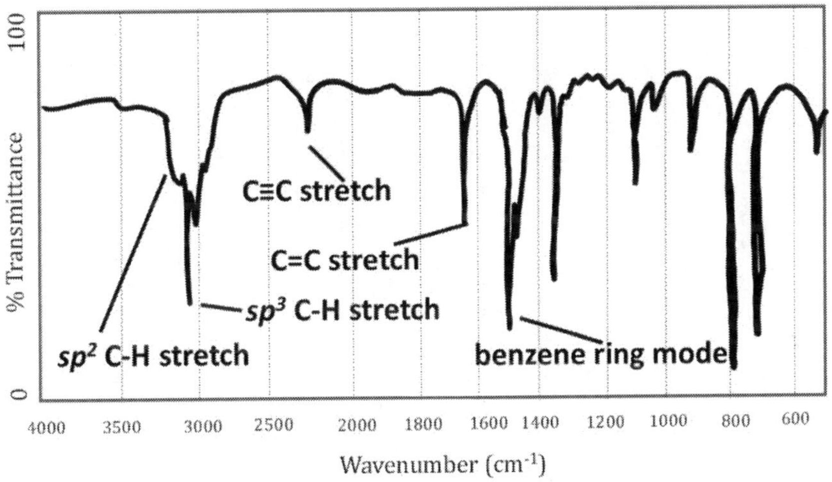

***The absence of an *sp* C-H stretch at 3300 cm^{-1} rules out a terminal alkyne**

D)

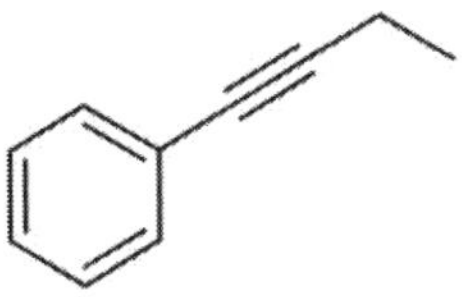

Answer to IR Spectroscopy Problem 9:

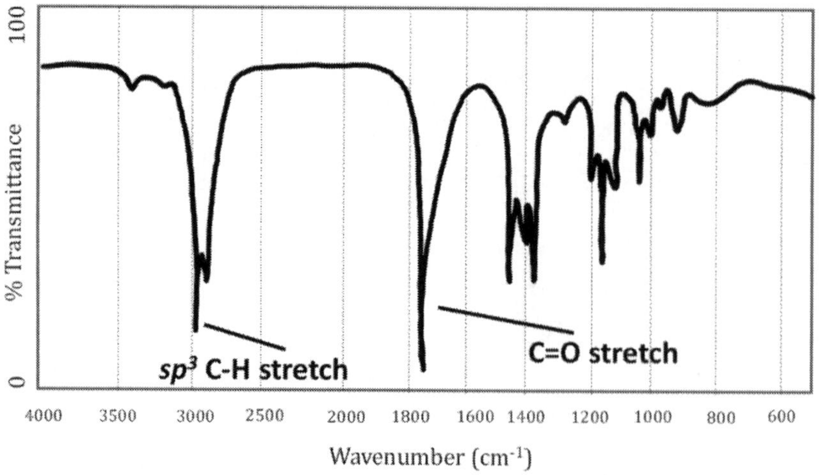

A)

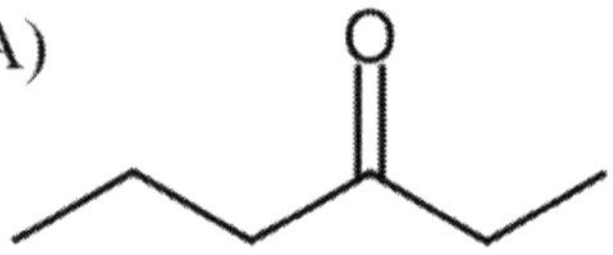

Answer to IR Spectroscopy Problem 10:

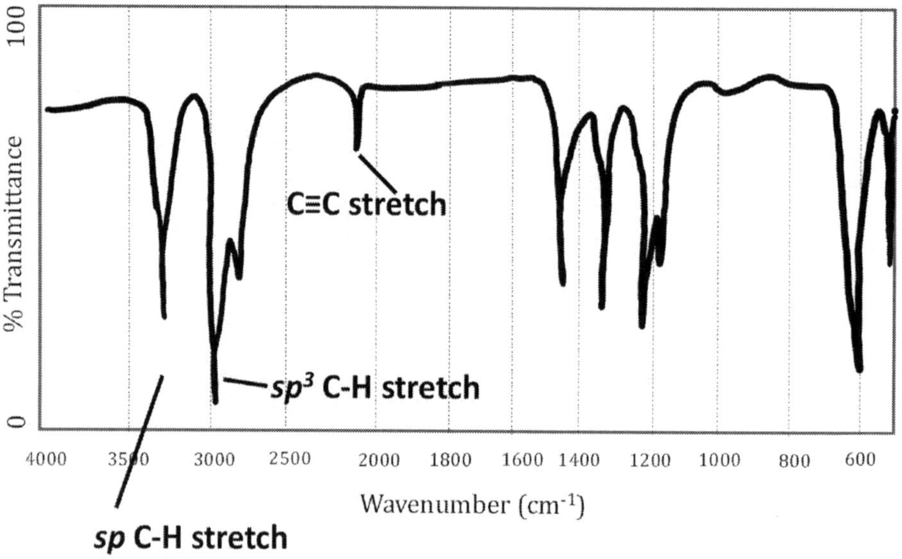

B)

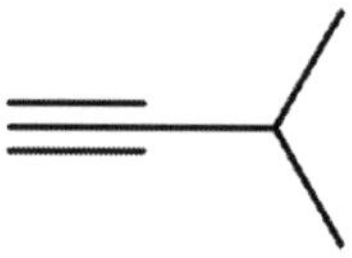

Answer to IR Spectroscopy Problem 11

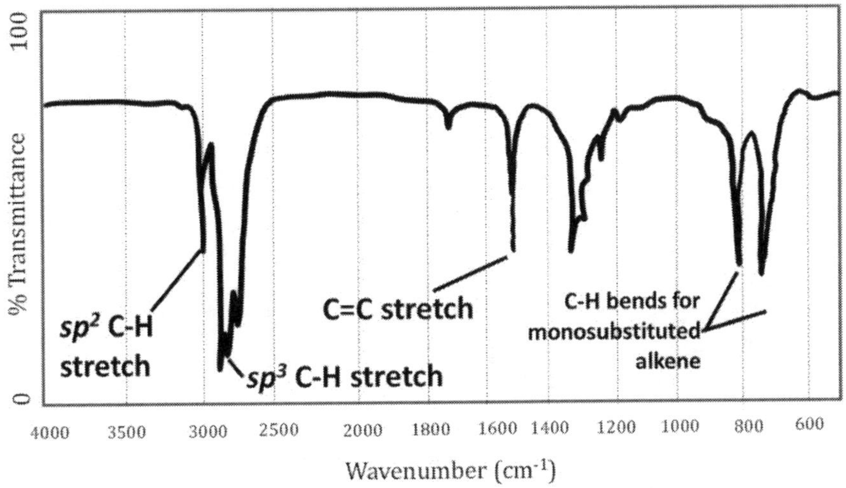

E)

Answer to IR Spectroscopy Problem 12

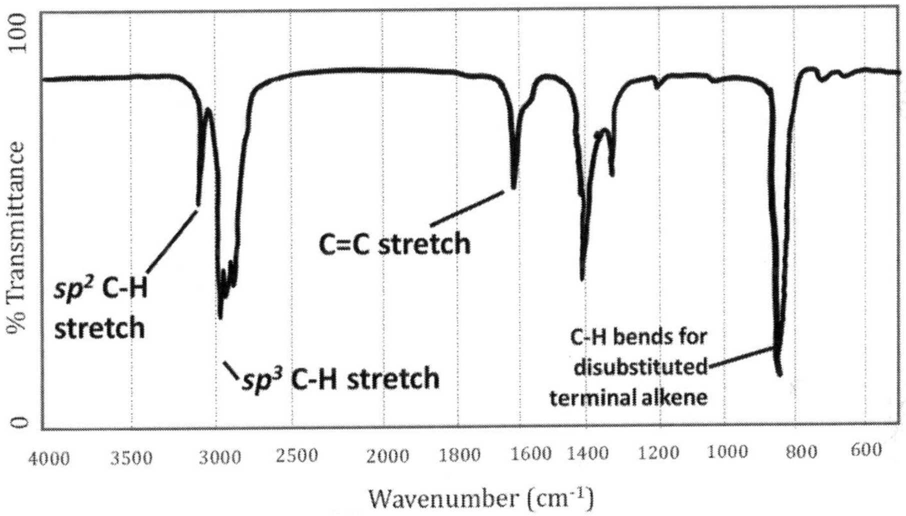

C)

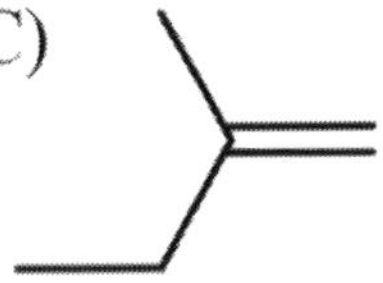

Answer to IR Spectroscopy Problem 13

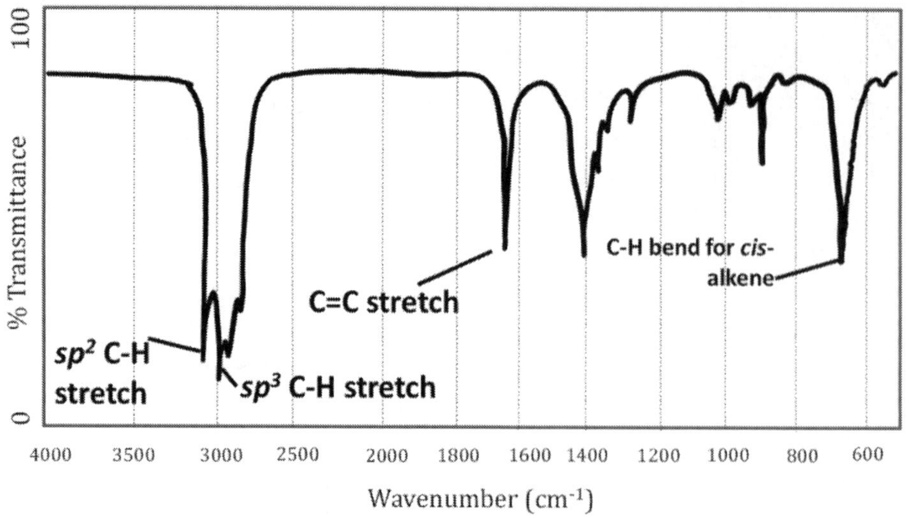

B)

Answer to IR Spectroscopy Problem 14

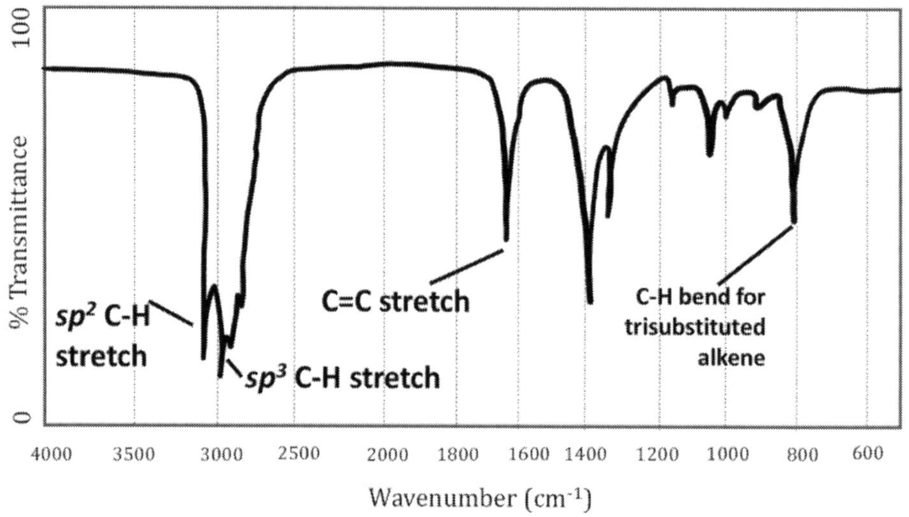

D)

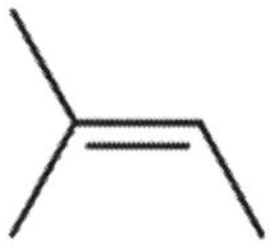

Answer to IR Spectroscopy Problem 15

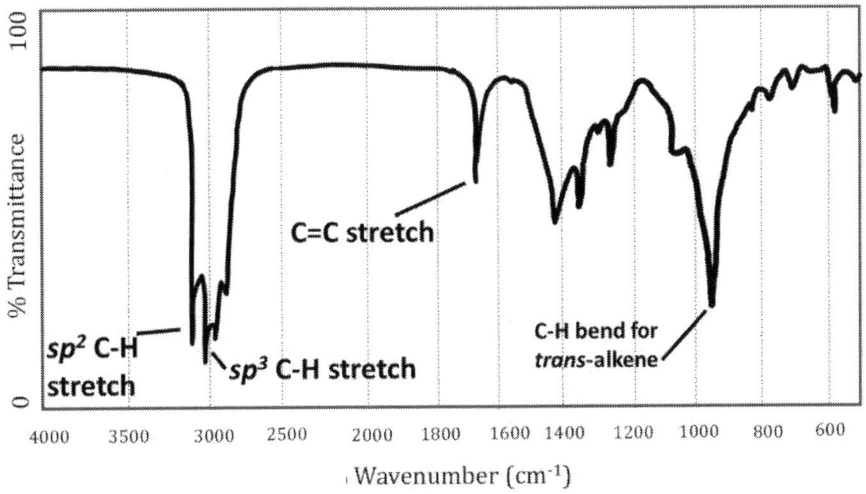

A)

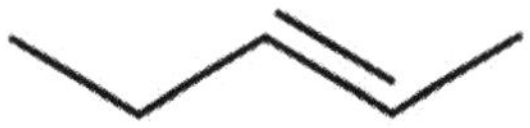

Lesson VIII.4. Proton NMR Practice Problems

For each of the following, deduce the structure having the provided molecular formula that is most likely to have produced the accompanying ^1H NMR spectrum.

Proton NMR Problem 1

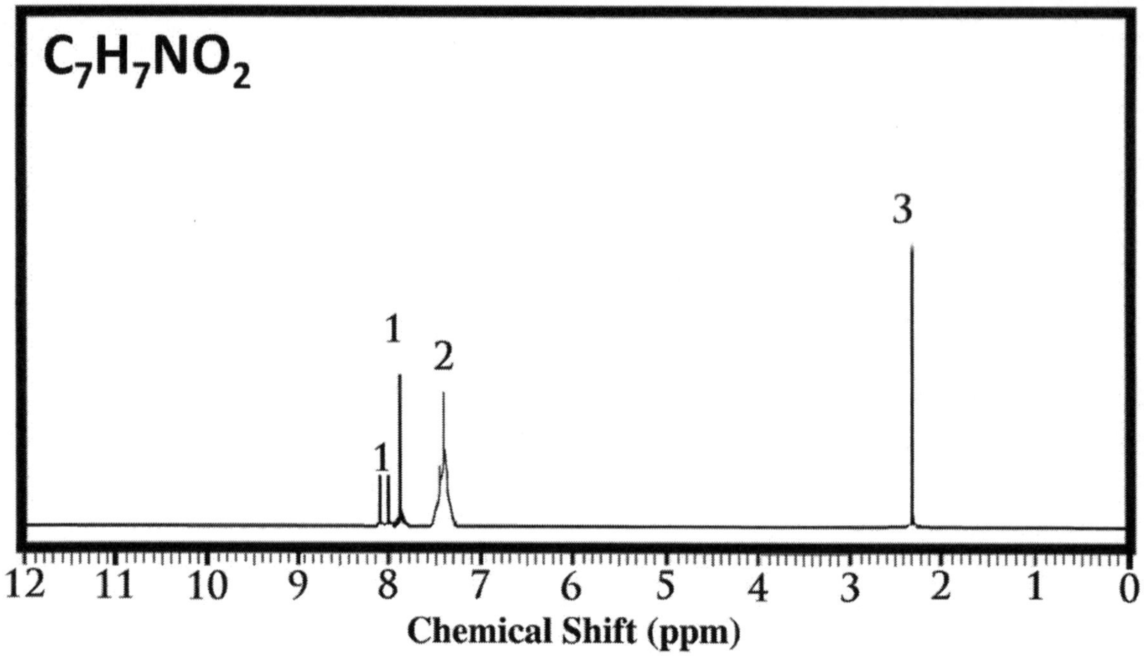

Proton NMR Problem 2

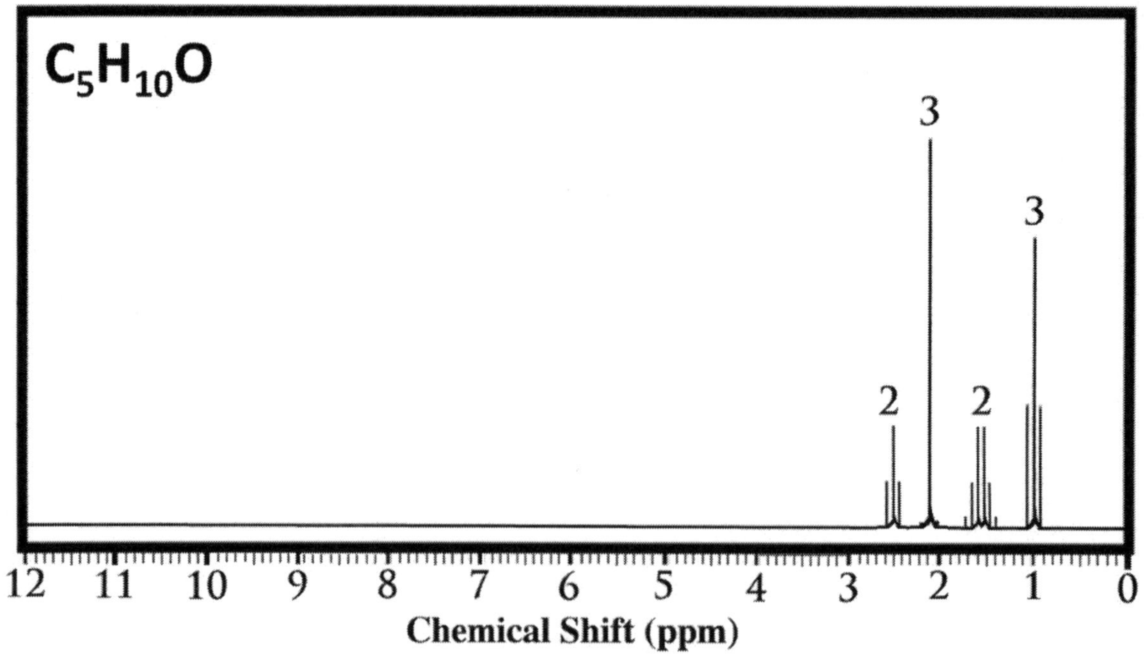

Proton NMR Problem 3

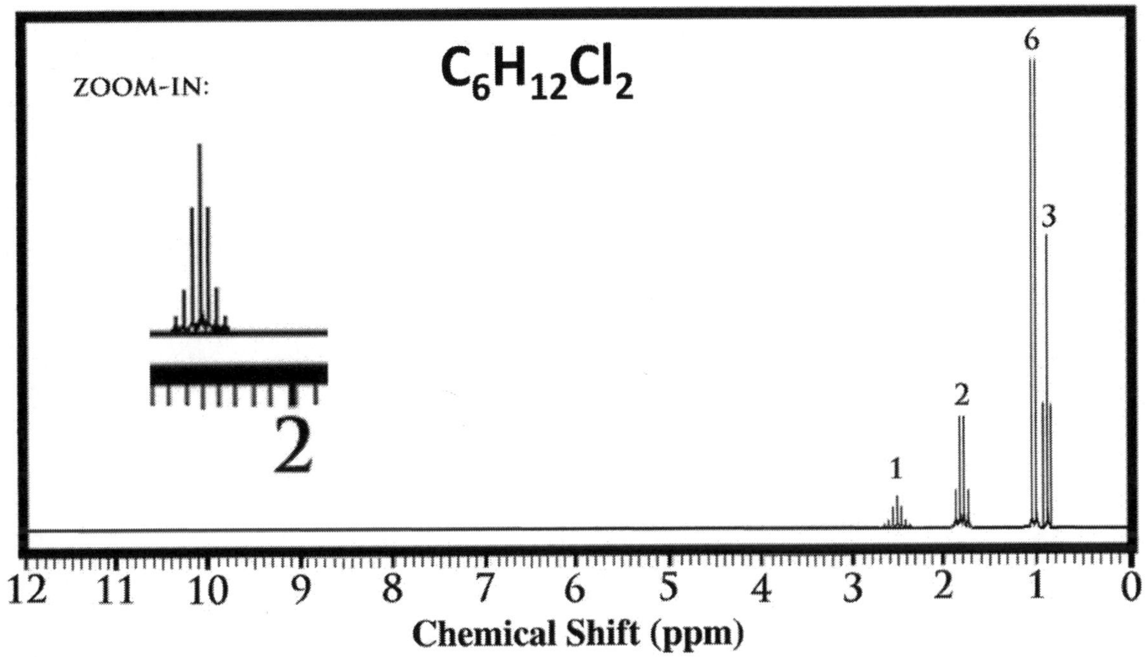

Proton NMR Problem 4

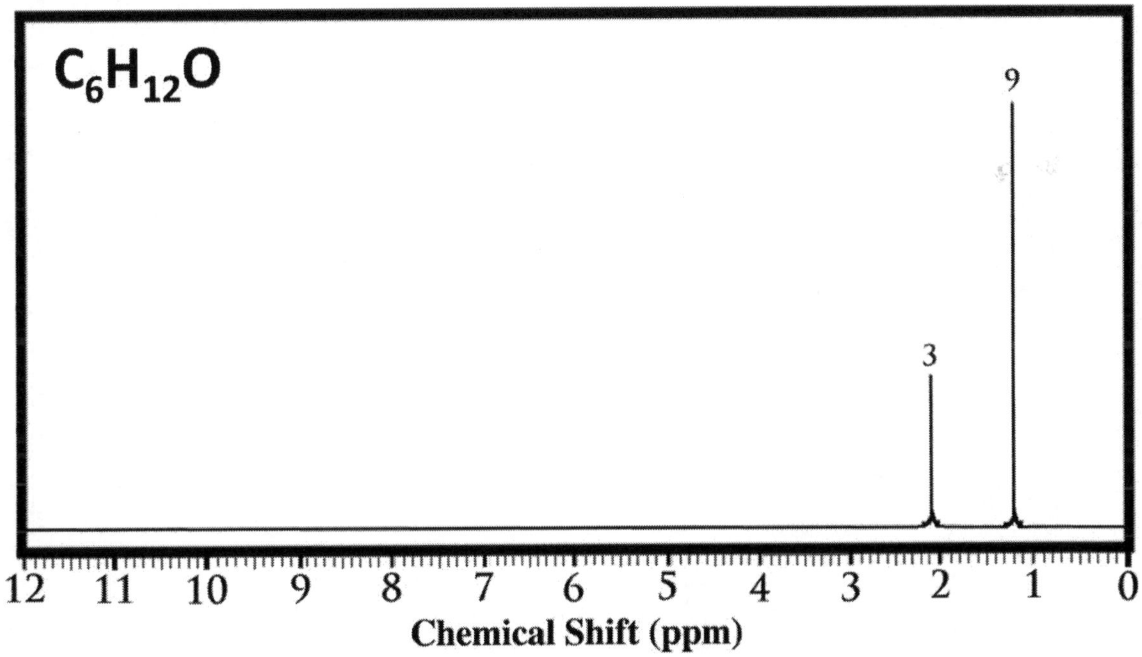

Proton NMR Problem 5

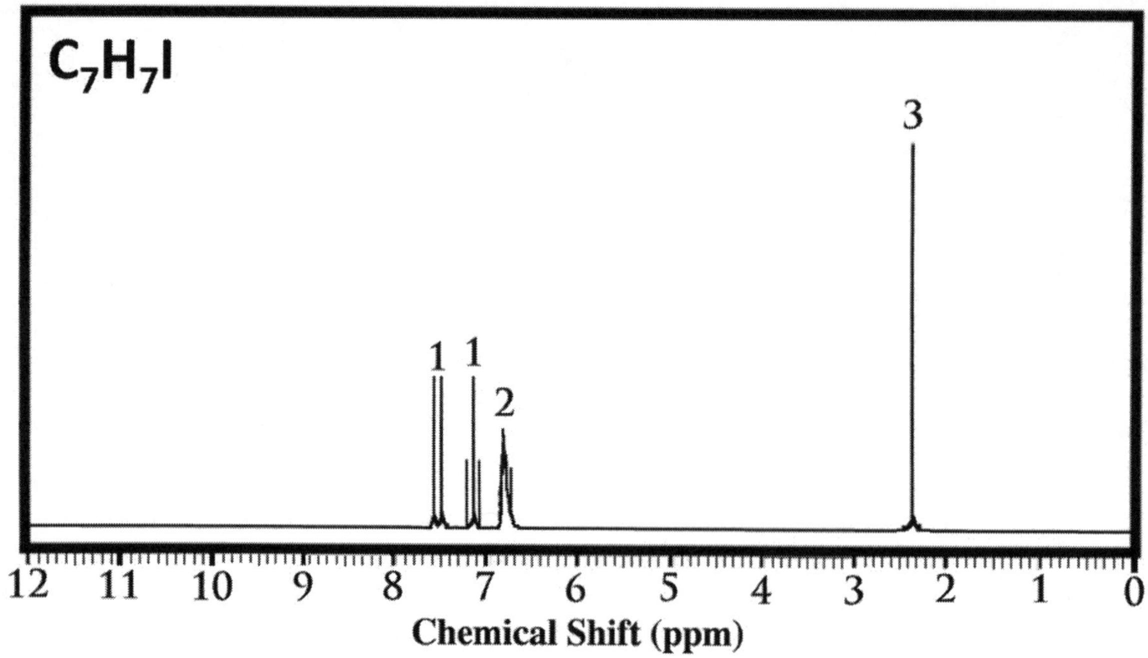

Proton NMR Problem 6

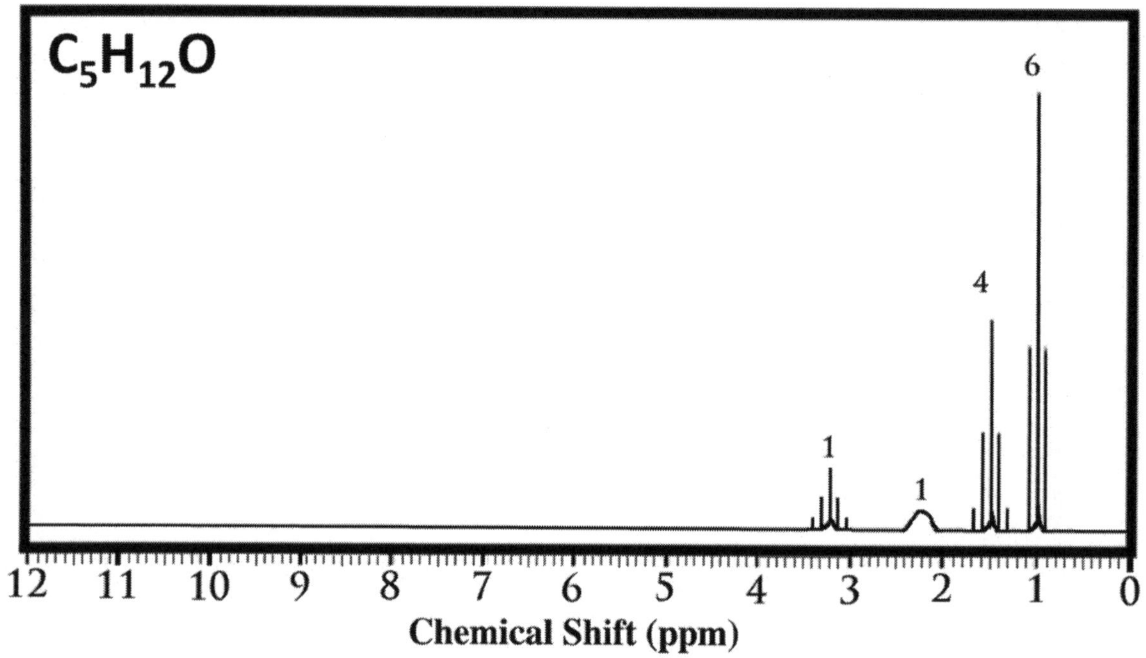

Proton NMR Problem 7

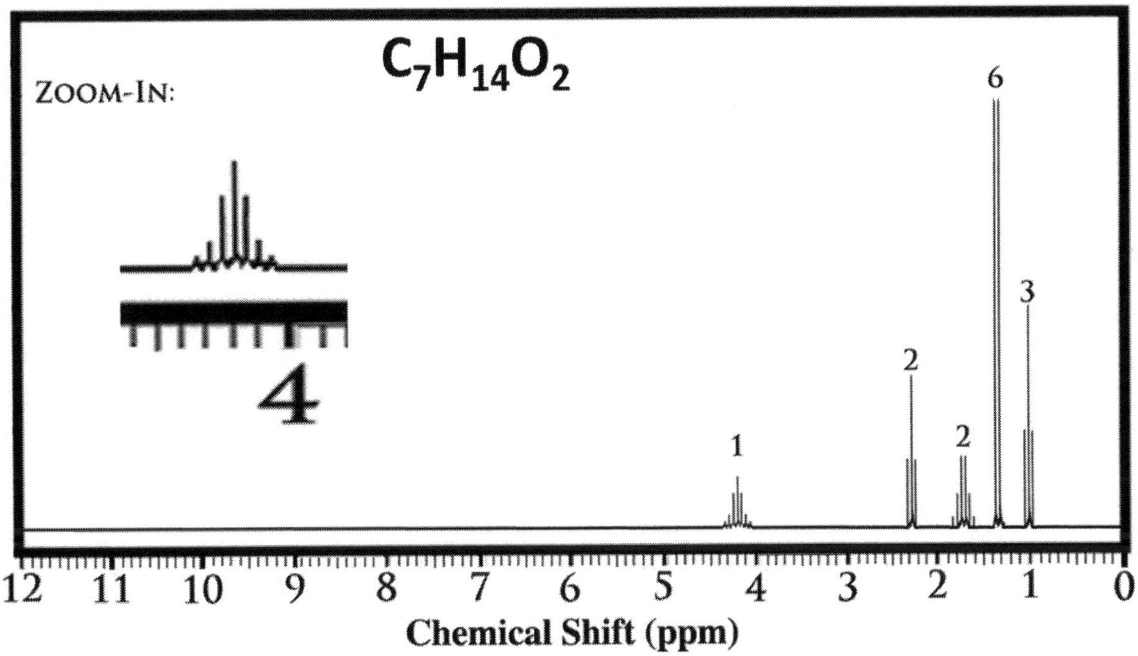

Proton NMR Problem 8

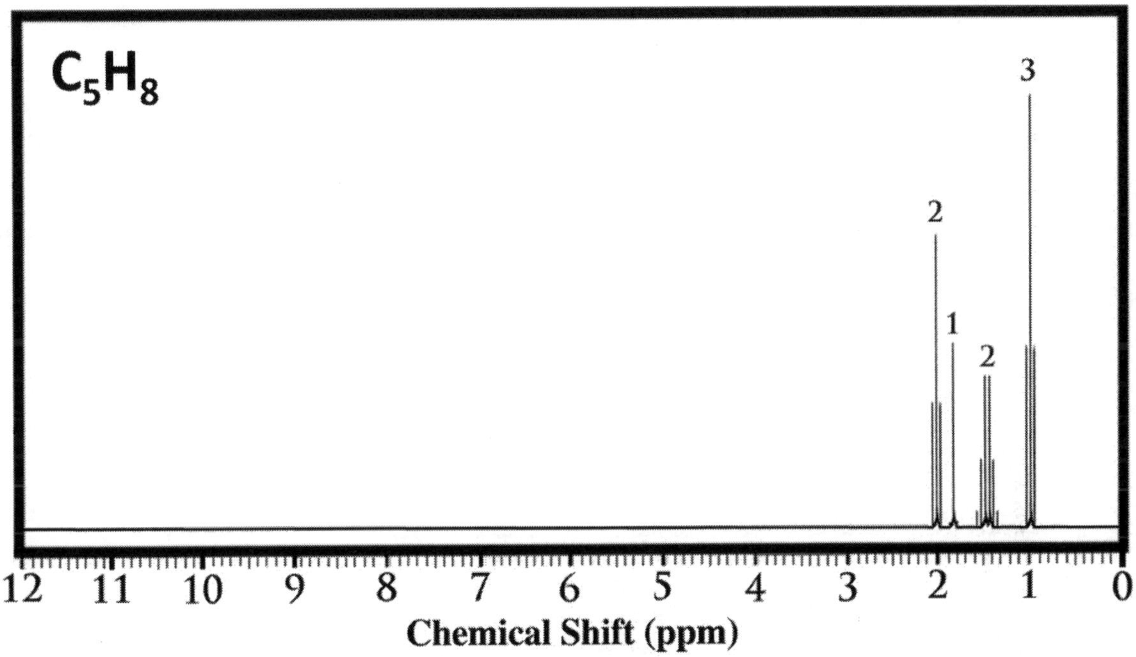

Proton NMR Problem 9

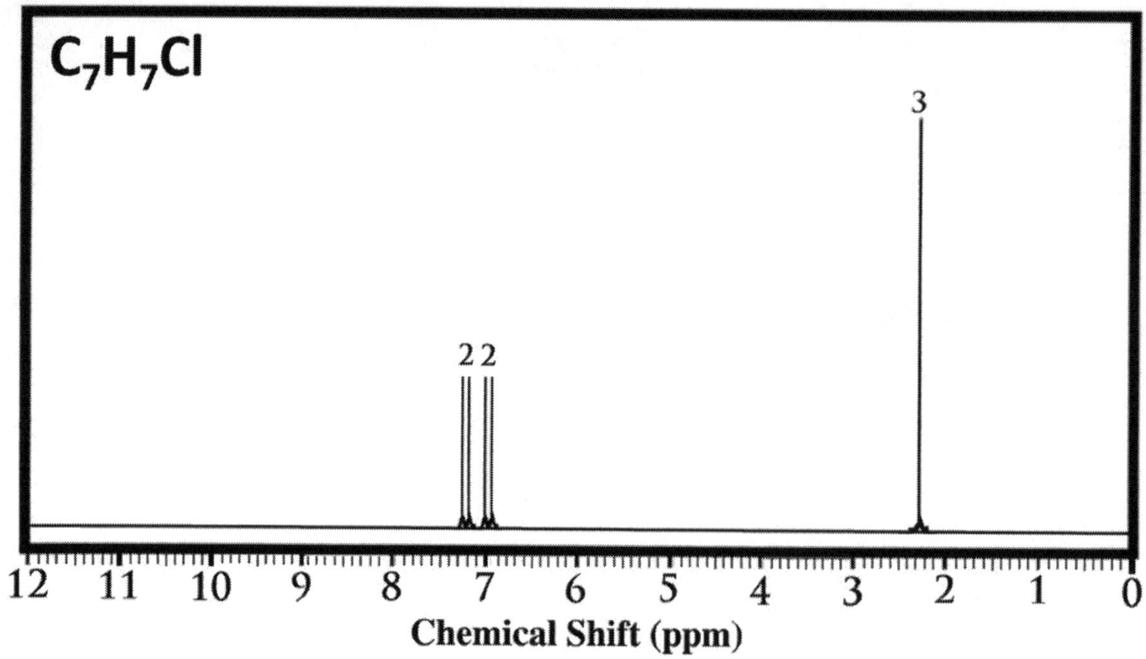

Proton NMR Problem 10

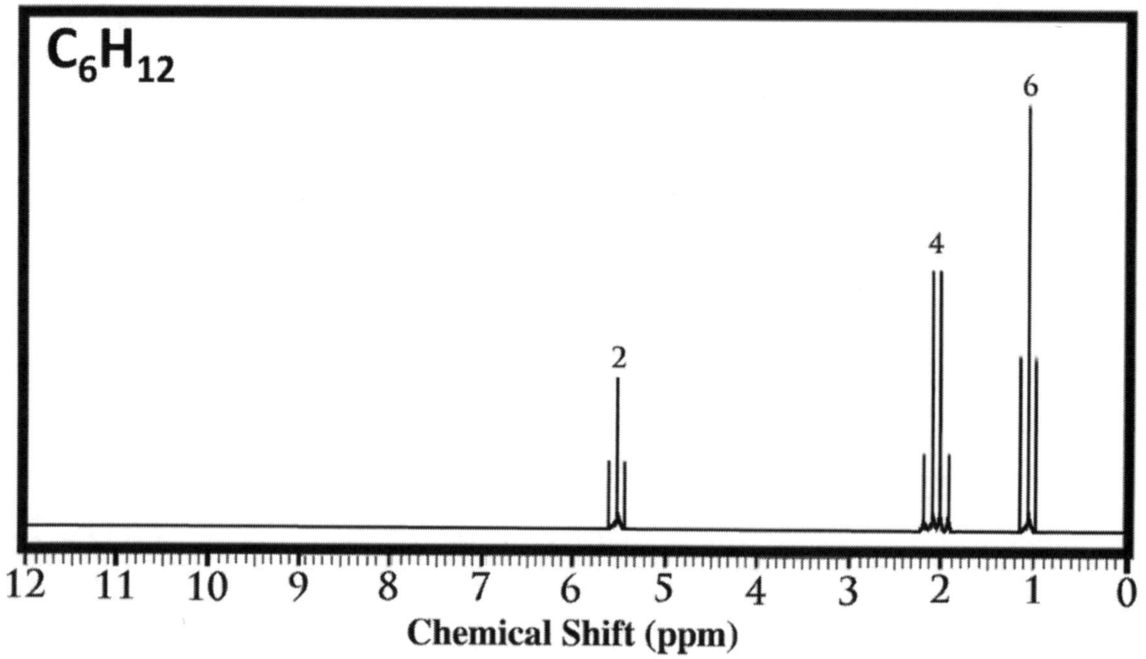

Proton NMR Problem 11

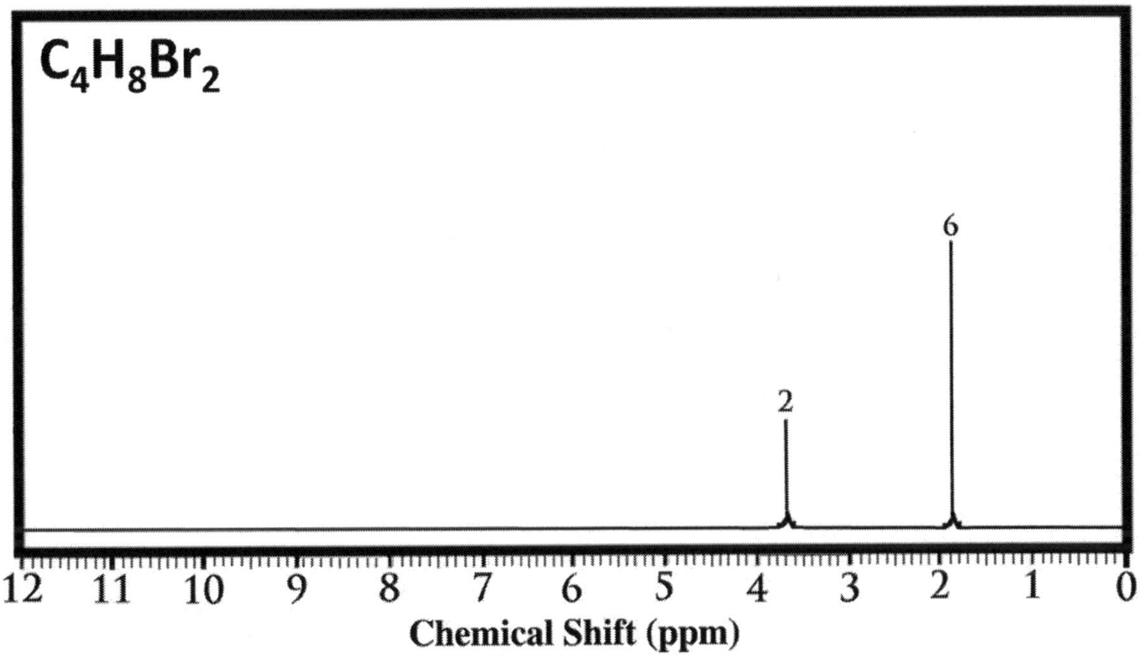

Proton NMR Problem 12

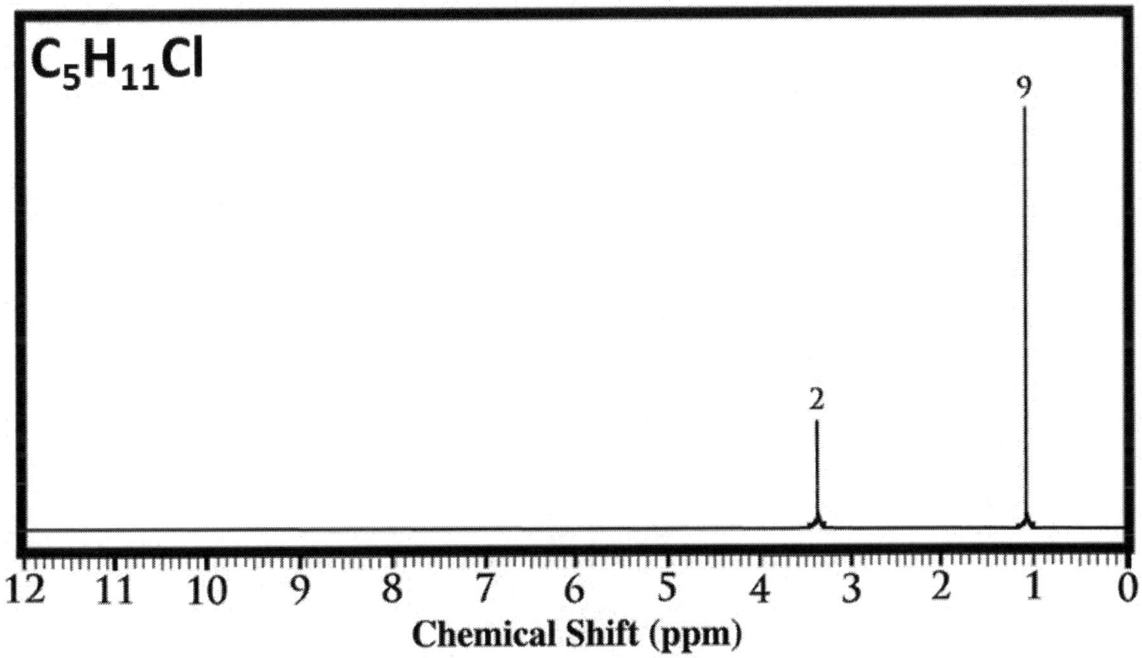

Proton NMR Problem 13

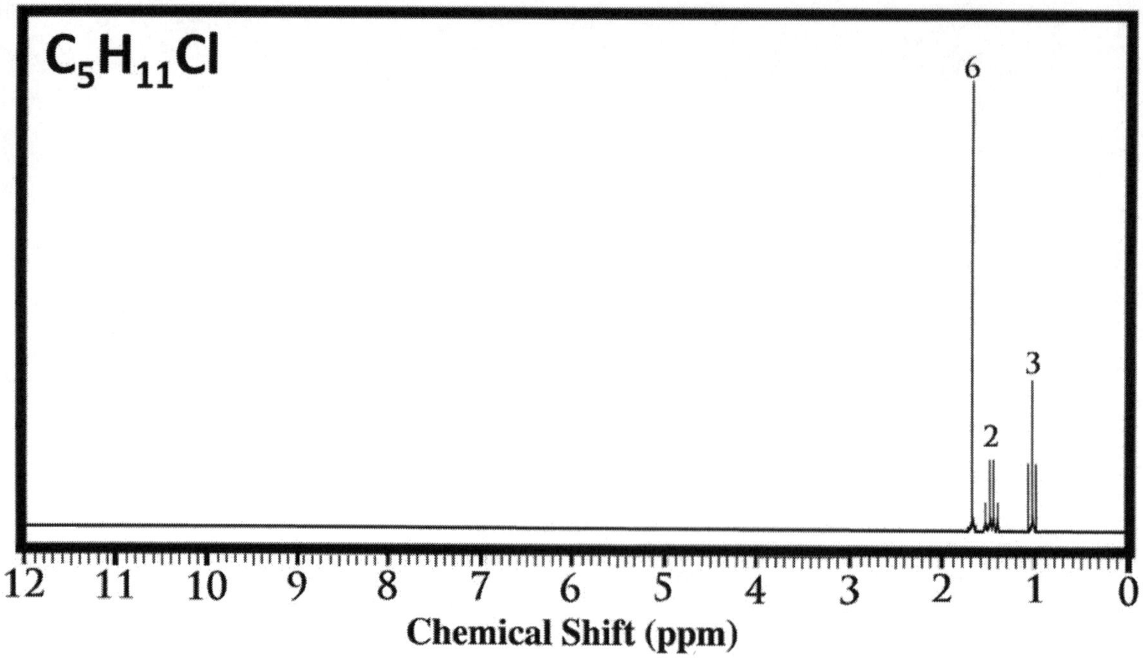

Proton NMR Problem 14

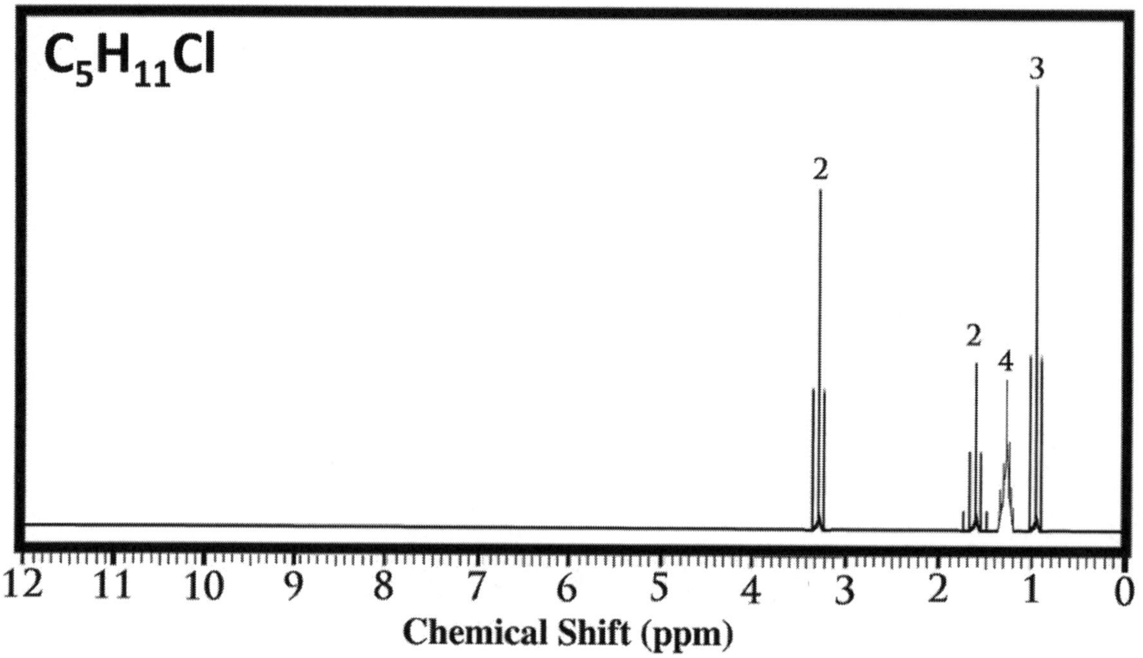

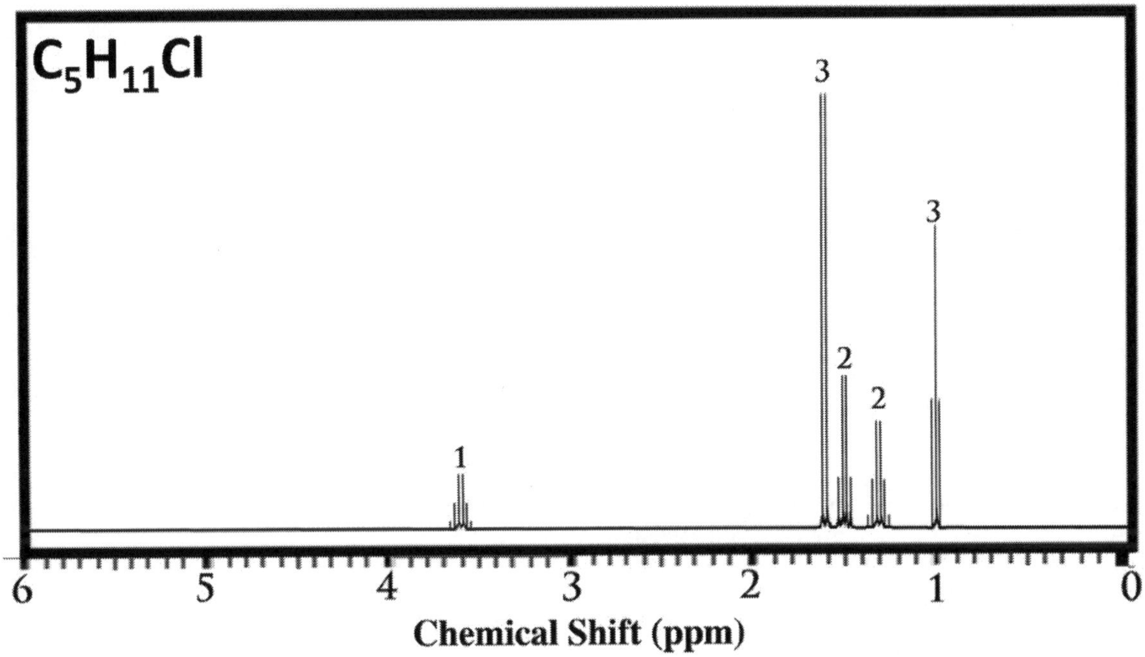
$C_5H_{11}Cl$
3
3
2
2
1
6
5
4
3
2
1
0
Chemical Shift (ppm)

Lesson VIII.5. Answers to Proton NMR Practice Problems

Answer to Proton NMR Problem 1

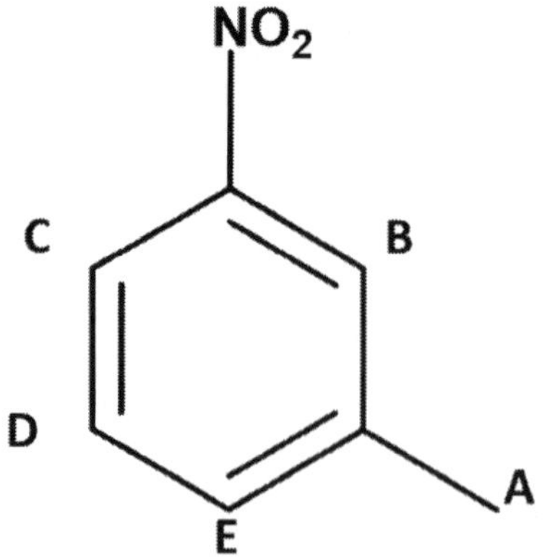

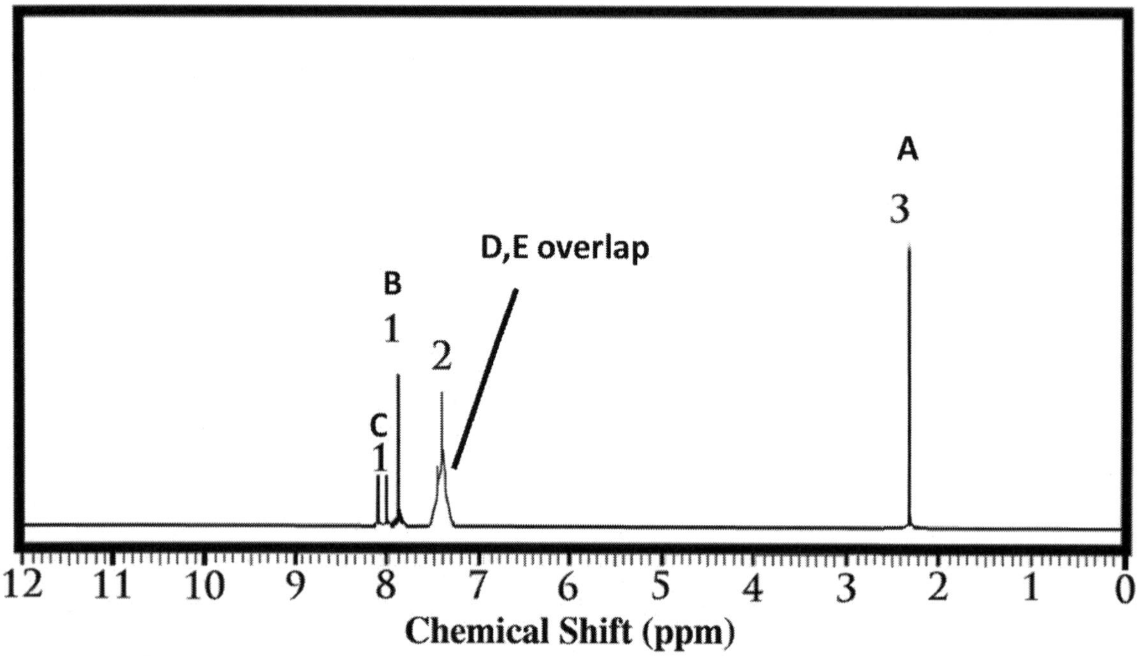

Answer to Proton NMR Problem 2

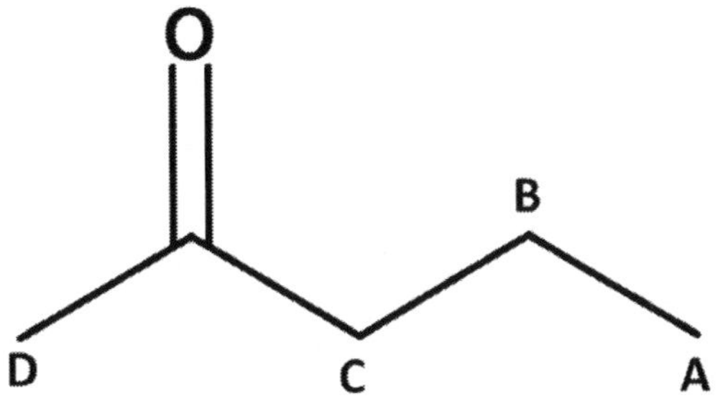

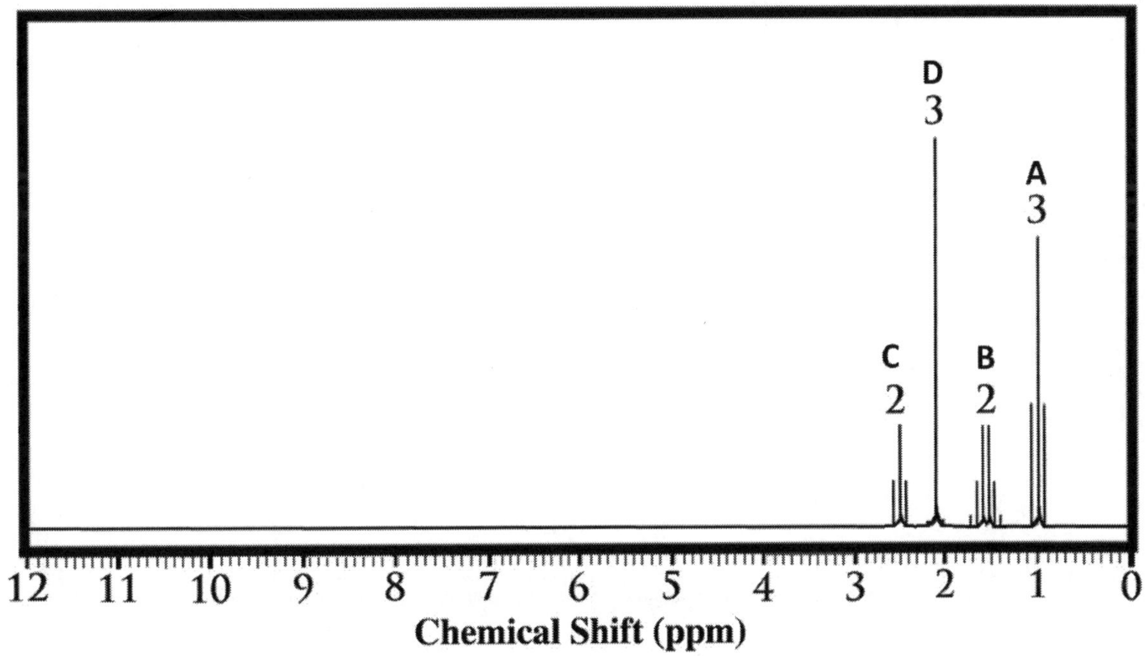

Answer to Proton NMR Problem 3

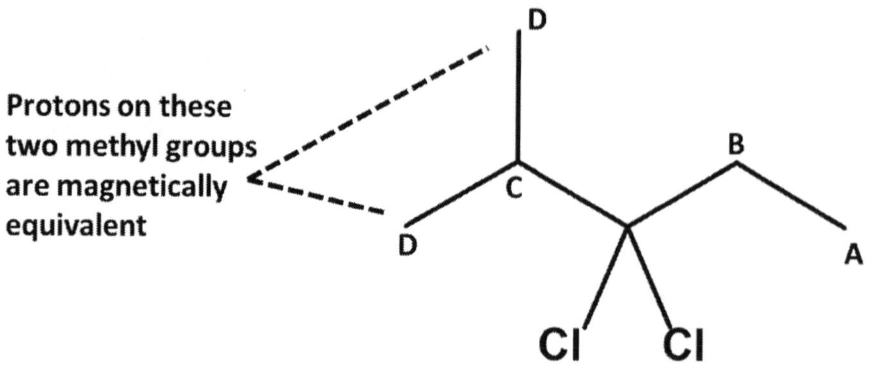

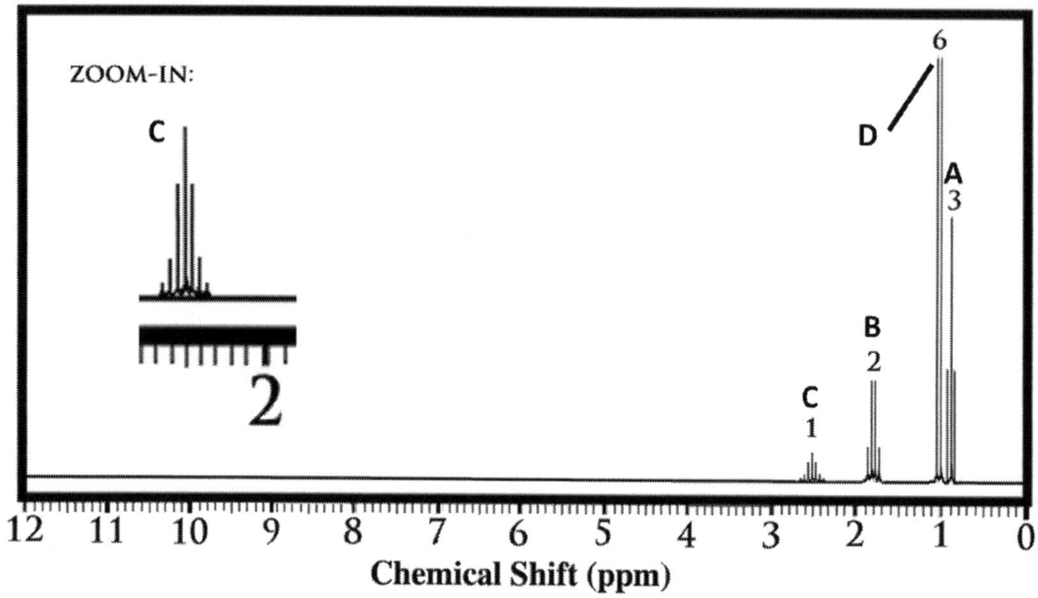

Answer to Proton NMR Problem 4

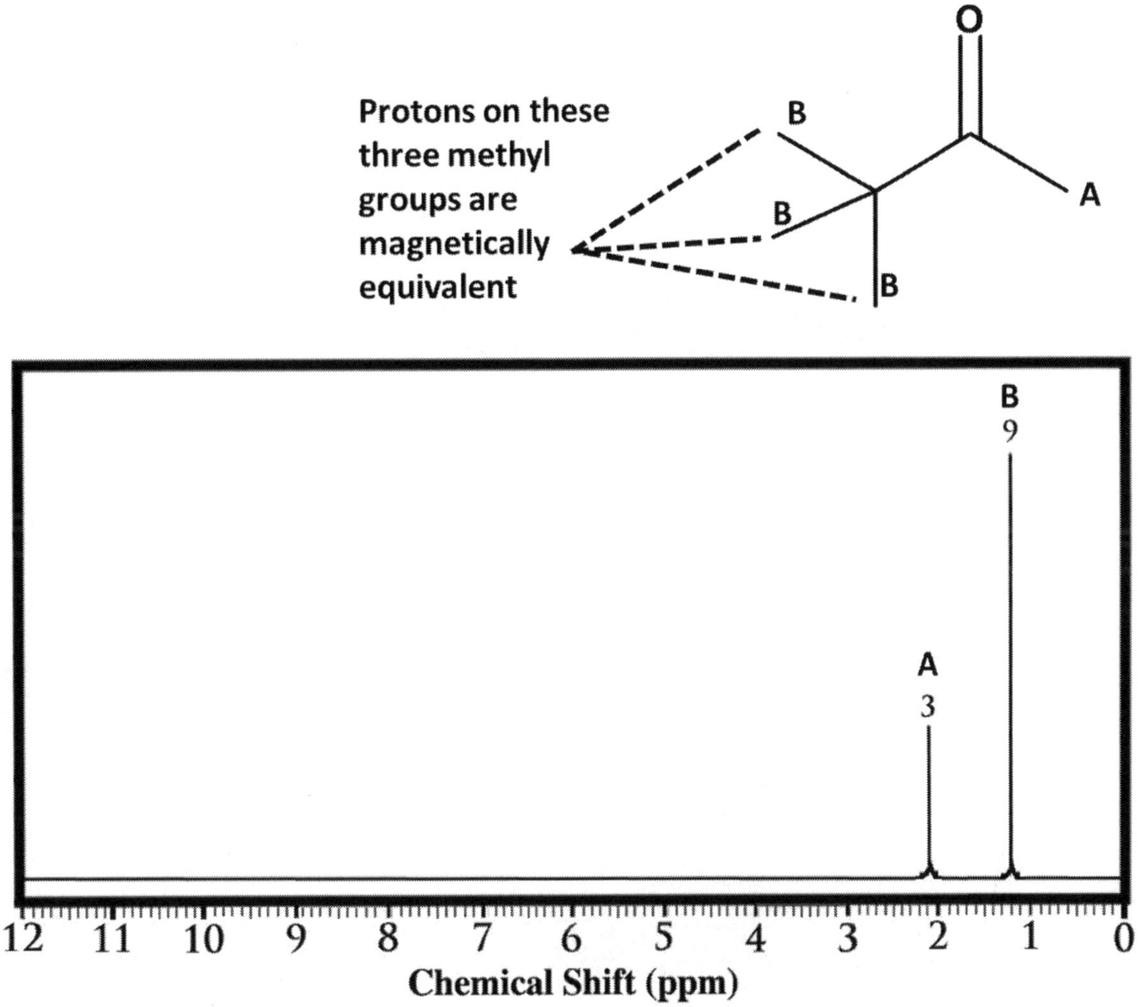

Answer to Proton NMR Problem 5

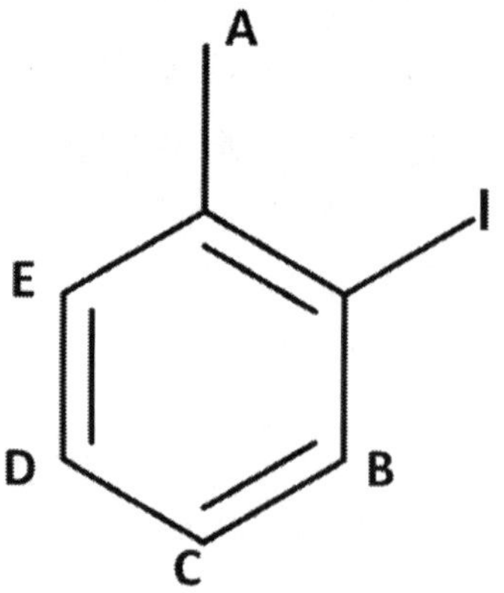

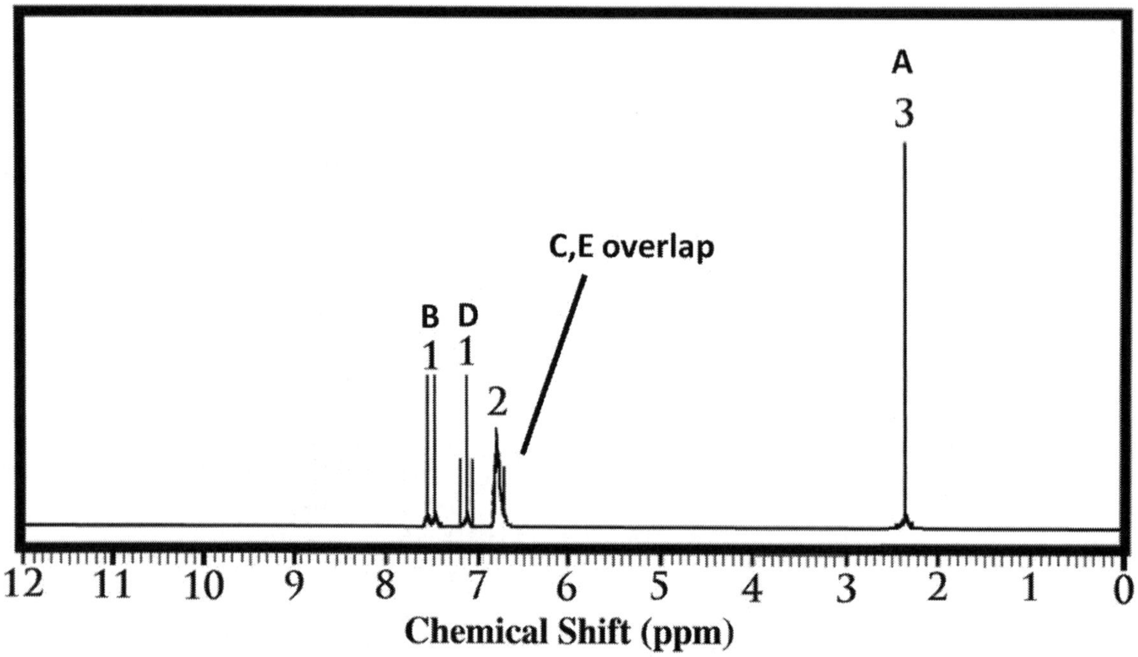

Answer to Proton NMR Problem 6

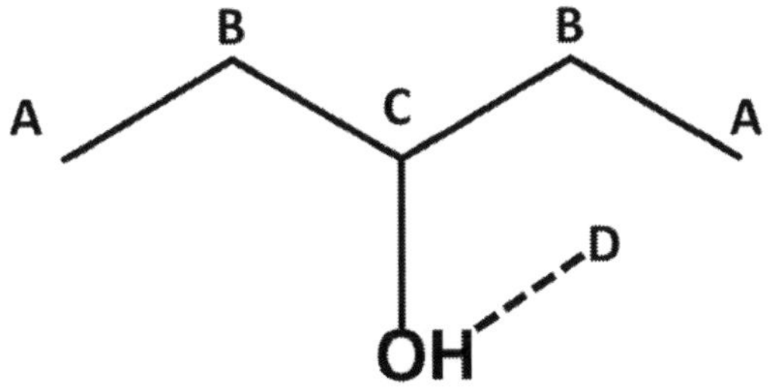

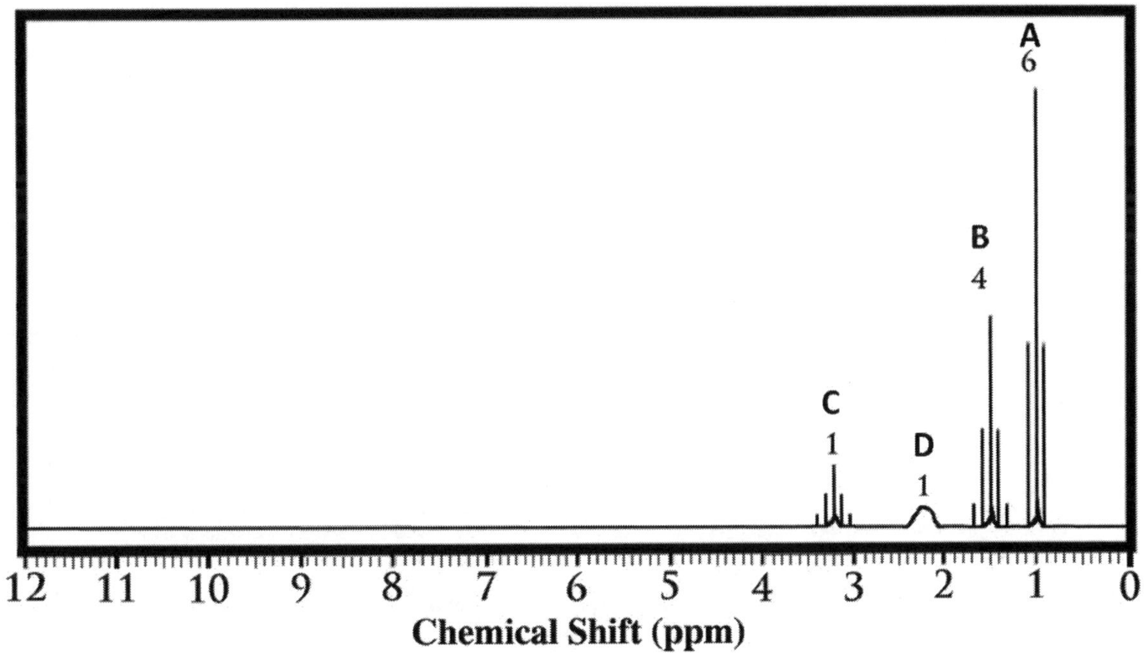

Answer to Proton NMR Problem 7

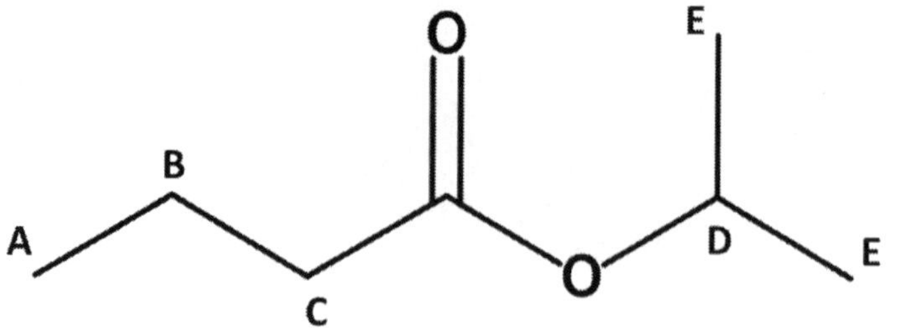

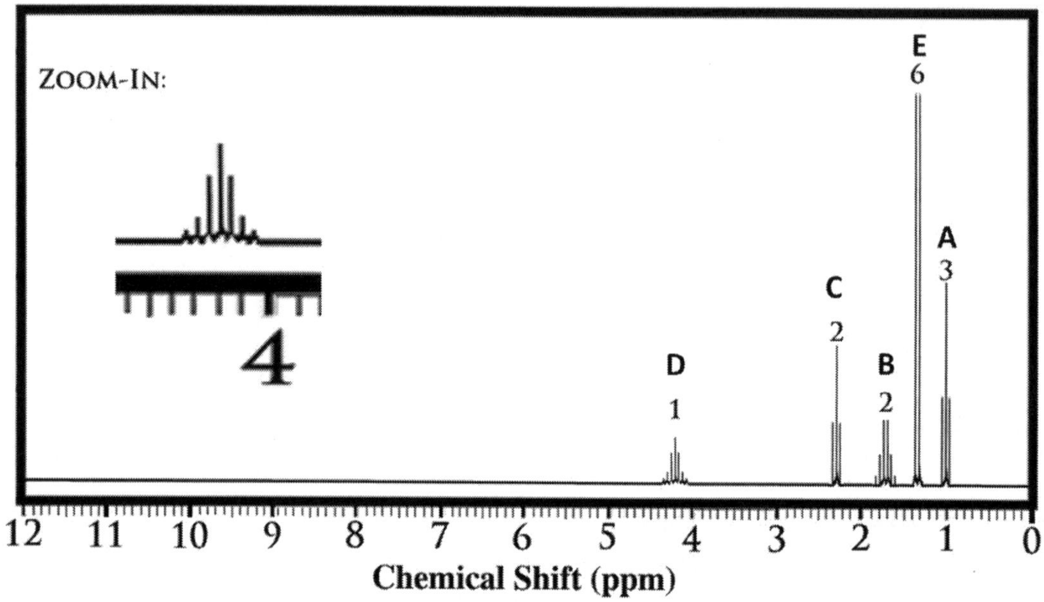

Answer to Proton NMR Problem 8

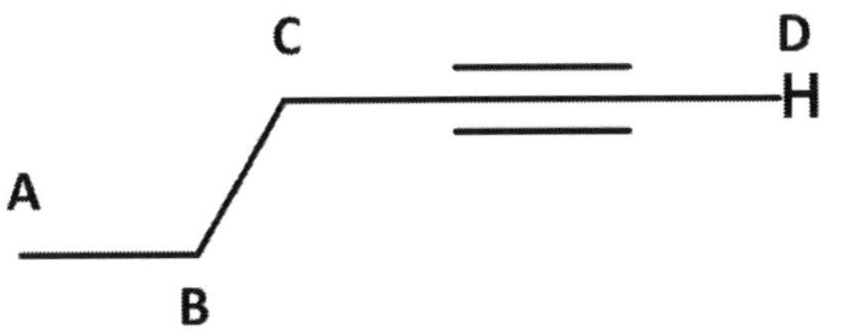

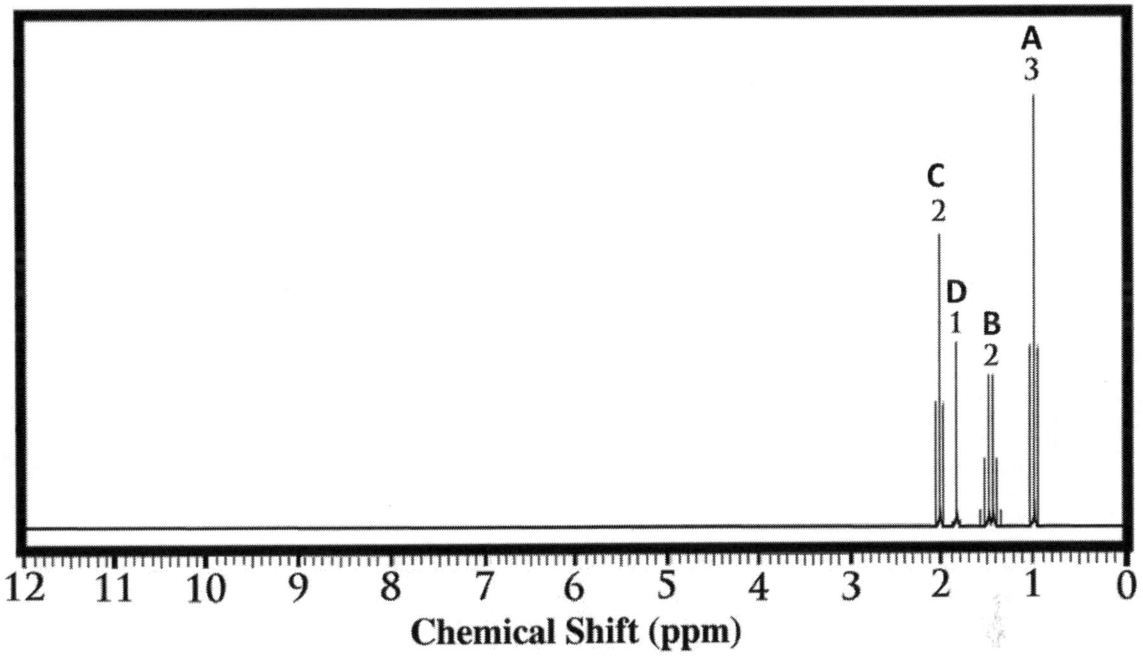

Answer to Proton NMR Problem 9

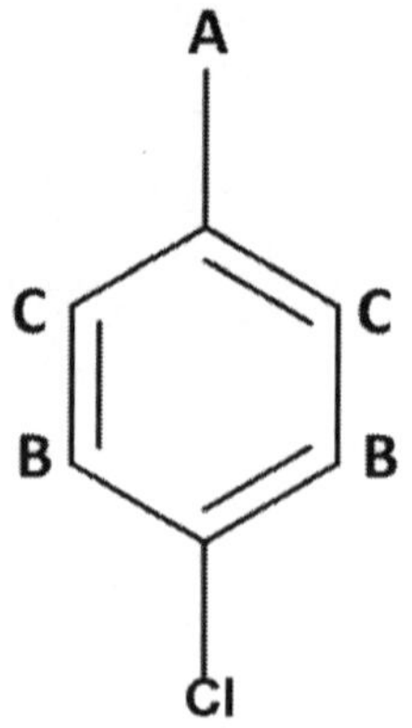

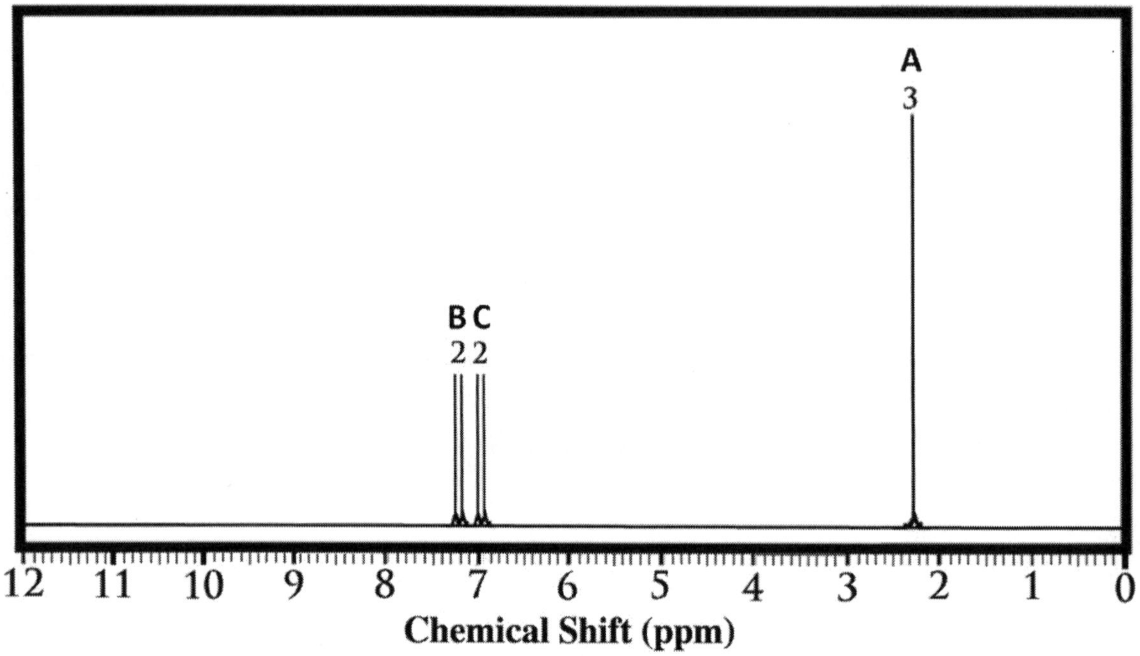

Answer to Proton NMR Problem 10

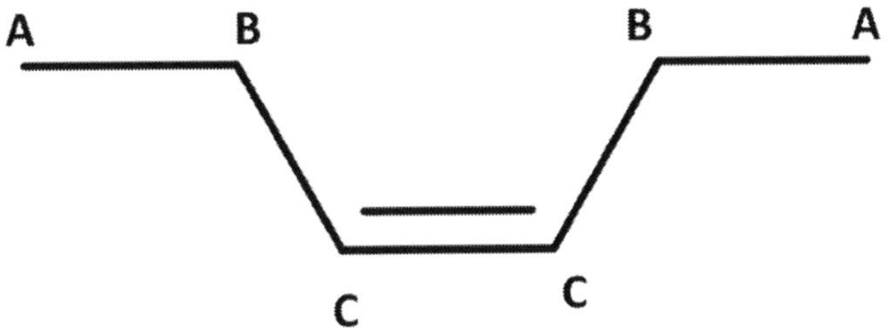

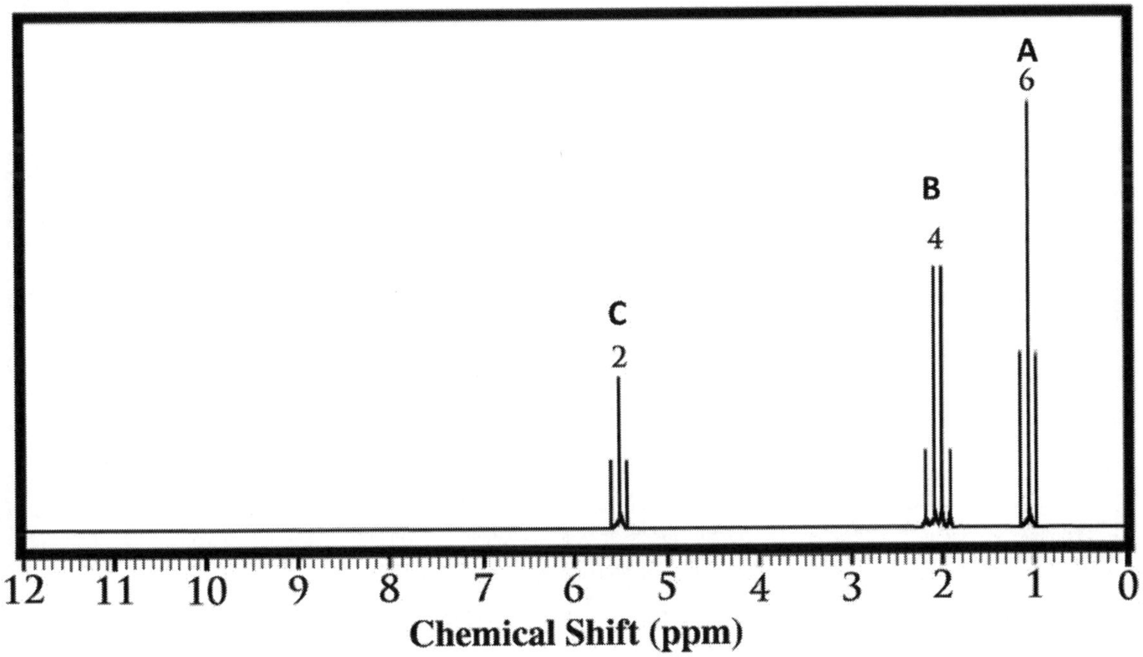

Answer to Proton NMR Problem 11

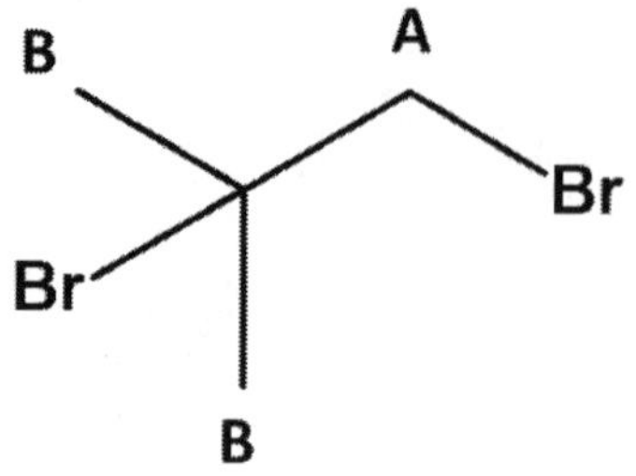

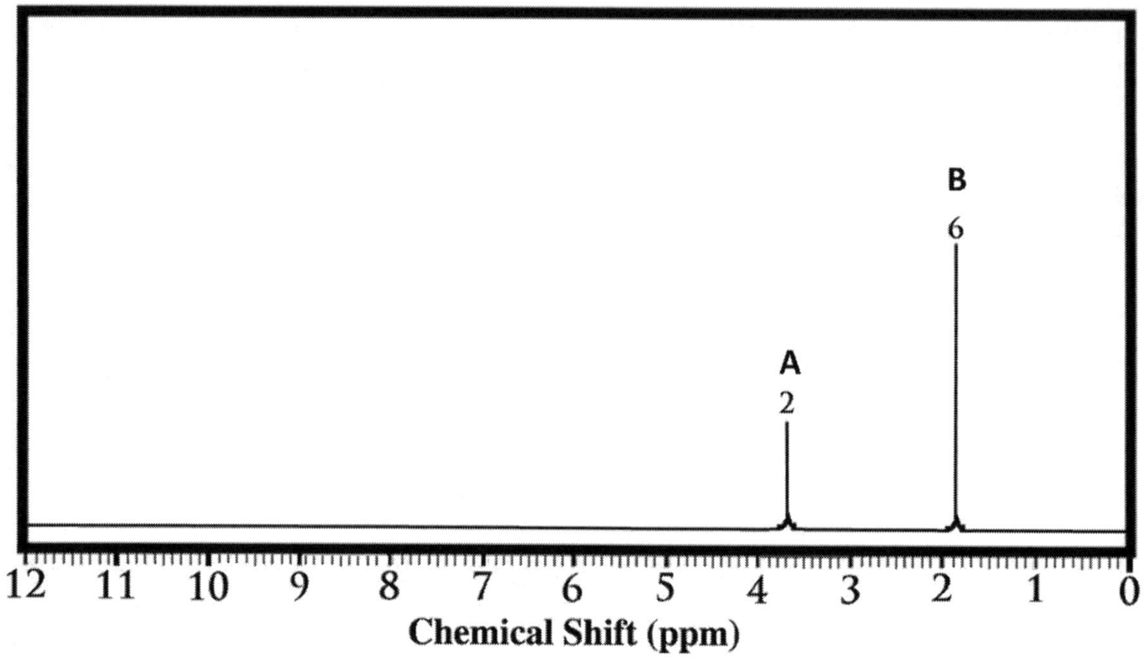

Answer to Proton NMR Problem 12

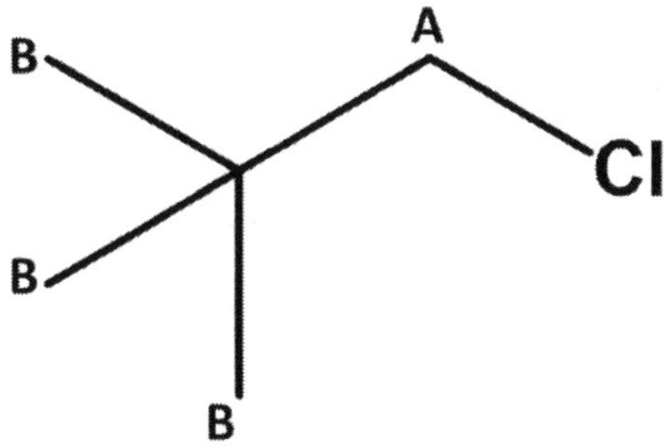

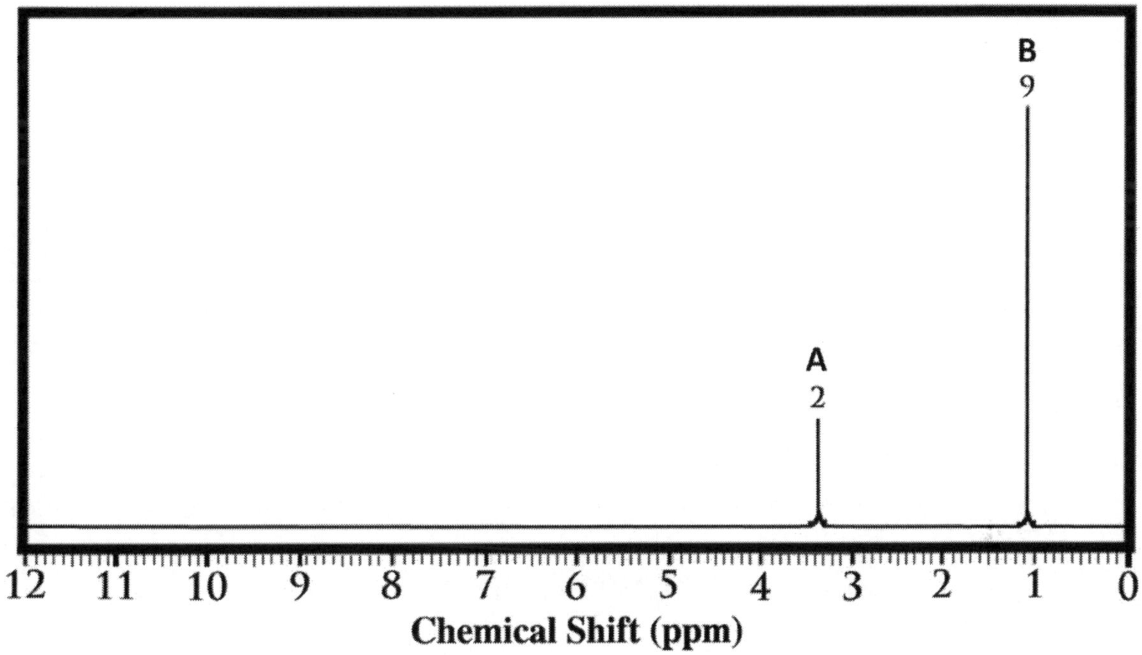

Answer to Proton NMR Problem 13

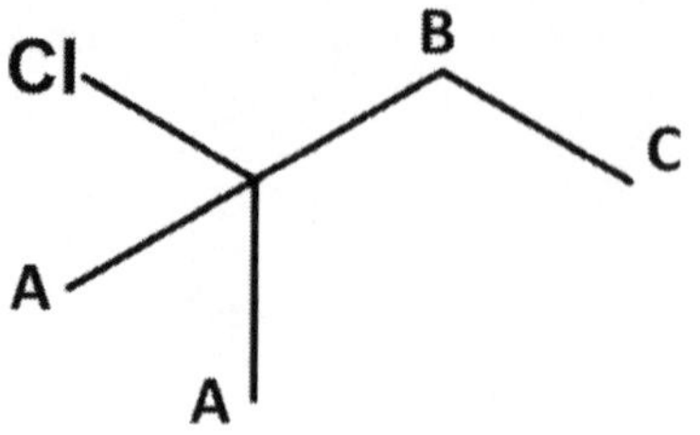

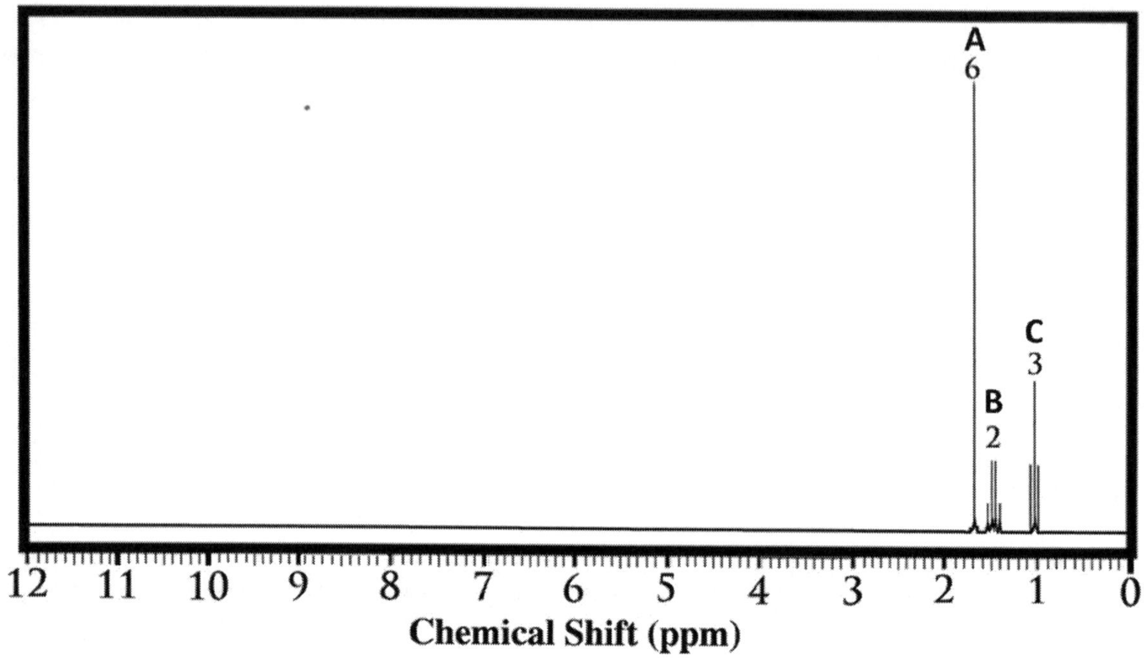

Answer to Proton NMR Problem 14

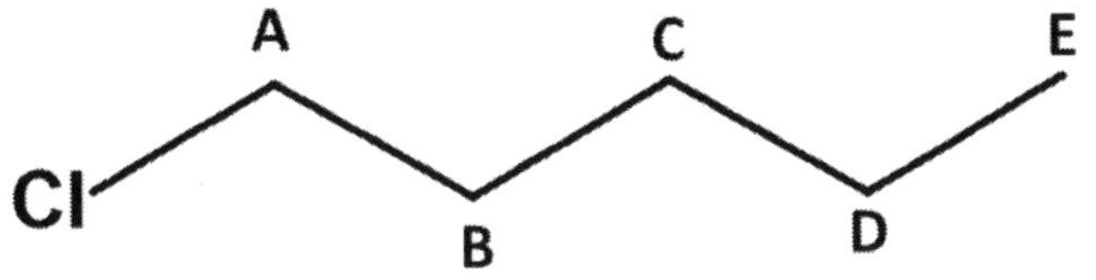

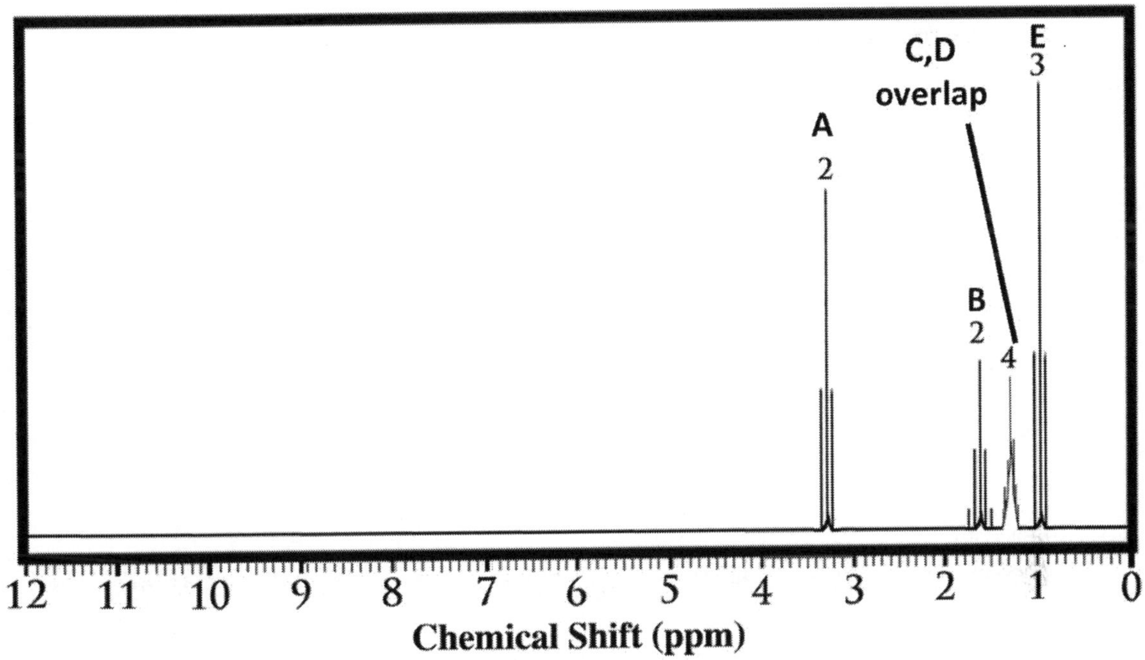

Answer to Proton NMR Problem 15

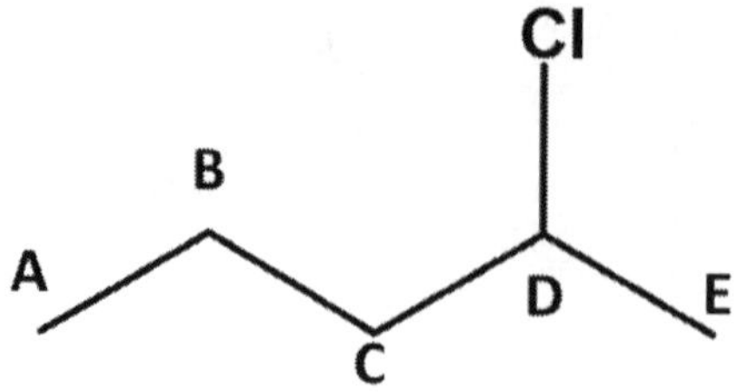

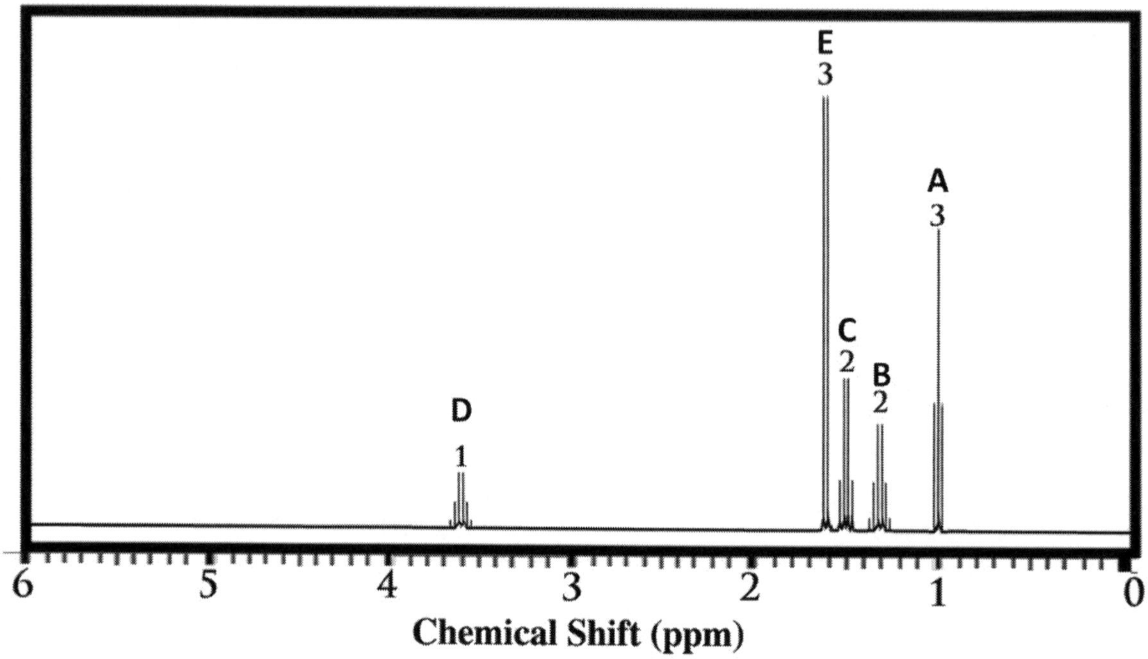

Lesson VIII.6. Carbon-13 NMR Practice Problems with Solutions

Carbon-13 NMR Question 1.

Which of the molecules **I-IV** would give rise to the ^{13}C NMR spectrum shown below?

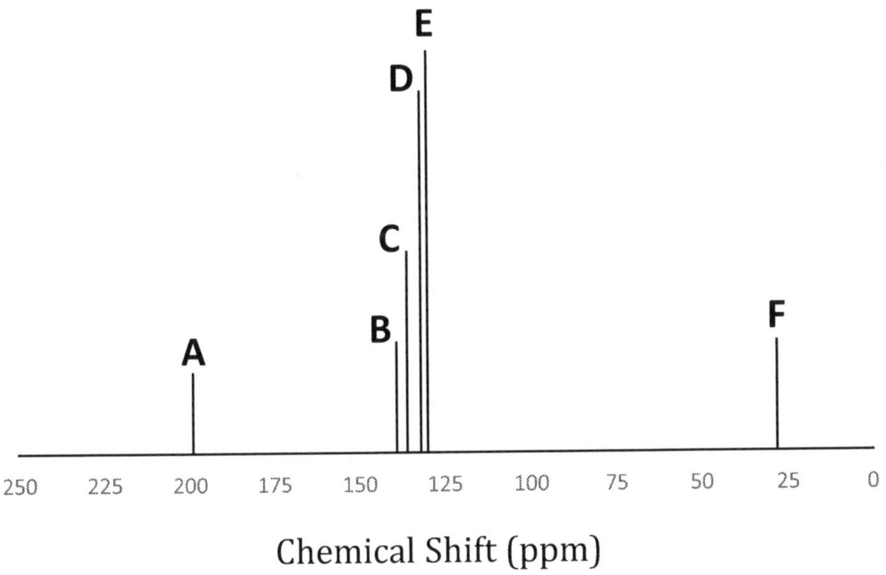

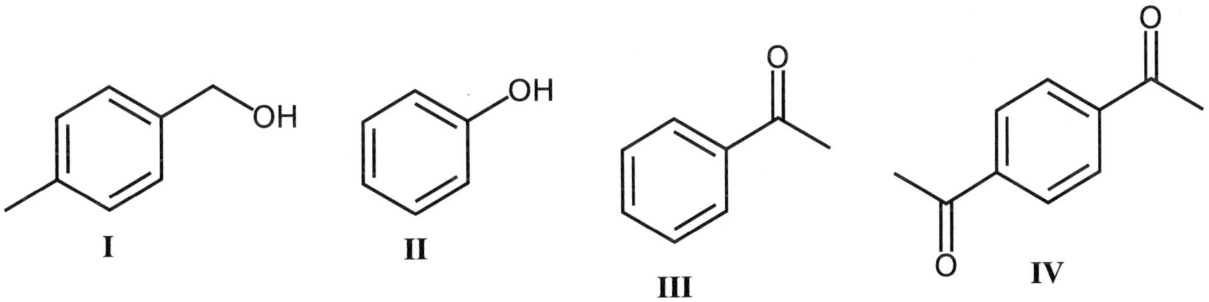

Carbon-13 NMR Answer 1.

Peaks

Carbonyls:	1
C=C:	4
sp/highly E.N.:	0
Alkyl:	1

Answer Choices:

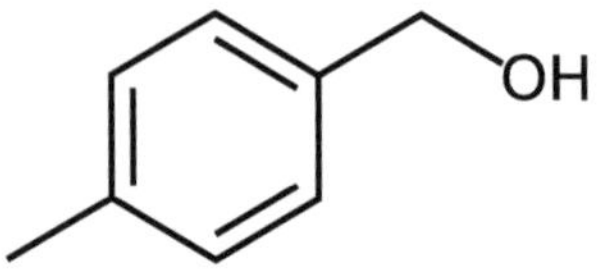

I

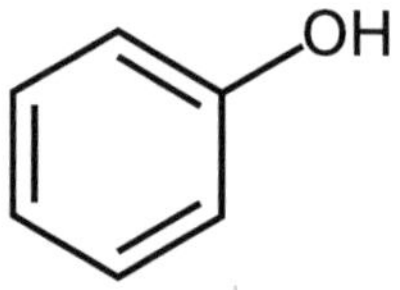

II

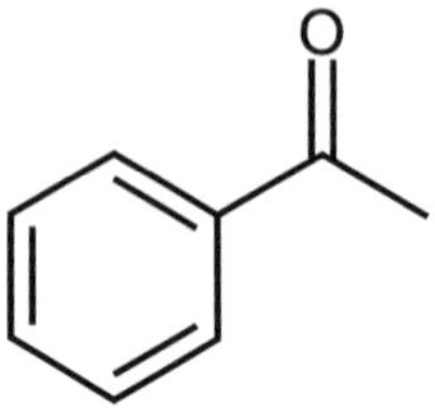

III

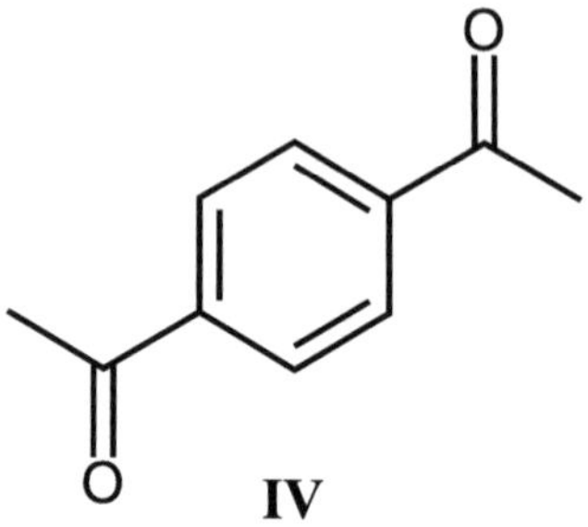

IV

I and **II** do not have any carbonyl-based functional groups and would thus not exhibit any peaks in the **Carbonyls** region. **IV** has multiple mirror planes and would only exhibit 2 peaks in the **C=C** region. **III** is the only molecule consistent with the ^{13}C NMR spectrum, and the peak assignments are shown below:

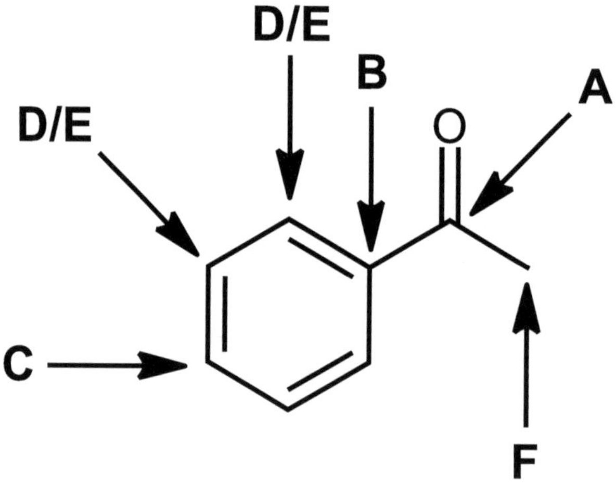

Carbon-13 NMR Question 2.

Which of the molecules **I-IV** would give rise to the ^{13}C NMR spectrum shown below?

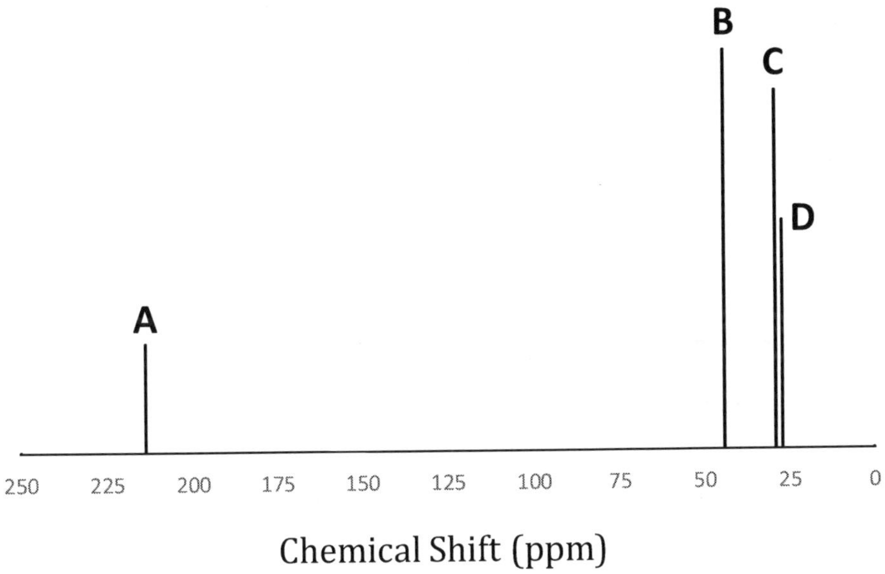

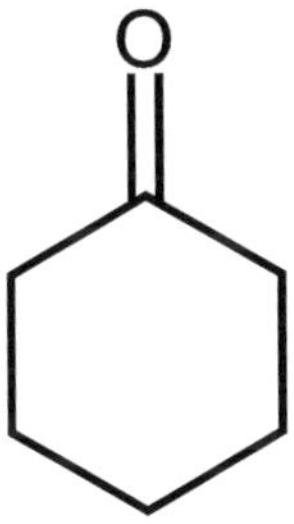

I

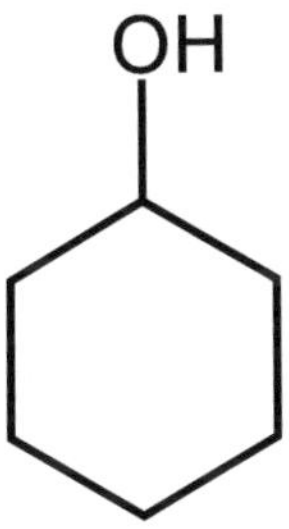

II

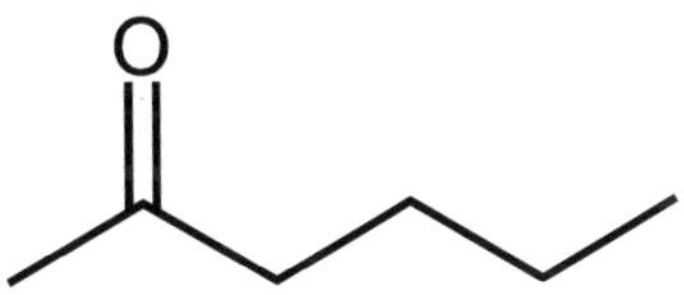

III

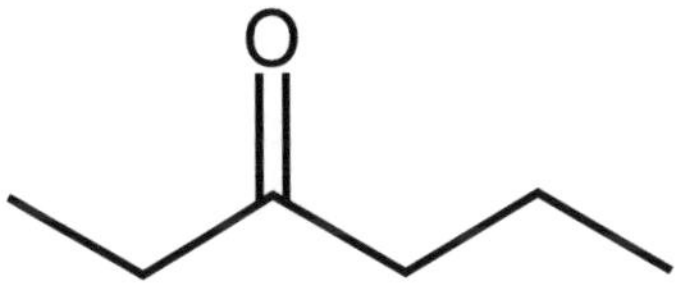

IV

Carbon-13 NMR Answer 2.

Peaks

Carbonyls: 1

C=C: 0

sp/highly E.N.: 1 (borderline)

Alkyl: 2

Answer Choices:

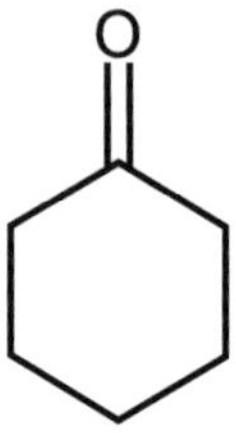

I

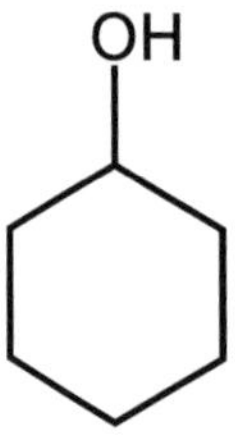

II

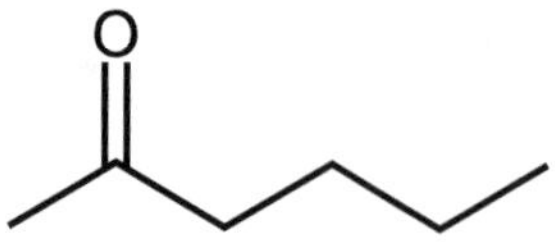

III

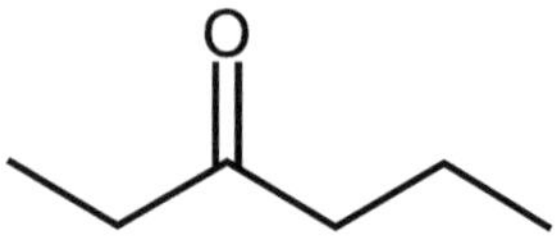

IV

II does not have any carbonyl-based functional groups and would thus not exhibit any peaks in the **Carbonyls** region. All of the carbons in **III** and **IV** are inequivalent, so both would exhibit 6 total peaks in the combined **sp/highly E.N.** and **Alkyl** regions. **I** is the only molecule consistent with the ^{13}C NMR spectrum, and the peak assignments are shown below:

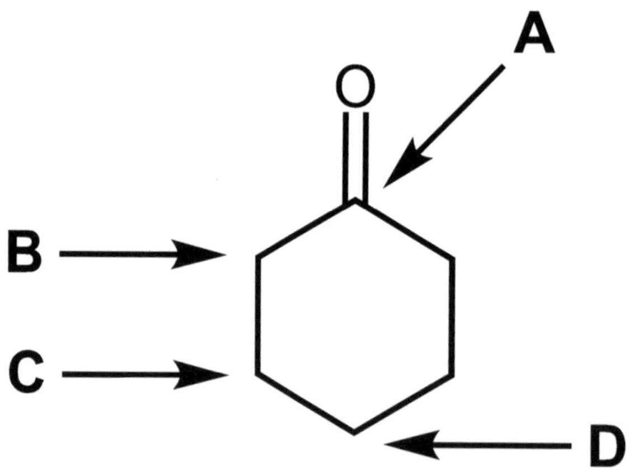

Carbon-13 NMR Question 3.

Which of the molecules **I-IV** would give rise to the ^{13}C NMR spectrum shown below?

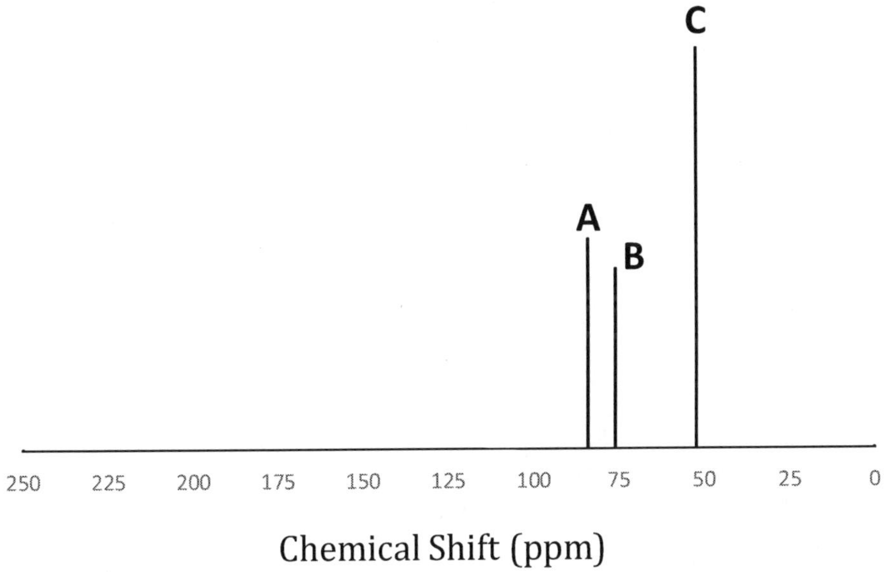

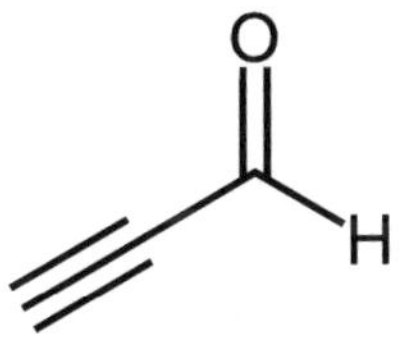

I

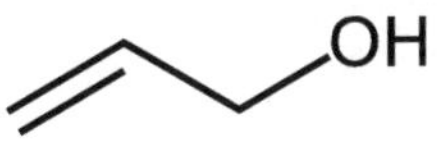

II

III

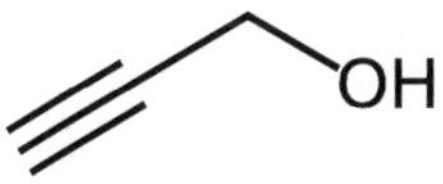

IV

Carbon-13 NMR Answer 3.

Peaks

Carbonyls:	0
C=C:	0
sp/highly E.N.:	3
Alkyl:	0

Answer Choices:

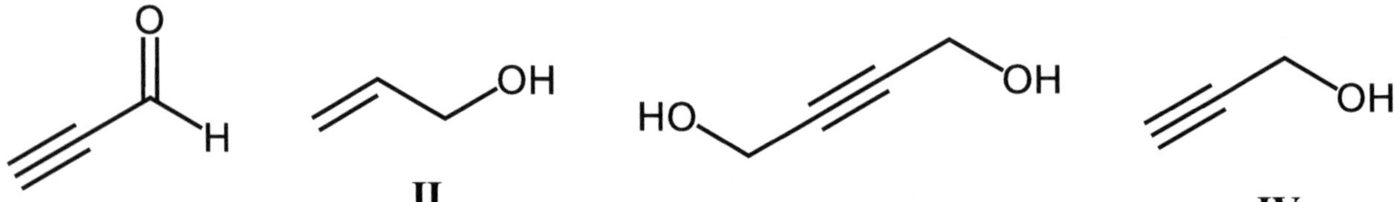

I II III IV

I has a carbonyl-based functional group and would thus exhibit a peak in the **Carbonyls** region. **II** has an alkene and would thus exhibit 2 peaks in the **C=C** region. **III** contains a mirror plane and would thus only exhibit 2 peaks in the **sp/highly E.N.** region. **IV** is the only molecule consistent with the ^{13}C NMR spectrum, and the peak assignments are shown below:

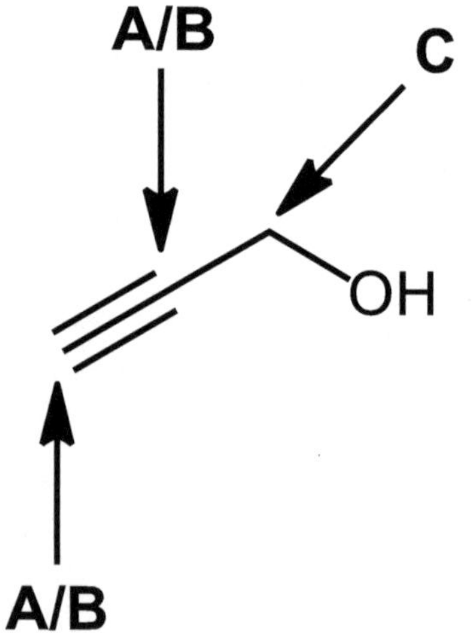

Carbon-13 NMR Question 4.

Which of the molecules **I-IV** would give rise to the ^{13}C NMR spectrum shown below?

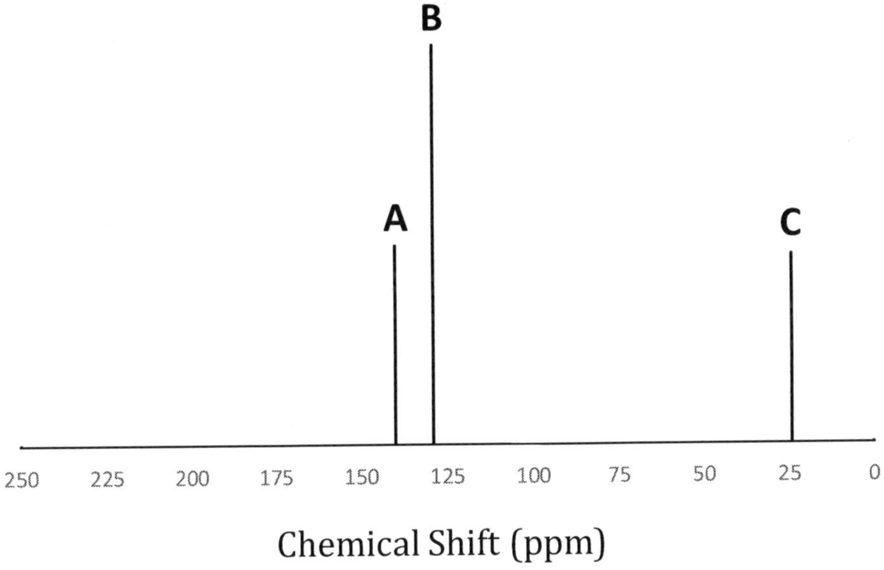

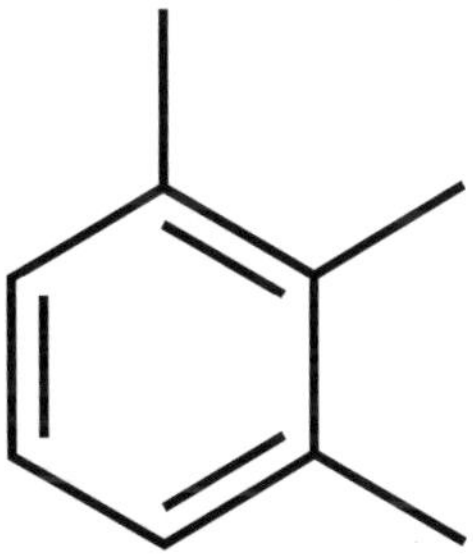

I

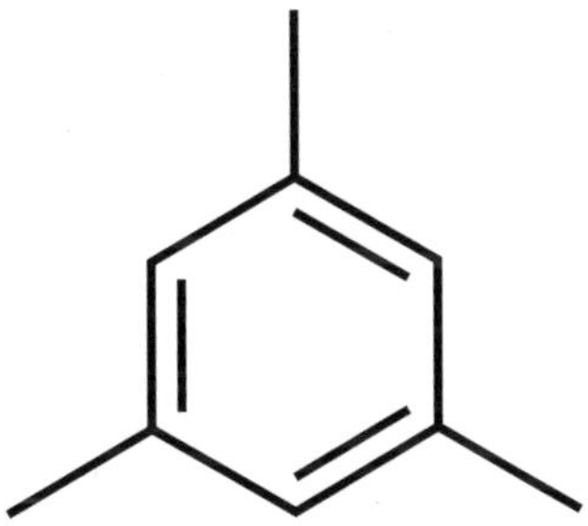

II

III

IV

Carbon-13 NMR Answer 4.

Peaks

Carbonyls:	0
C=C:	2
sp/highly E.N.:	0
Alkyl:	1

Answer Choices:

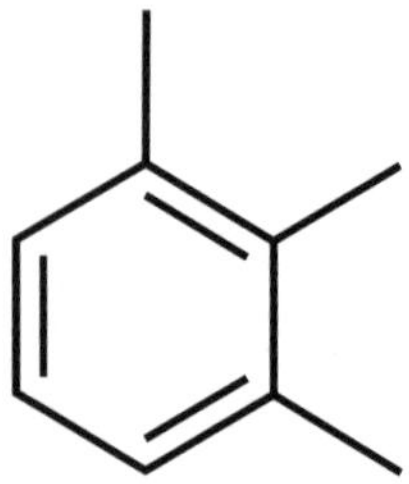

I

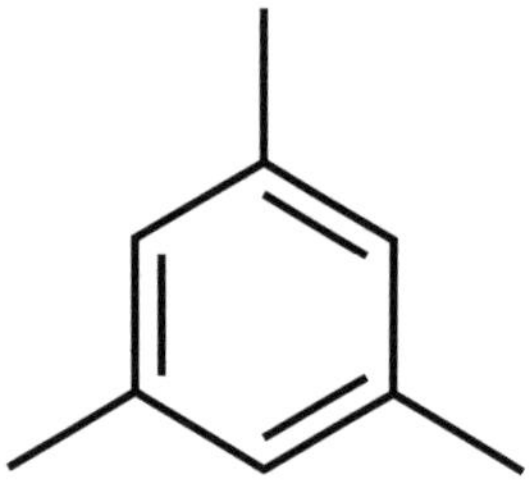

II

III

IV

I and **IV** have a mirror planes but each would still have 2 inequivalent *sp*^{3}-carbons (without highly electronegative substituents), so each would exhibit 2 peaks in the **Alkyl** region. **III** has a mirror plane that would make the 2 alkene carbons equivalent, so it would exhibit only 1 peak in the **C=C** region. **II** is the only molecule consistent with the ^{13}C NMR spectrum, and the peak assignments are shown below (the carbons with H-substituents will relax faster):

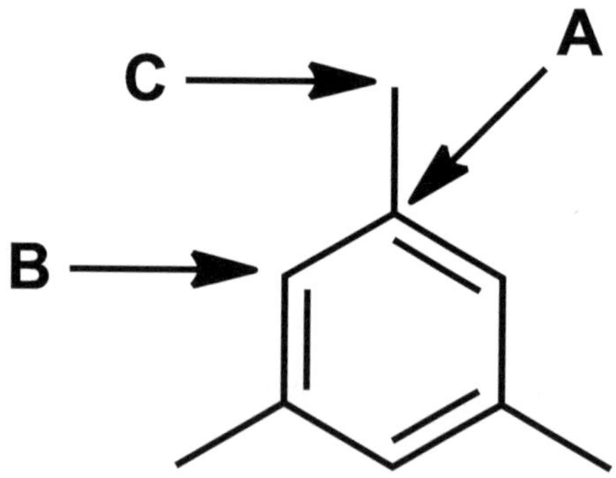

Carbon-13 NMR Question 5.

Which of the molecules **I-IV** would give rise to the ^{13}C NMR spectrum shown below?

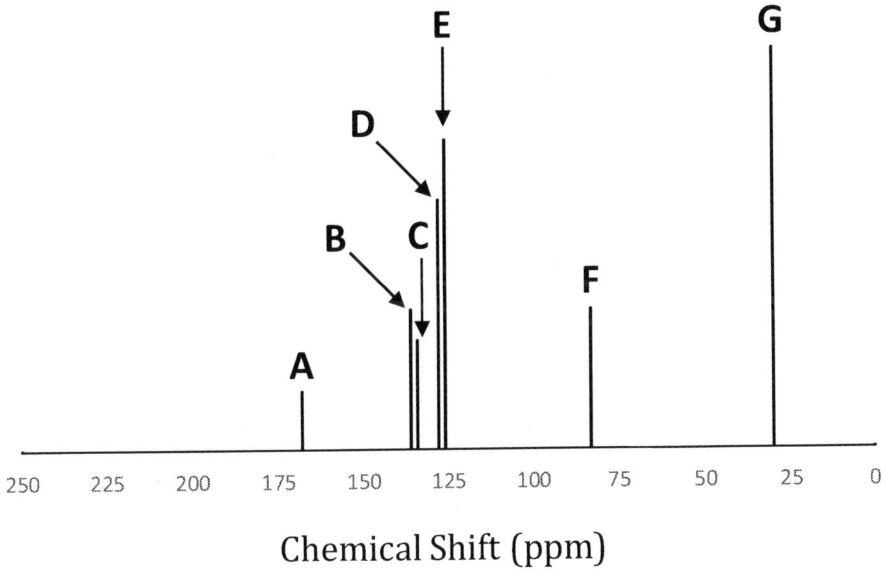

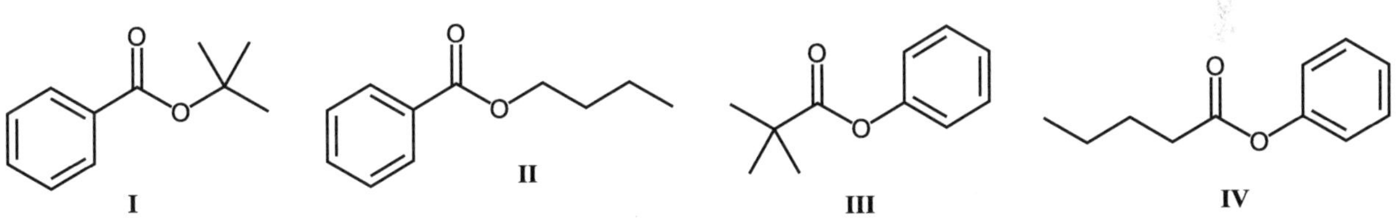

Carbon-13 NMR Answer 5.

Peaks

Carbonyls:	1
C=C:	4
sp/highly E.N.:	1
Alkyl:	1

Answer Choices:

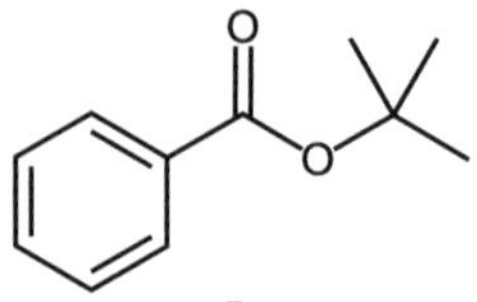

I

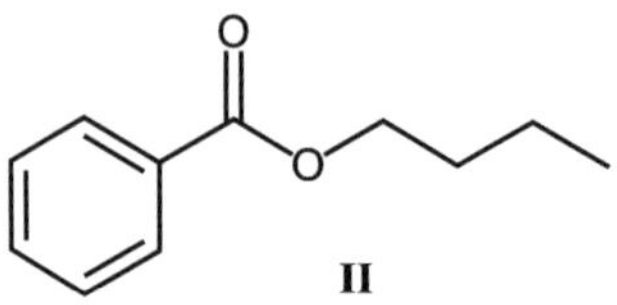

II

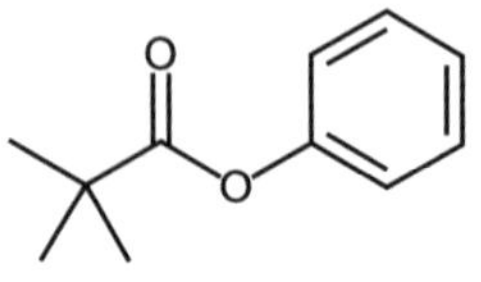

III

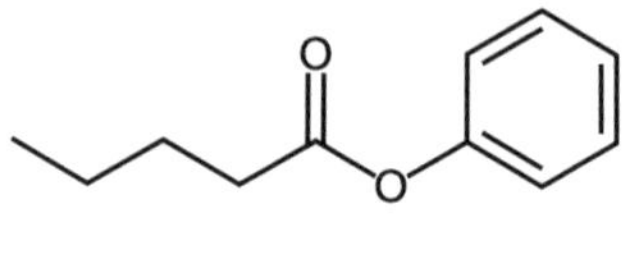

IV

III and **IV** do not have any sp^3-carbons with highly electronegative substituents, so neither molecule would exhibit any peaks in the **sp/highly E.N.** region. **IV** has 1 sp^3-carbon with a highly electronegative substituent and 3 sp^3-carbons without highly electronegative substituents, so it would exhibit 1 and 3 peaks in the **sp/highly E.N.** and **Alkyl** regions, respectively. **I** is the only molecule consistent with the ^{13}C NMR spectrum, and the peak assignments are shown below (remember that carbons with H-substituents will relax faster and more equivalent carbons will give stronger peak intensities):

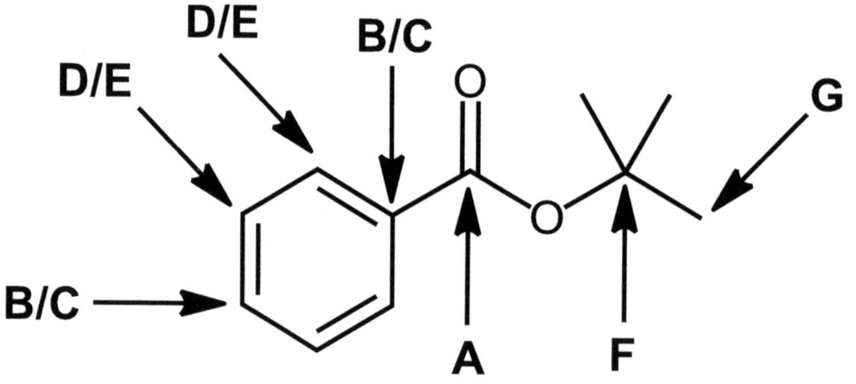

Carbon-13 NMR Question 6.

Which of the molecules **I-IV** would give rise to the ^{13}C NMR spectrum shown below?

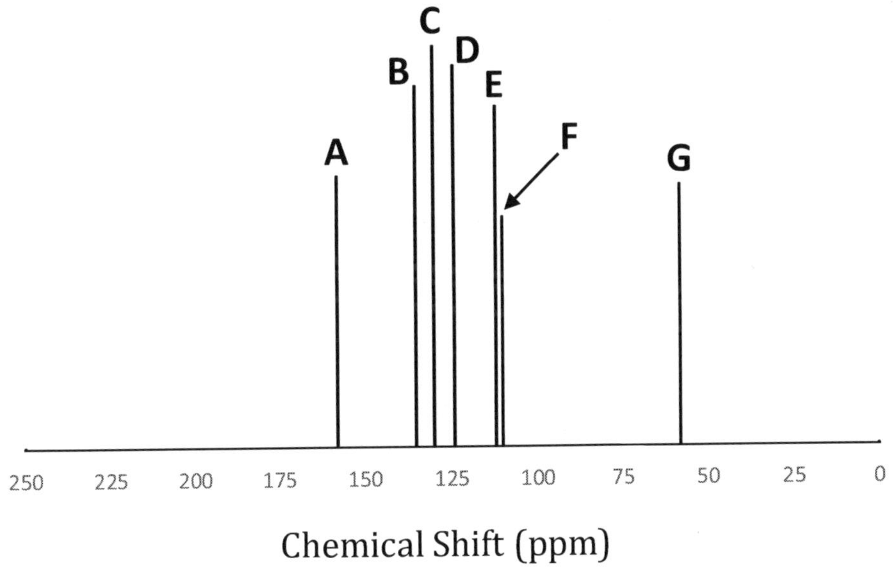

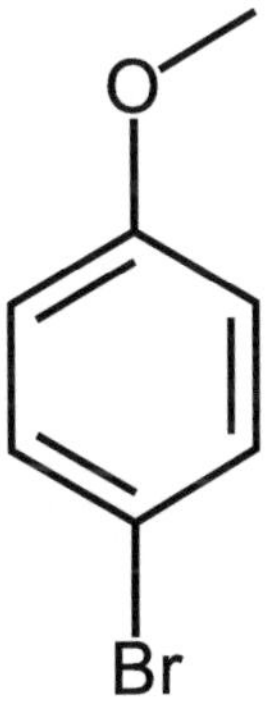

I

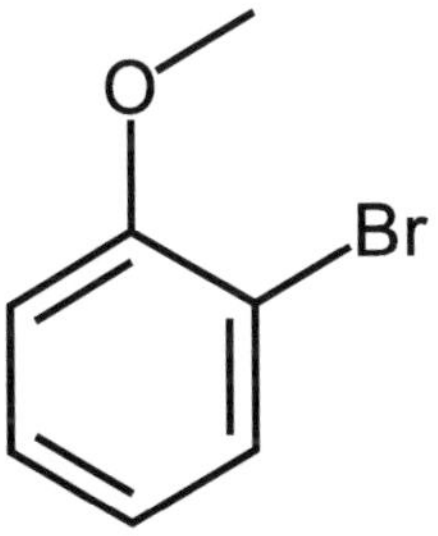

II

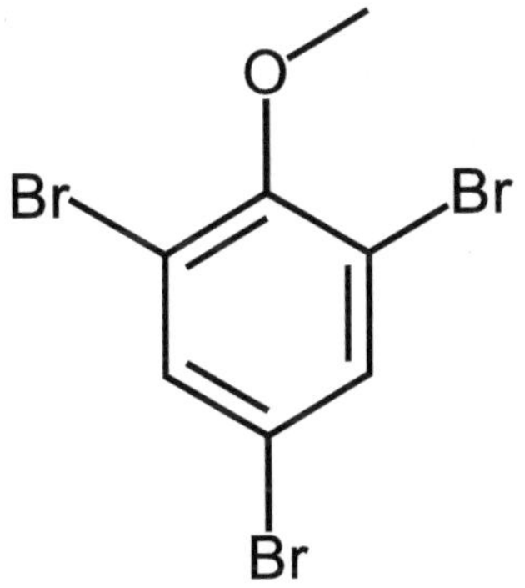

III

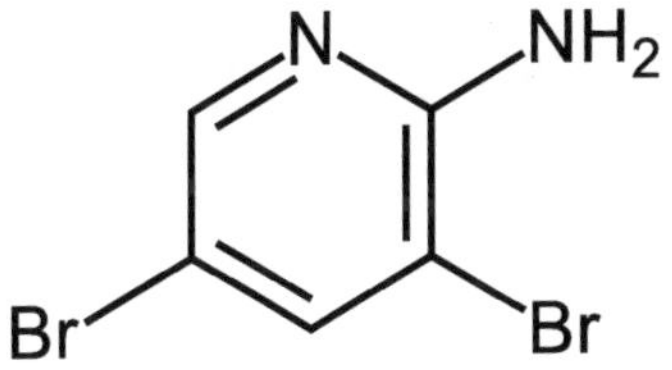

IV

Carbon-13 NMR Answer 6.

Peaks

Carbonyls: 0

C=C: 6 (1 is borderline with **Carbonyls**)

sp/highly E.N.: 1

Alkyl: 0

Answer Choices:

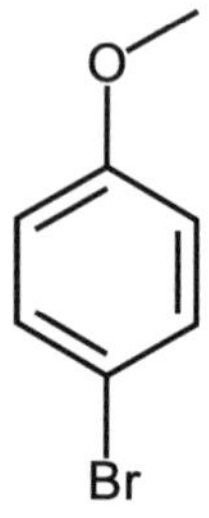

I

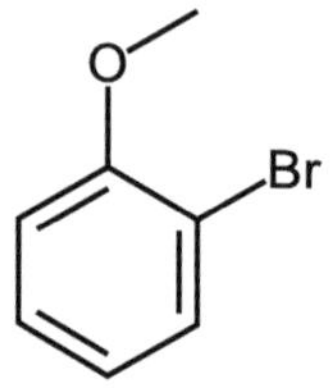

II

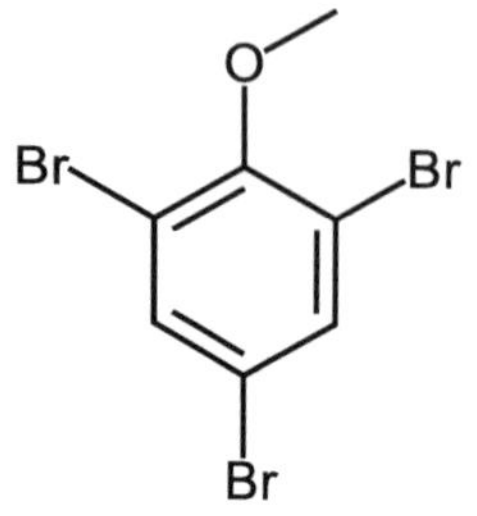

III

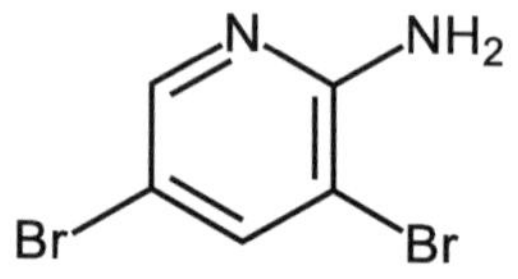

IV

I contains a mirror plane and would thus exhibit only 5 peaks in total, and only 4 peaks in the **C=C** region. Likewise, **III** contains a mirror plane and would thus exhibit only 5 peaks in total, and only 4 peaks in the **C=C** region (although its peaks would be shifted more downfield). **IV** only has 5 carbons in total and no heteroatoms that commonly cause peak splitting, so there is no way it could exhibit 7 total peaks in its ^{13}C NMR spectrum. II is the only molecule consistent with this ^{13}C NMR spectrum, and the peak assignments are shown below (remember that oxygen is a strong *ortho*/*para*-director, so it increases electron density at the positions *ortho* and, to a lesser extent, *para* relative to it!):

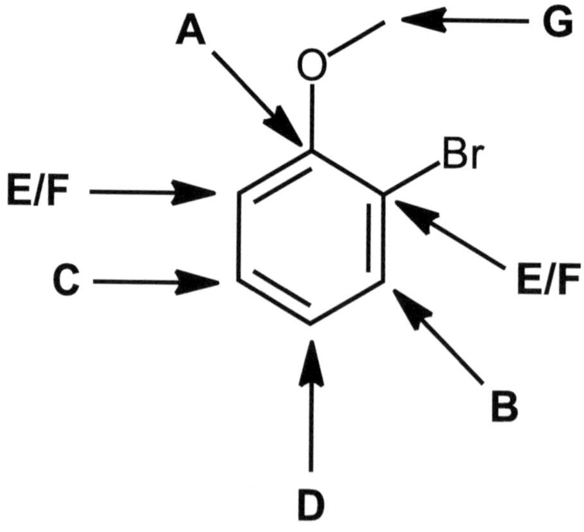

Carbon-13 NMR Question 7. Which of the molecules **I-IV** would give rise to the ^{13}C NMR spectrum shown below?

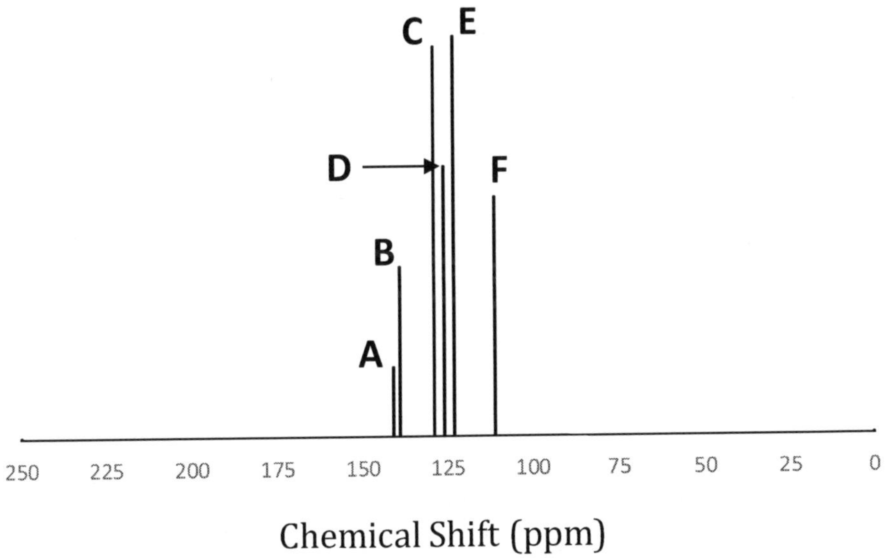

I

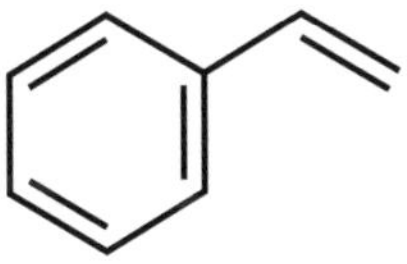

II

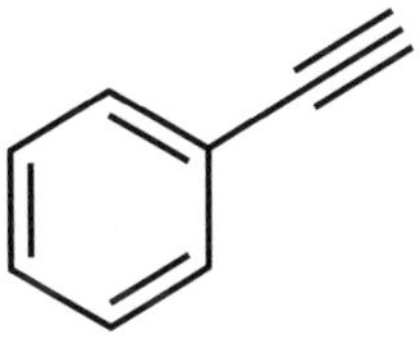

III

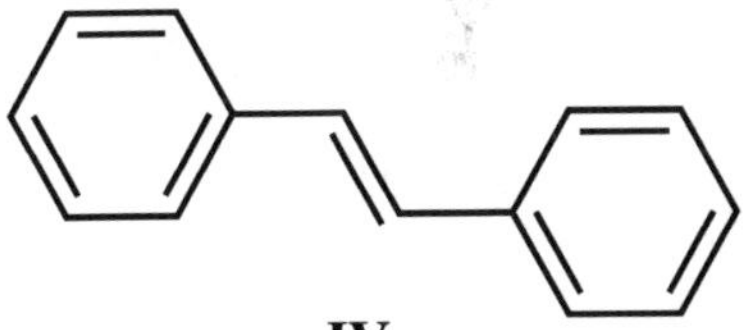

IV

Carbon-13 NMR Answer 7.

Peaks

Carbonyls:	0
C=C:	6
sp/highly E.N.:	0
Alkyl:	0

Answer Choices:

I

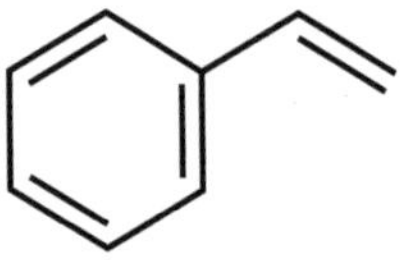

II

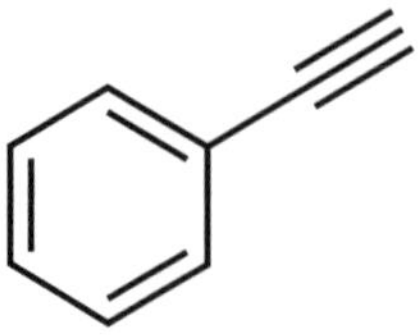

III

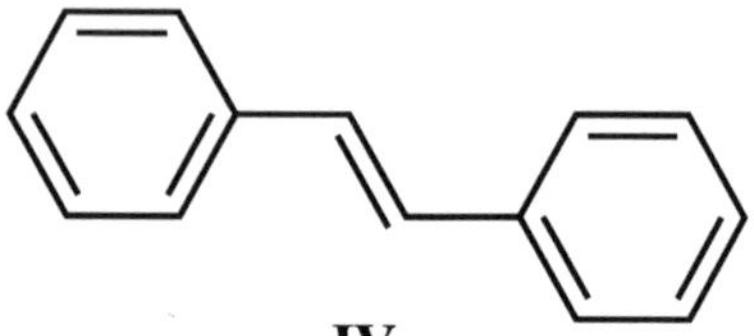

IV

III has 2 inequivalent *sp*-carbons and would thus exhibit 2 peaks in the **sp/highly E.N.** region. **I** has only 3 inequivalent sp^2-carbons (rotate molecule by 180º perpendicular to plane of page to see why), so it would only exhibit 3 peaks in the **C=C** region. Similarly, **IV** has only 5 inequivalent sp^2-carbons (rotate molecule by 180º perpendicular to plane of page to see why, and remember C–C single bonds can rotate), so it would only exhibit 5 peaks in the **C=C** region. **II** is the only molecule consistent with the ^{13}C NMR spectrum. Because all of the carbons are sp^2-hybridized with no heteroatom substituents, there is not enough information to be able to make any peak assignments.

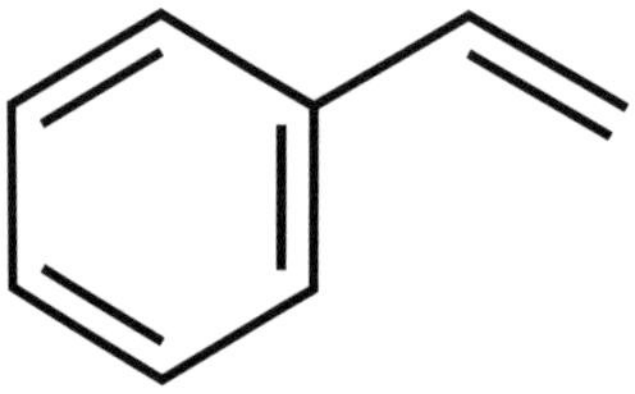

Carbon-13 NMR Question 8.

Which of the molecules **I-IV** would give rise to the ^{13}C NMR spectrum shown below?

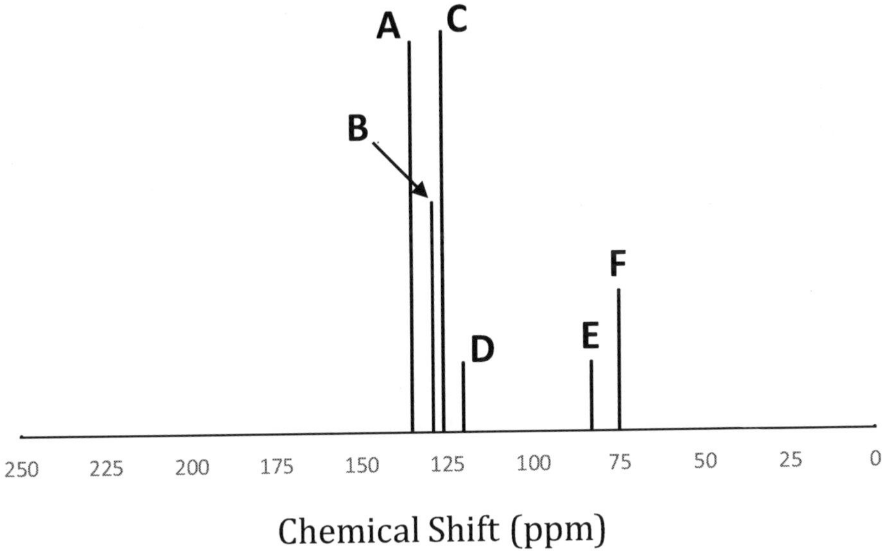

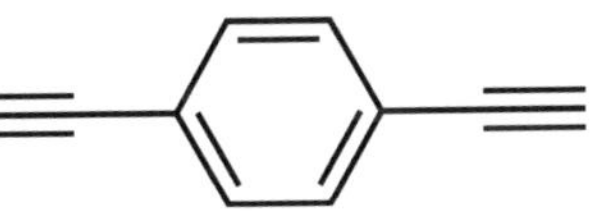

I

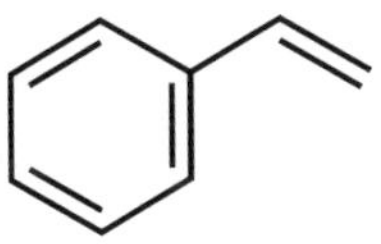

II

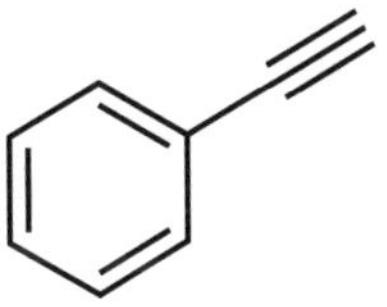

III

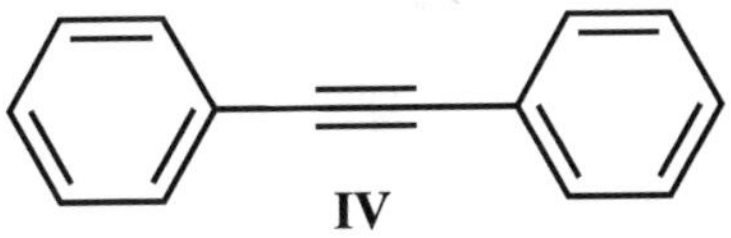

IV

Carbon-13 NMR Answer 8.

Peaks

Carbonyls:	0
C=C:	4
sp/highly E.N.:	2
Alkyl:	0

Answer Choices:

I

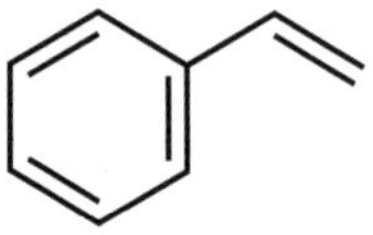
II

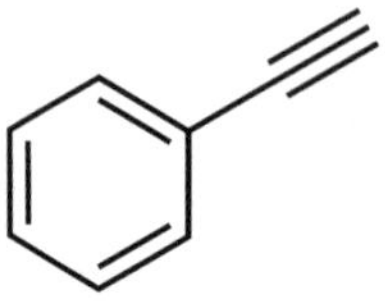
III

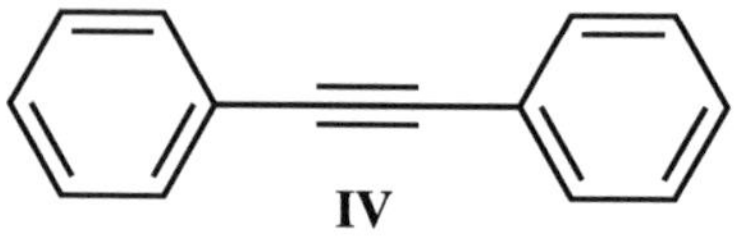
IV

II only has *sp*^{2}-carbons, so it would not exhibit any peaks in the **sp/highly E.N.** region. **I** has multiple mirror planes and would thus only exhibit 2 peaks in the **C=C** region. **IV** also has multiple mirror planes and would only exhibit 1 peak in the **sp/highly E.N.** region. **III** is the only molecule consistent with the ^{13}C NMR spectrum, and the peak assignments are shown below (remember that carbons with H-substituents will relax faster and more equivalent carbons will give stronger peak intensities):

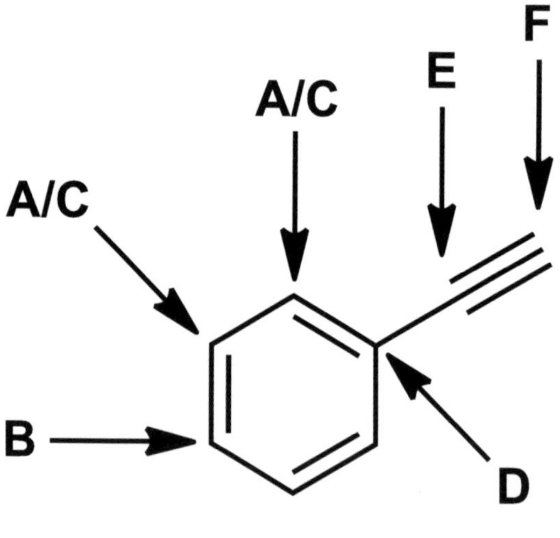

Carbon-13 NMR Question 9.

Which of the molecules **I-IV** would give rise to the ^{13}C NMR spectrum shown below?

Hint: **Peak B is in the region for a C directly attached to a highly electronegative atom.**

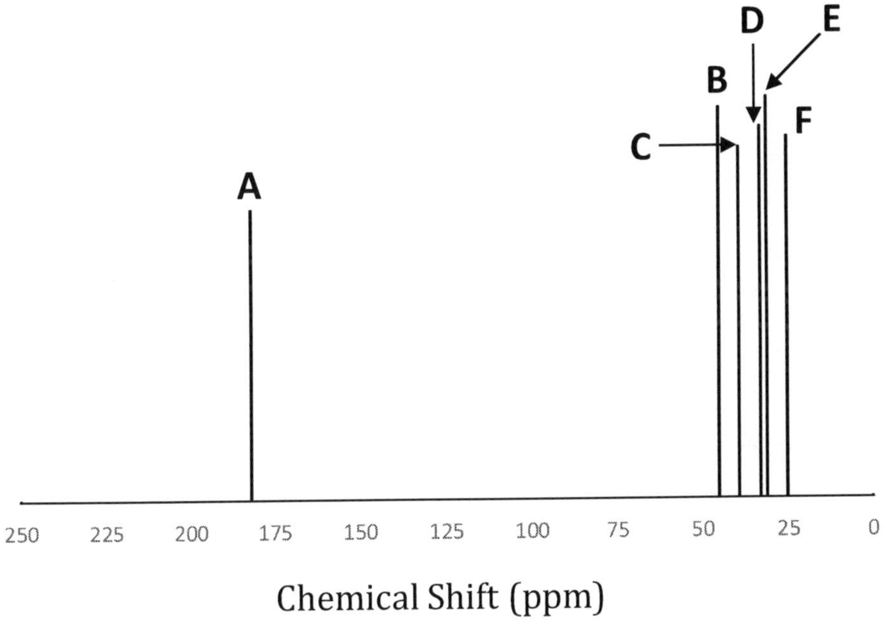

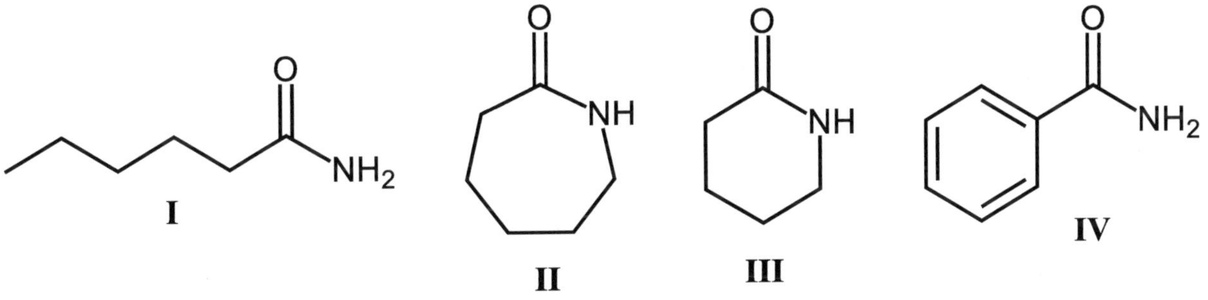

Carbon-13 NMR Answer 9.

Peaks

Carbonyls:	1
C=C:	0
sp/highly E.N.:	1
Alkyl:	4

Answer Choices:

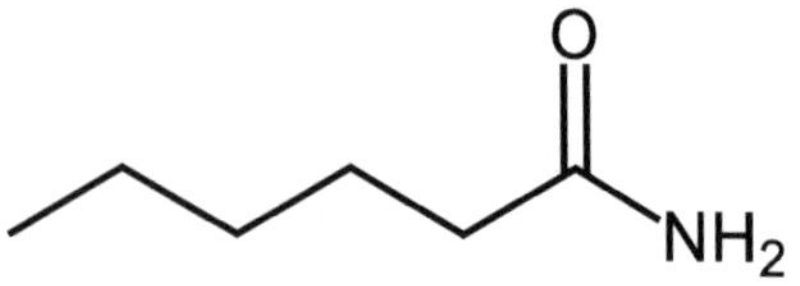

I

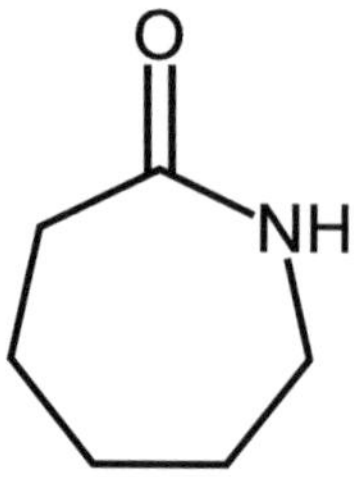

II

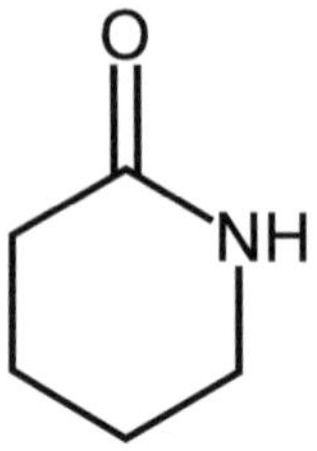

III

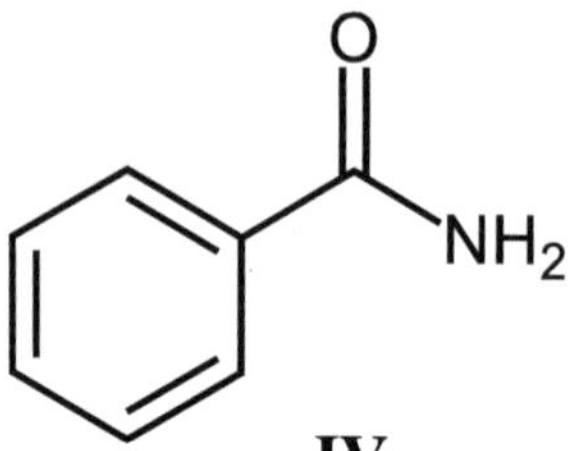

IV

I has no substituents on the nitrogen (other than the carbonyl C that we will count in the carbonyl region peaks), so it would not exhibit any peaks in the **sp/highly E.N.** region. **IV** has a phenyl ring, so it would exhibit 4 peaks in the **C=C** region. **III** only has 5 carbons, so it could not give rise to more than 5 peaks total in the spectrum. **II** is the only molecule consistent with the ^{13}C NMR spectrum. We can easily assign peaks **A** (only 1 carbonyl in the structure) and **B** (only one carbon attached to a highly-electronegative substituent), but we do not have enough information to assign peaks **C-F** (this is a non-trivial exercise even for advanced graduate students).

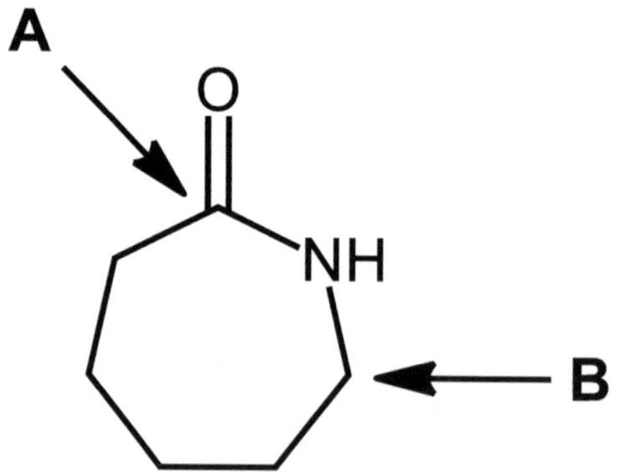

Carbon-13 NMR Question 10.

Which of the molecules **I-IV** would give rise to the ^{13}C NMR spectrum shown below?

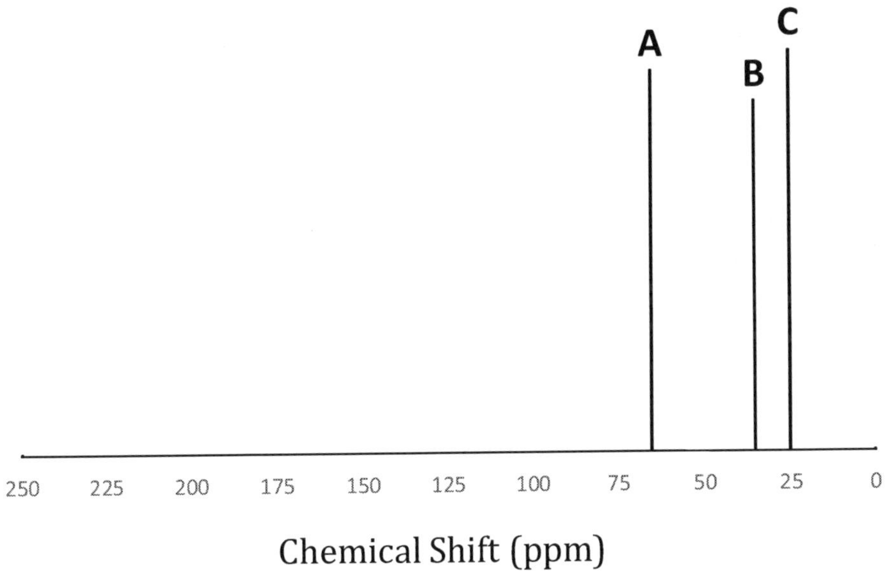

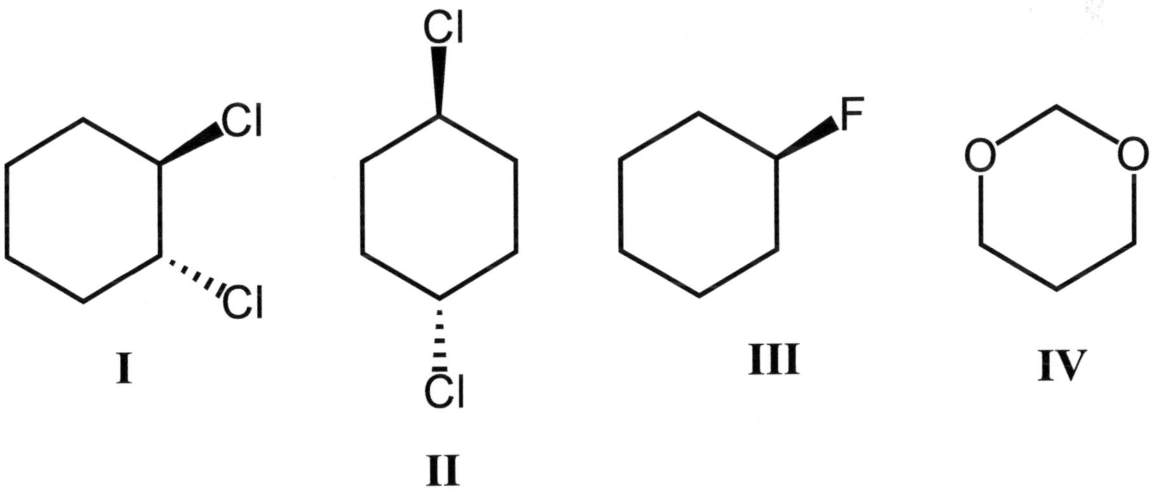

Carbon-13 NMR Answer 10.

Peaks

Carbonyls: 0

C=C: 0

sp/highly E.N.: 1

Alkyl: 2

Answer Choices:

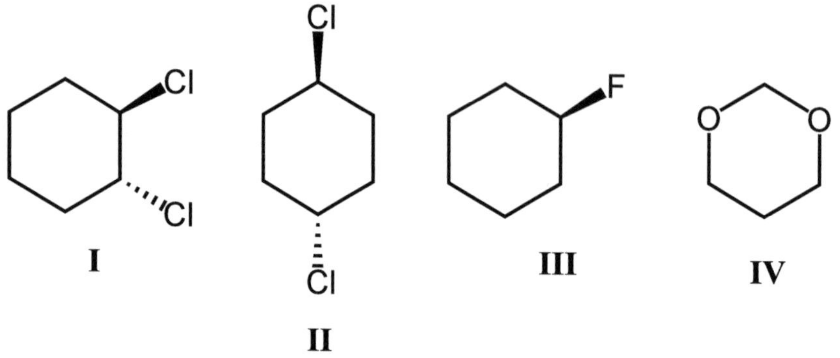

II only contains 2 unique carbons total (there is a mirror plane bisecting the carbons with the Cl substituents, and you can imagine rotating 180º on an axis containing the plane of the page to see why), and would thus exhibit only 1 peak in the **sp/highly E.N.** region and 1 peak in the **Alkyl** region. **III** contains 4 inequivalent carbons. The mirror plane in **IV** leaves 3 inequivalent carbons, one of which is between two oxygens (and gets shifted up to the **C=C** region), so **IV** would exhibit 1 peak in the **C=C** region, 1 peak in the **sp/highly E.N.** region, and 1 peak in the **Alkyl** region. **I** is the only molecule consistent with this ^{13}C NMR spectrum, and the peak assignments are shown below:

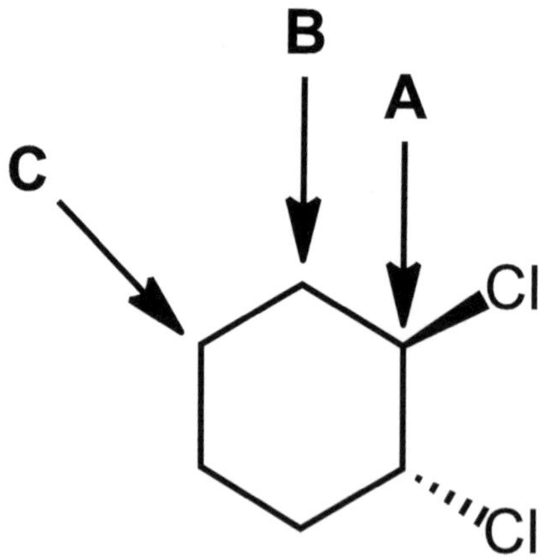

Carbon-13 NMR Question 11.

Which of the molecules **I-IV** would give rise to the ^{13}C NMR spectrum shown below?

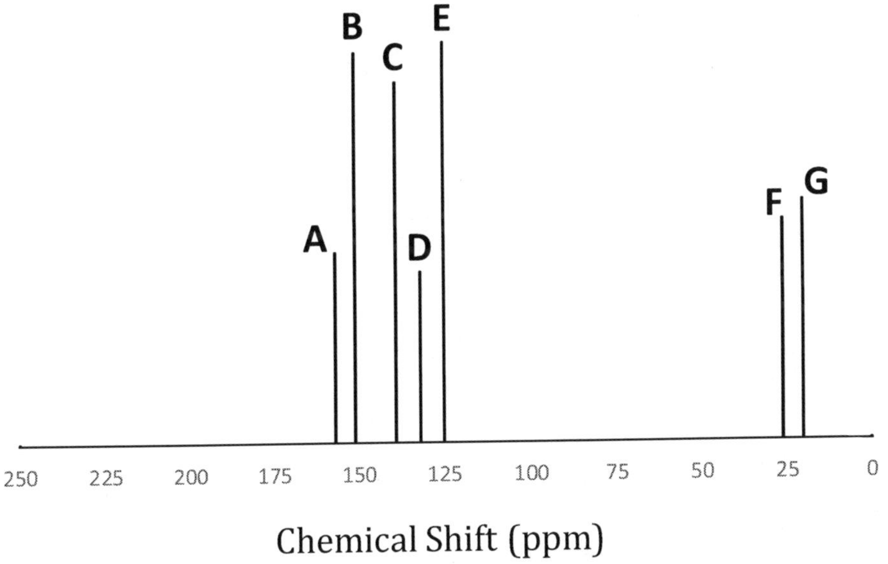

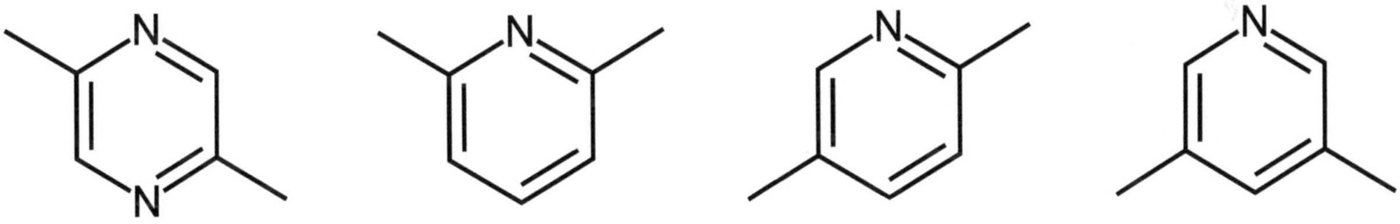

I II III IV

Carbon-13 NMR Answer 11.

Peaks

Carbonyls:	0
C=C:	5 (2 are borderline with **Carbonyls**)
sp/highly E.N.:	0
Alkyl:	2

Answer Choices:

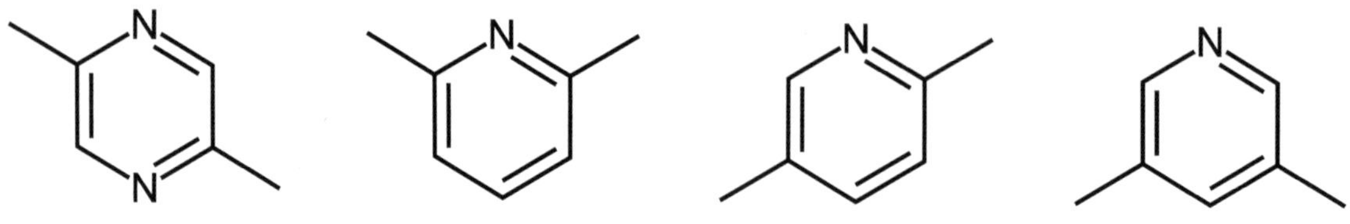

I II III IV

I only contains 3 inequivalent carbons (imagine rotating the molecule 180º perpendicular to the plane of the page to see why), so it would only exhibit 2 peaks in the **C=C** region and 1 in the **Alkyl** region. **II** and **IV** each contain mirror planes which means that each molecule only has 4 inequivalent carbons in total. **II** and **IV** would thus exhibit only 3 peaks in the **C=C** region and 1 peak in the **Alkyl** region. **III** is the only molecule consistent with this ^{13}C NMR spectrum, and the peak assignments are shown below (carbons that have H substituents will give more intense peaks than carbons without H substituents, and carbons next to a highly-electronegative element will be more downfield than those not next to one):

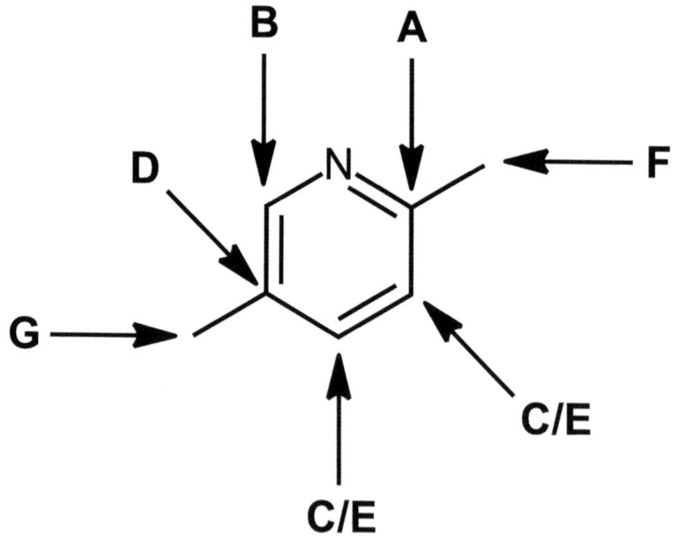

Carbon-13 NMR Question 12.

Which of the molecules **I-IV** would give rise to the ^{13}C NMR spectrum shown below?

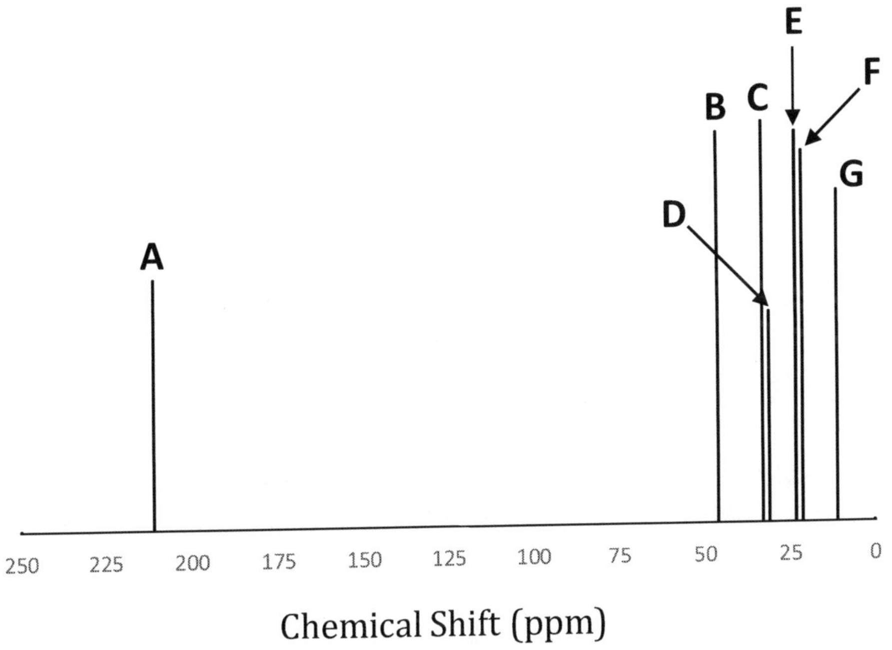

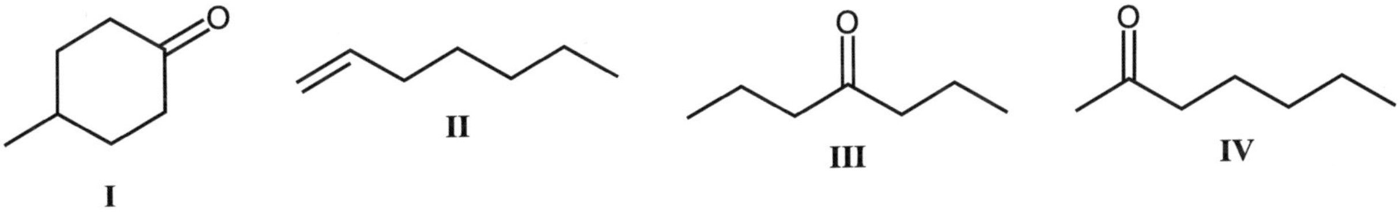

Carbon-13 NMR Answer 12.

Peaks

Carbonyls: 1

C=C: 0

sp/highly E.N.: 0

Alkyl: 6 (1 is borderline with **sp/highly E.N.**)

Answer Choices:

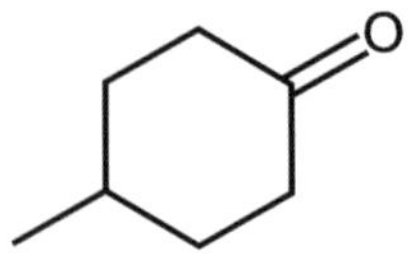

I

II

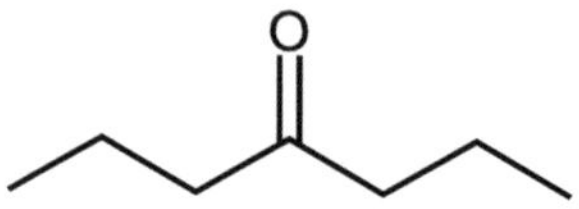

III

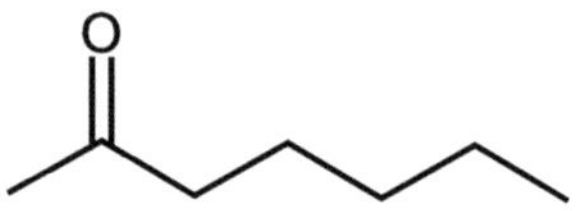

IV

II has 7 inequivalent carbons, but it would not exhibit any peaks in the **Carbonyls** region. **I** and **III** would each exhibit 1 peak in the **Carbonyls** region, but the presence of a mirror plane in each molecule would cause **I** to only exhibit 5 peaks in the **Alkyl** region, and **III** to only exhibit 4 peaks in the **Alkyl** region. **IV** is the only molecule consistent with the ^{13}C NMR spectrum. We can easily assign peak **A** (only 1 carbonyl in the structure), but we do not have enough information to assign peaks **B**-**G** (this is a non-trivial exercise even for advanced graduate students).

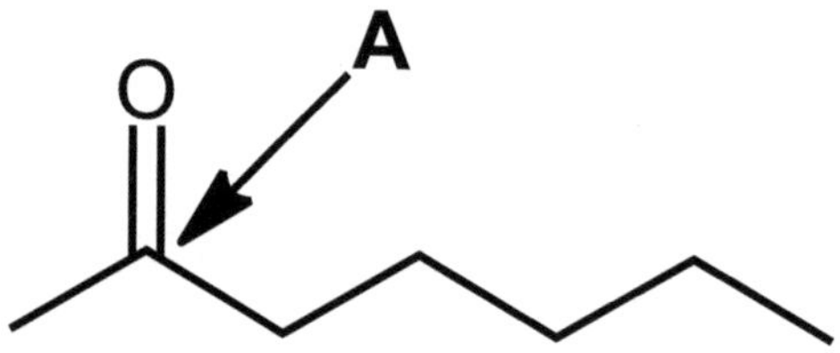

Carbon-13 NMR Question 13.

Which of the molecules **I-IV** would give rise to the ^{13}C NMR spectrum shown below?

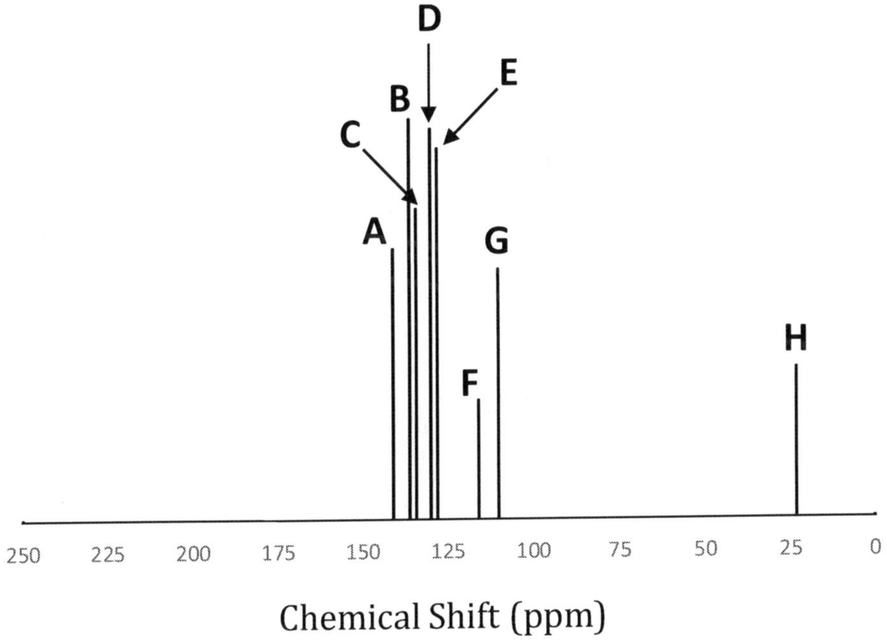

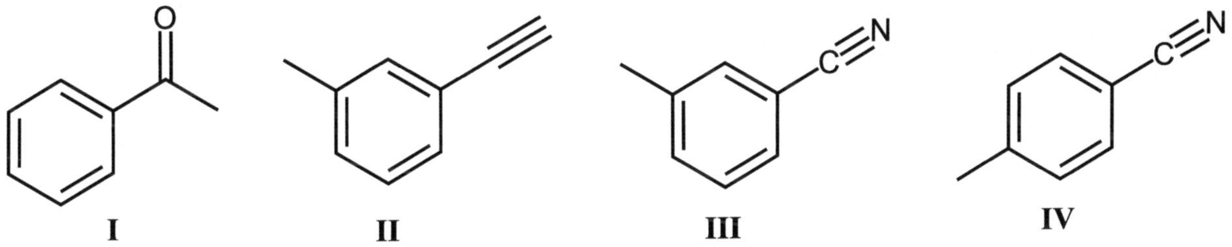

Carbon-13 NMR Answer 13.

Peaks

Carbonyls:	0
C=C:	7
sp/highly E.N.:	0
Alkyl:	1

Answer Choices:

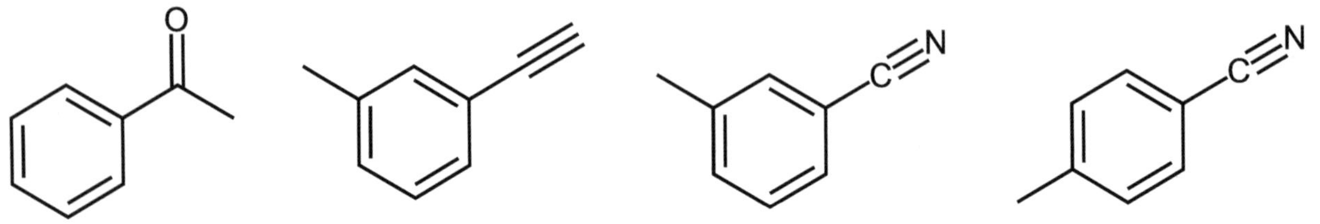

I II III IV

I would exhibit 1 peak in the **Carbonyls** region and can be eliminated. **II** would exhibit 2 peaks in the **sp/highly E.N.** region and can be eliminated. **III** and **IV** each have an *sp*-carbon attached to a highly-electronegative element, which shifts its signal into the **C=C** region. However, **IV** has a mirror plane and would thus have only 4 inequivalent aryl carbons, so it would exhibit only 5 peaks (4 aryl + 1 nitrile) in the **C=C** region. **III** is the only molecule consistent with the ^{13}C NMR spectrum. We can easily assign peak **A** (only 1 sp^3-carbon with no highly-electronegative substituents in this molecule), but we do not have enough information to assign peaks **B**-**H** (this is a non-trivial exercise even for advanced graduate students).

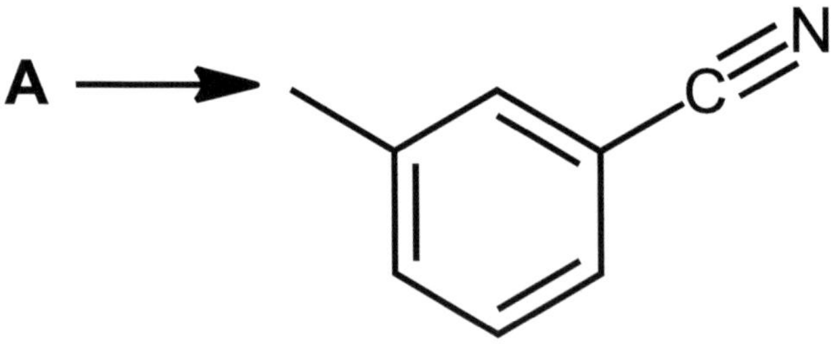

Carbon-13 NMR Question 14.

Which of the molecules **I-IV** would give rise to the ^{13}C NMR spectrum shown below?

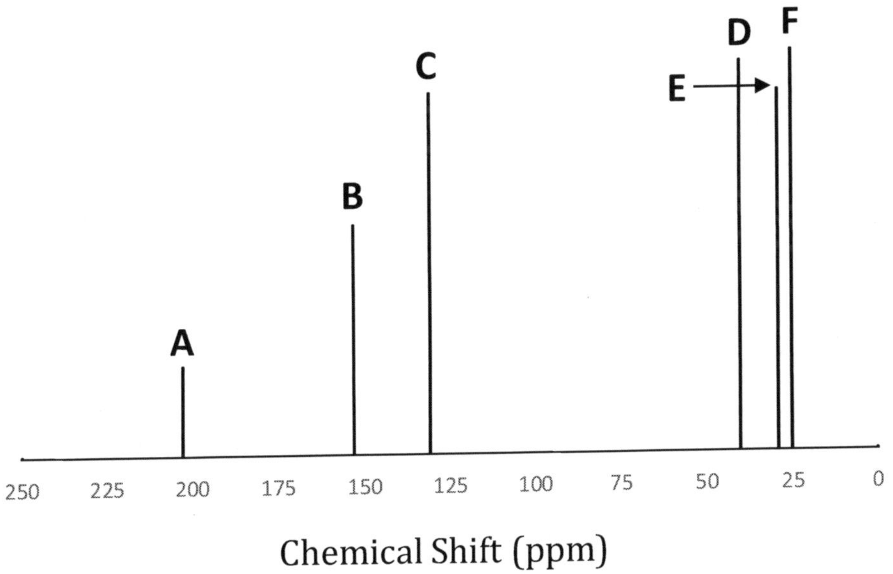

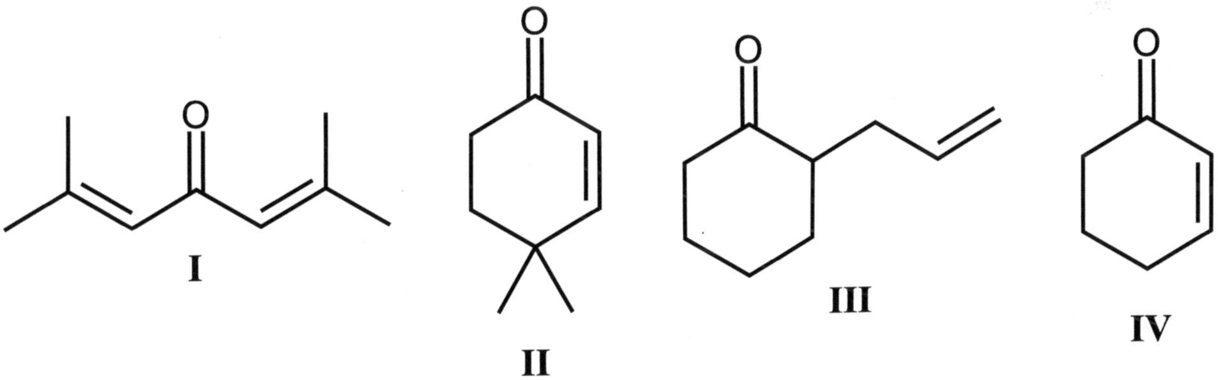

Carbon-13 NMR Answer 14.

Peaks

Carbonyls:	1
C=C:	2 (1 is borderline with **Carbonyls**)
sp/highly E.N.:	0
Alkyl:	3 (1 is borderline with **Sp/highly E.N.**)

Answer Choices:

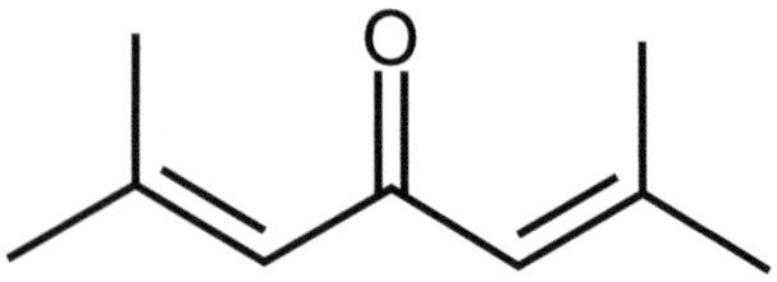

I

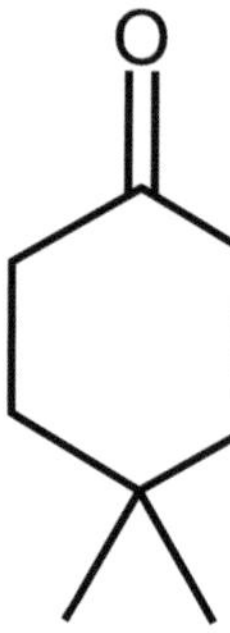

II

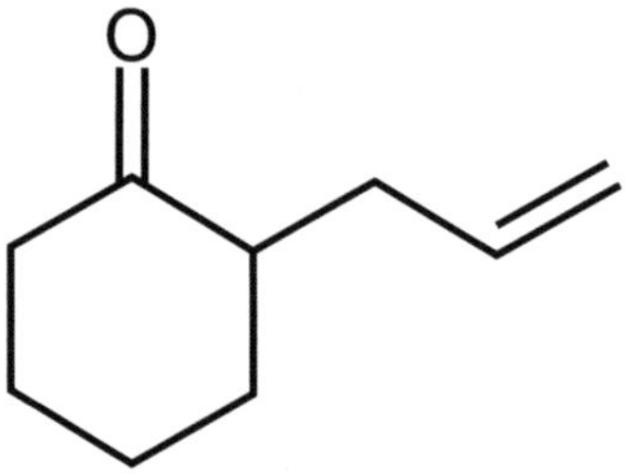

III

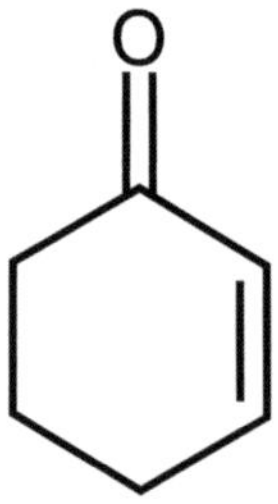

IV

I has a mirror plane and would thus exhibit only 2 peaks in the **Alkyl** region. The 2 methyl groups in **II** are equivalent, so this compound would exhibit 4 peaks in the **Alkyl** region. **III** would exhibit 6 peaks in the **Alkyl** region. **IV** is the only molecule consistent with the ^{13}C NMR spectrum. We can easily assign peak **A** (only 1 carbonyl carbon in this molecule) and peaks **B-C** (only one alkene in this molecule), but we do not have enough information to assign peaks **D-F** (this is a non-trivial exercise even for advanced graduate students).

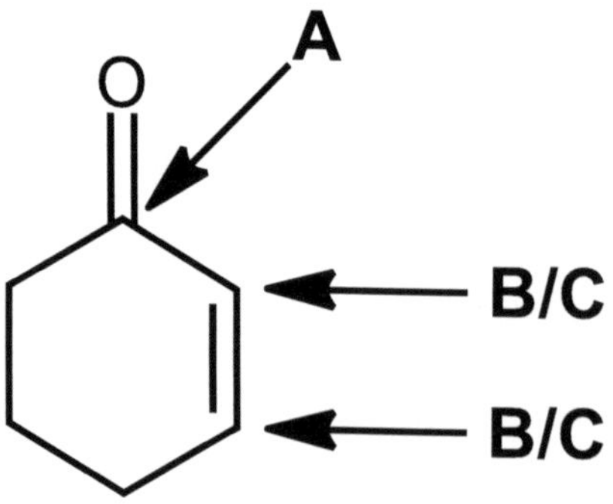

Carbon-13 NMR Question 15.

Which of the molecules **I-IV** would give rise to the ^{13}C NMR spectrum shown below?

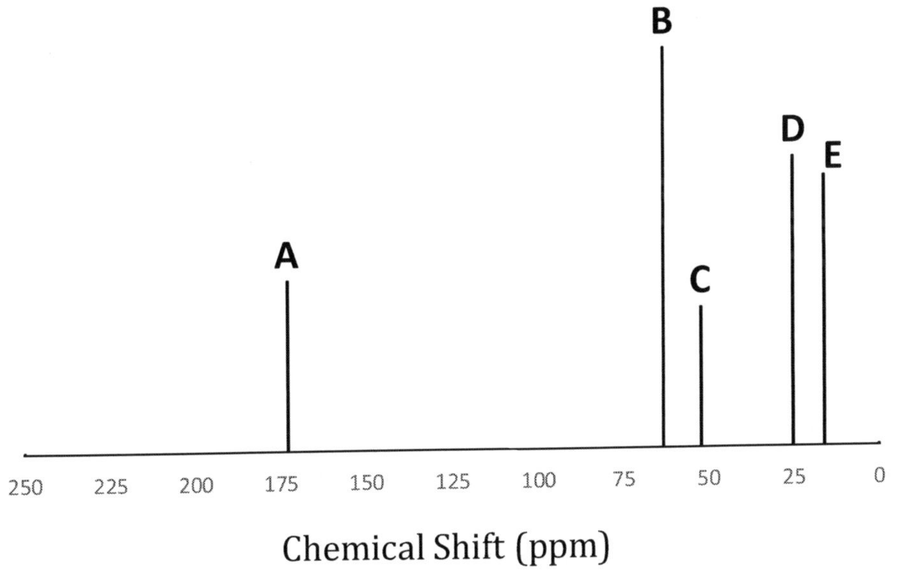

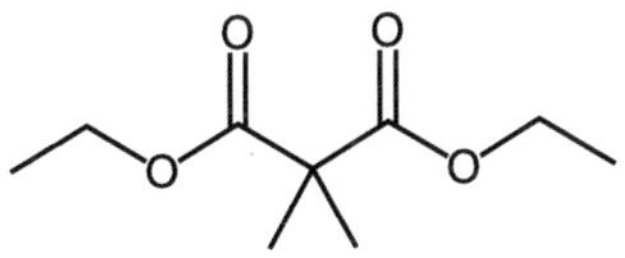

I

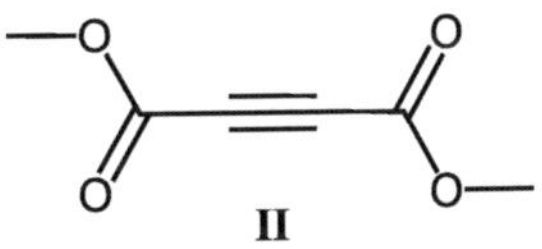

II

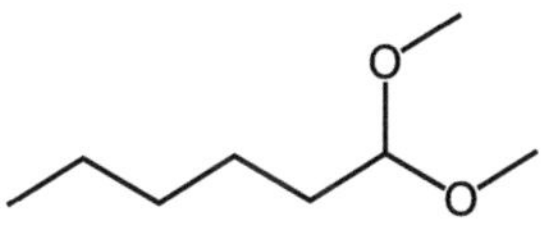

III

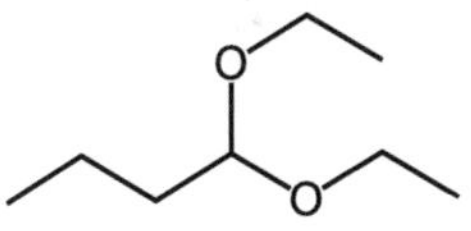

IV

Carbon-13 NMR Answer 15.

Peaks

Carbonyls: 1

C=C: 0

sp/highly E.N.: 2 (1 is borderline with **Alkyl**)

Alkyl: 2

Answer Choices:

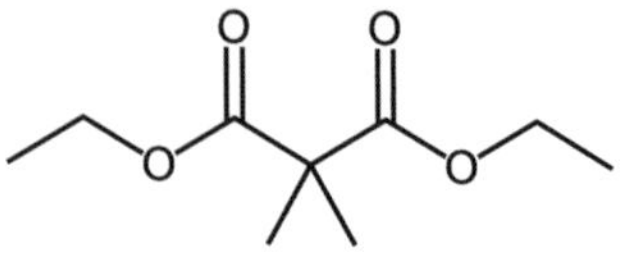

I

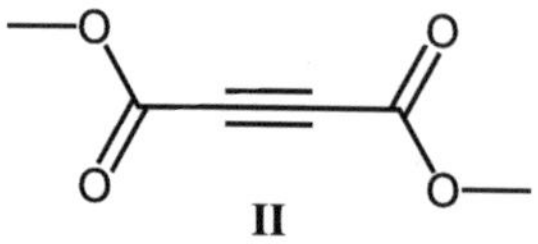

II

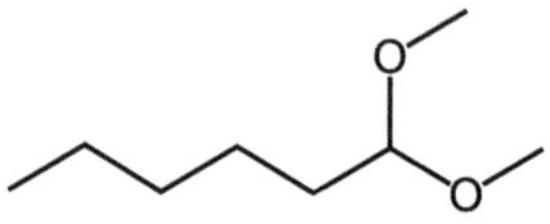

III

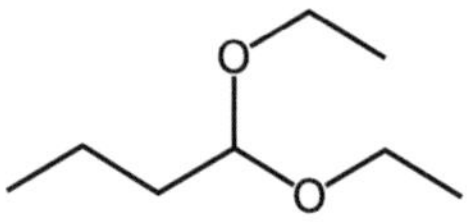

IV

II would exhibit 1 peak in the **Carbonyls** region, as well as 2 peaks in the **Sp/highly E.N.** region because the 2 methoxy carbons and 2 alkyne carbons are equivalent (imagine rotating the molecule 180º perpendicular to the plane), but it would not exhibit any peaks in the **Alkyl** region. **III** would exhibit 5 peaks in the **Alkyl** region, **IV** would exhibit 4 peaks in the **Alkyl** region, and neither molecule would exhibit any peaks in the **Carbonyls** region. **I** is the only molecule consistent with the ^{13}C NMR spectrum, and the peak assignments are shown below (a carbon between 2 carbonyl groups can get shifted into the **sp/highly E.N.** region because it is beside **two** $C\delta+$ atoms!):

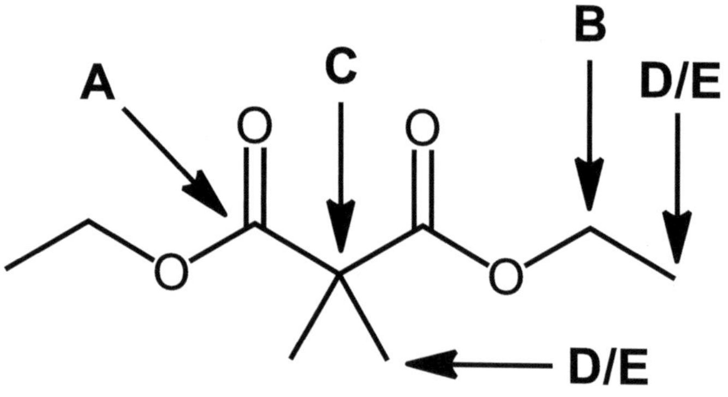

Carbon-13 NMR Question 16.

Which of the molecules **I-IV** would give rise to the ^{13}C NMR spectrum shown below?

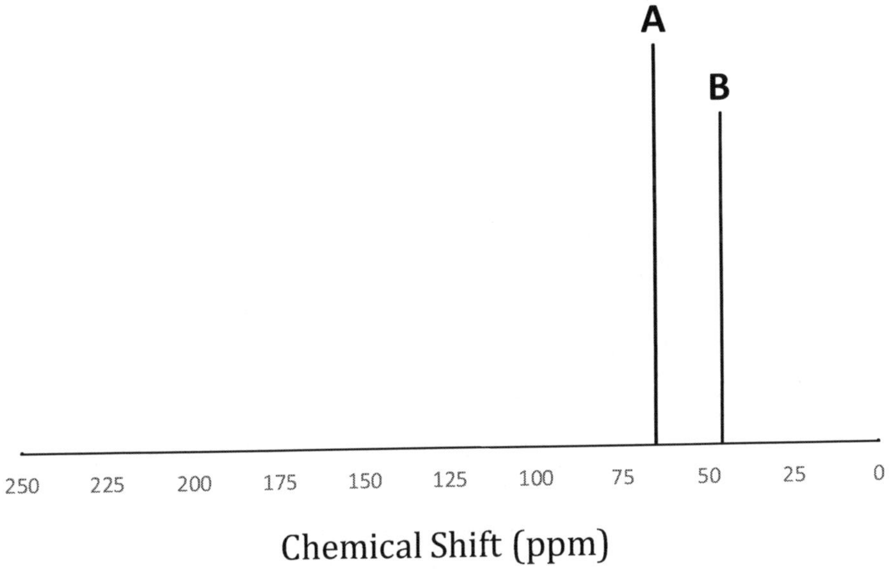

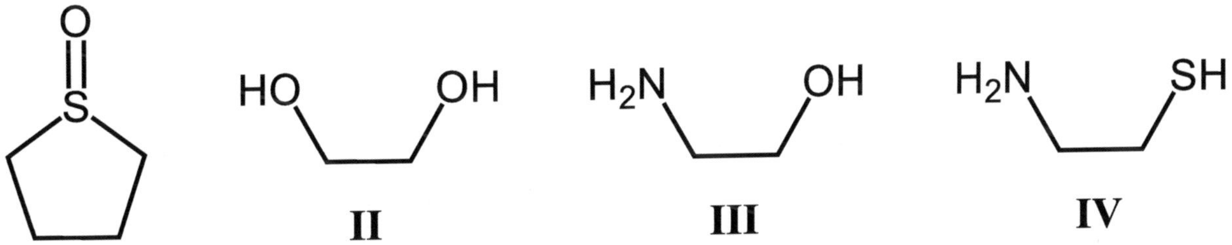

I

II

III

IV

Carbon-13 NMR Answer 16.

Peaks

Carbonyls:	0
C=C:	0
sp/highly E.N.:	2
Alkyl:	0

Answer Choices:

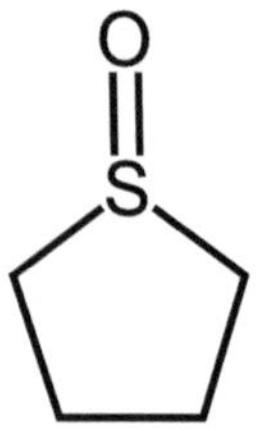

I

II

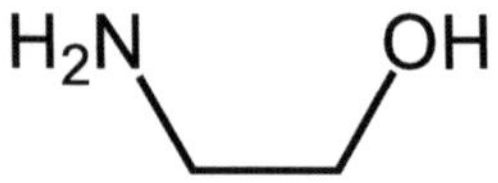

III

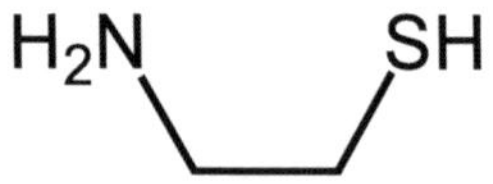

IV

I would exhibit 1 peak in the **Alkyl** region, and not anywhere near borderline. The mirror plane in **II** makes the 2 carbons equivalent, thus it would exhibit only 1 peak total and it would be in the **sp/highly E.N.** region. Sulfur is not a highly-electronegative element, therefore **IV** would exhibit 1 peak in the **sp/highly E.N.** region and 1 peak in the **Alkyl** region. **III** is the only molecule consistent with this ^{13}C NMR spectrum, and the peak assignments are shown below (oxygen is more electronegative than nitrogen):

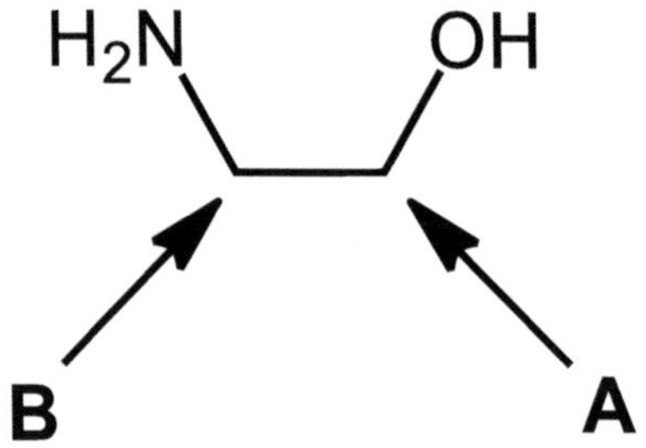

INDEX

F

G

H

I

J

K

L

M

N

U

V

W

Z

Made in United States
Orlando, FL
04 November 2026

90201520R00173